Edmund Turney Allen
Test Pilot - Aviation Pioneer
Aeronautical Engineer

1896 - 1943

Letters From My Father
By Turney (Allen) Oswald

Edited By Terry Kent Oswald

"Boeing was fortunate that a leader arrived at the right time to set a course for success by not only pioneering the organization that continues today as Boeing Test & Evaluation but also insisting the company build its own wind tunnel. The Edmund T. Allen Memorial Aeronautical Laboratories are named in his honor."

"His idea has helped shape many innovations - its a widespread belief without Allen's wind tunnel, Boeing would not be the company it is today. ….. The chances are that one of our competitors — one of them would have built America's first jet bomber and probably have gone on to build America's first commercial jet ….and Boeing would have just faded away.
I firmly believe Boeing would not be here today if not for that wind tunnel."

    Michael J. Lombardi - Boeing Corporate Historian   October 2013

"My personal enthusiasm for the man may well have distorted my appraisal of his worth to America. But my experience in life and in aviation tells me Eddie Allen was the most important figure in America's aviation firmament since the inception of the aeronautical science. He was second only to the Wright Brothers."

    Thomas Collison - Aviation Author 1944

Foreword

This is the personal story of a true aviation pioneer told through his own letters and published writings. His story begins as young student in 1910 and continues until his death in 1943. The letters (more than 200 in number) are written mainly to his family and represent a diary of his life. His published writings are numerous and include many aviation magazine articles for the general public as well as technical reports for the aviation community. He tells an aviation story that covers 26 years, involving aircraft from the Curtiss "Jenny" bi-plane to the B-29 Superfortress, perhaps the most sophisticated aircraft of WW II. He tells of his trials in learning to fly the "Jenny", of becoming an expert in "stunting", spins, stalls and the trauma of the airplane crashes so common in WW I flying. His stories as an instructor in training young student pilots and their patriotic fervor to help the French are vivid reminders of history forgotten today. Returning to the U.S. he flew as the first NACA civilian pilot at Langley Field during the summers while a student at MIT. As an aeronautical engineering student, he participates in glider competitions with many of the same French and German aviators that he trained to fly with and against a few years before. Eddie survived early attempted "helicopter" flights (although the rudimentary aircraft would be difficult to call credible helicopters), and in the late 1920's, he flew the airmail over the most challenging routes over the Rocky Mountains. Following this he began 10 years of independent flight testing in the 1930's, initiating the first flight on more than 29 new aircraft from 9 different manufacturers. He will be recognized as the foremost test pilot of large multi-engine aircraft in the U.S. He will tell you of his personal experiences visiting the German Air Ministry in the mid-thirties and he clearly forecast the impending tragedy that would become WW II. Finally in his last years he was the Director of Flight Test and Aerodynamics for the Boeing Company delivering the Boeing Clipper, the Stratoliner, the B-17 and its many improved versions, and the first flights of the B-29.

Throughout the letters is a family story through four decades, a look at social and economic issues, politics, literature, philosophy, religion, and marital issues — now a hundred years in the past yet somehow relevant today.

And finally the art of articulate letter writing is on display -a lost skill in our world of the internet, texting, and the general reduction of emphasis on language arts in our educational systems today.

Edmund Turney Allen

1896 - 1943

Table of Contents

Foreword	5
Acknowledgements	8
Introduction	11
Overview of Life	14
Childhood	16
Dairy Farm	26
University of Illinois 1916-1917	44
Army Basic Training/ Pre-Flight Training 1917	50
Flight Training and Flight Instruction 1917-1918	74
Flight Instructor Role 1918	141
Instructors Survey - England / McCook Field 1918-1919	173
NACA/ University of Illinois 1919-1920	193
MIT Course Work/ Glider Competitions in Europe 1920-1922	220
Washington D.C. / Dayton, Ohio 1923-1925	282
Air Mail Cheyenne / Salt Lake City 1925-1927	291
Independent Flight Test and Consulting 1928 -1929	336
Aircraft Radio Communication Study	349
Independent Flight Test and Consulting 1930-1939	363
Douglas Aircraft DC-1, DC-2 and DC-3	386
German Air Ministry 1935	390
Testing/ Demonstration in the South American Market 1936	400
Eddie Allen and Vought Aircraft	403
Edmund T. Allen and the Boeing Company	405
Trauma and Celebration - March to July 1939	414
The Final Flight	422
Most Unforgettable Character - Readers Digest February 1965	432
Memorable Quotations	437
Honors and Awards	439
Bibliography	440
References	443
Appendices	444

Acknowledgements

Numerous people have researched and written about the life of Edmund Turney Allen over the last 75 years. At least three books have been attempted and as far as I know none has been successfully completed, but valuable draft research information was collected and much of it was provided to our family over the years.

This document is centered on the personal letters of my father. I have attempted to weave a story of his life using these letters and taking advantage of the unfinished work of several key people. For their efforts they must be recognized.

Margaret Reynolds Broussard (MRB) - the keeper of the letters over many years. Margaret is Eddie Allen's niece, the daughter of his sister, Margaret Allen Reynolds (MAR) who was a key participant in this story. Margaret Broussard diligently digitized the more than 200 letters and provided valuable Allen family history. She also provided her mother's personal history and boxes of historical background material.

Richard K. Smith PhD (RKS) - A Smithsonian Aviation historian, Richard first approached our family in 1985 and spent significant time reviewing the trunks of data we had stored related to my father. Over the next 10 years we communicated often and Richard provided draft research material with always a little "nudge" to encourage us to find more. Most of the aviation chronology in this document is based upon Richard's work. Sadly, Richard's illness and his death in 2001 prevented completion of his book. He did leave several volumes of research material which are included in the references to this document.

Thomas Collison - Aviation author whose friendship with Eddie Allen began in 1925 while Allen was a pilot for the Mountain Division of the U.S. Postal airmail service in Cheyenne, Wyoming, and continued until the final crash at Boeing Field in Seattle in 1943. Thomas Collison drafted the most complete biography of Eddie Allen in 1944. As he says "My personal enthusiasm for the man may well have distorted my appraisal of his worth to America. But my experience in life and in aviation tells me Eddie Allen was the most important figure in America's aviation firmament since the inception of the aeronautical science. He was second only to the Wright Brothers."

Several of the stories written by aviation pioneers such as Kelly Johnson and Frank Hawks, have been included because they referred to their direct interaction with my father. During the years we have visited with friends of Eddie Allen: Bob Robbins,

Harry Changnon, Bob Lamson, and Ed Wells I remember well. They all gave a consistent story of their respect and affection for my father. Bob Robbins offered us a ride in the B-29 "Fifi" during the 50th Anniversary Event. I regret not accepting. Harry Changnon left us a valuable copy of his draft biography of my father. Also included are several remembrances written by William H. Cook, chief aerodynamicist at Boeing and an appreciative proponent of the decision to build the Edmund T. Allen Memorial Wind Tunnel Facility.

Terry Kent Oswald (TKO) - And finally I must thank my husband for editing the story of my father. My husband is a retired aerospace engineer with 33 years experience at Boeing and Douglas Aircraft Companies. His enthusiasm for the story was contagious.

Introduction

"Eddie" Allen died in the crash of the prototype B-29 bomber in the early years of World War II. This traumatic crash endangered the development of the aircraft which the Department of Defense would rely upon for success in the war in the Pacific. The progress of the war was very much in doubt in February 1943 and all across the country the mobilization to provide the aviation weaponry to win the war was underway. The pressure to bring the B-29 to the war meant the acceptance of development risks that would not have been taken during peacetime. But this was not peacetime. Flight testing continued even when critical developmental problems occurred. More than 1600 aircraft were on order even without a completed prototype program. Contracts had been let to build construction facilities at Wichita, Kansas; Marietta, Georgia and Omaha, Nebraska, as well as Seattle. Thus, when engine fires became a reoccurring problem, the flight test program attempted to work through the problems.. In 23 flight tests they had to replace 16 engines and could achieve only 27 hours of flight time at the time of Eddie Allen's final flight on February 18, 1943.

Throughout his life Eddie Allen had been the most conservative of engineering test pilots. He was sought by aviation companies for "first flight" duties at the insistence of insurance companies. In the weeks before his death he had been requested by Lockheed for the "first flight" of their new transport aircraft, the Lockheed Constellation, even though he was Director of Flight Test for the Boeing Company. It was with this background that the crash occurred which resulted in the early death of my father.

I was not yet 3 years old at the time of his death so my younger brother and I never knew Eddie Allen as a father nor as a person. His achievements as a test pilot and aeronautical engineer have been well documented and celebrated in various aviation articles and awards (many are listed in the references for this document). Some articles such as those written by his associates give views of his personality and have helped us appreciate what a unique person he was. Many of these were written following his death and thus may have been appropriately gratuitous. Certainly from my meetings with his friends and working associates, I gained an appreciation for my father's personality and his unique expertise as a pilot, engineer, and inspiration to so many people.

In recent months a cousin (Margaret R. Broussard, the daughter of Margaret Allen, Eddie Allen's sister), has sent me a treasure trove of letters written by my father to his family along with a family history written by Margaret Allen. The letters were written over many years (from his boyhood experiences away from home in 1910 to shortly before his death in 1943).

Reading the letters has been fascinating, certainly from my perspective as his daughter, but also I expect to those interested in aviation history because of the times and events in which he was a participant.

Most of the early aviators were not "writers". They were in many instances risk takers, sometimes exciting personalities. Generally, someone else would describe their aviation achievements. My father was a prolific writer. In the letters included herein, he describes his experiences learning how to fly the Curtis JN4D "Jenny", the Thomas-Morse Scout, the Standard J and others. He describes the maneuvers, the "stunting", the risks of learning to fly. He wrote about his difficulties in learning to fly and his doubts that he would succeed. He described his experiences flying gliders in early meets in Europe in 1922, and his experiences testing an early helicopter prototype in 1923. He describes the challenges of flying air mail routes over the Rockies in 1925 to 1927. He wrote about his experiences meeting with German aeronautical engineers in 1935, visiting their facilities and even flying their aircraft. He clearly warned of the potential impending catastrophe that would be WW II. And he wrote about his experiences as a test pilot who conducted "first" flight tests for more than 29 different aircraft from at least 9 different companies.

Throughout his life he documented his experiences. Most often he wrote articles in Aviation Magazine, but also technical journals, and other aviation magazines. During some periods he was more active in writing to his family and I have included these letters as written. These include current events, political issues, religion, philosophy, early aviation technology and personal issues. The personal letters regarding his first wife and family issues associated with that marriage are left unchanged and frankly do no harm at this time. They show how much we as a society have changed in the last 100 years. Issues that today are met with acceptance might split a family apart in those previous years.

The value of written, articulate communication is also on display. Today's email and texts are no substitute for the detail and thoroughness of the written letter. How many of us can say we have freely expressed within our families the range of ideas that are articulated in these letters? Our abilities to communicate seem lacking today compared with this family of a 100 years ago. Perhaps this is an art lost with the advent of television, the personal computer, smart phones, and the internet and the resulting lost hours of personal communication.

Following the death of his mother in 1931, there were fewer personal letters saved by the family although there is interesting correspondence with key aviation figures of the time. I have included excerpts written by my father from his many articles as well as anecdotes written by friends and associates. Both give an interesting look at the personality of my father.

Onboard the Boeing Clipper 1939

Overview of Life

In order to appreciate the context of the letters, an overview of his life is necessary. The letters weave through the major events of our sometimes traumatic history.

The United States joined the British and French allies in WW I in April of 1917 in a war that had been raging since July of 1914. Following his freshman year at the University of Illinois, Eddie volunteered and joined the Army in the spring of 1917. Because of his height and weight he was a candidate for the Army Air Corp. He became an accomplished pilot and with his expertise he became an instructor pilot. In the summer of 1918 he was sent to England to study the British pilot training program. The war concluded in November of 1918 and he returned to the U.S. where he was stationed at McCook Field, Ohio, the Army's aviation engineering and test center and Langley Field, Virginia, which was also an aviation research center. Following completion of his military service, Eddie became the first civilian test pilot for the NACA (National Advisory Committee for Aeronautics). The NACA ultimately became NASA in 1958. He worked for the NACA during his summers while pursuing his university education.

Following a second year at the University of Illinois, a rather interesting marriage to Allene Gregory in August of 1920 apparently shook up the family relations as Allene was a former professor at the University of Illinois and was Eddie's senior by 9 years. The marriage lasted only a few years but had a long term effect on family relations.

Eddie met his life-long mentor, Edward P. Warner, of NACA and MIT fame, who encouraged him to attend MIT. As a student at MIT (1920-1922) he built gliders which he flew in competitions in France and Germany and developed professional relationships that would prove important throughout his life. He flew an early helicopter prototype for Henry Adler Berliner in 1923 and survived the crash of this flawed prototype. He also built three different versions of his own light airplane design, two of which flew successfully although they were not a commercial success. And he flew postal service flights over the Rocky Mountains (1925-1927) and the early Boeing Air Transport (BAT) flights from San Francisco to Chicago after the postal department turned the service over to private contractors. During the 1930's he became a key consulting pilot/engineer for many of the aircraft built by Boeing, Douglas, Consolidated, Northrop, Lockheed, Sikorsky, Curtis Wright, Vought and North American. Finally, he became the Director of Flight Test and Aerodynamics for Boeing in 1939.

Eddie had a unique look into the German aviation culture after WW I when he participated in the glider competitions in Germany (1922) and in visits prior to WW II (in 1924, 1929 and 1935). He established relationships with key German aeronautical engineering people during these visits and as a result he was invited to speak at the newly built unified German Air Ministry in Berlin in 1935. He toured research and manufacturing facilities and flew German aircraft at a time when some looked at the "New Germany" as an economic wonder. His observations of the German aviation production and unification of resources, as well as the nationalism and anti-semitism, in retrospect, foretold of more sinister outcomes.

Finally his early death along with his entire flight test crew in 1943 in the crash of the prototype B-29 was a major blow to the development of the aircraft that would eventually end the war in the Pacific.

The letters are grouped into different periods of his life. I have attempted to summarize each period in terms of the events in his life as well as the background history. The letters relevant to that period follow.

Childhood

Eddie was born on January 6, 1896, and was the son of religious parents who valued education and the practice of medicine. His father, Edmund Turney Allen Sr., was an accomplished physician specializing and teaching ophthalmology and writing two medical books which even today are recognized as medical text classics.

The practice of medicine was financially rewarding in Chicago but Edmund Sr. was not a skilled investor and apparently, combined with his generous nature, the family struggled financially. His kids remember a father unable to say "no" to any person in need. Also his restless and impatient nature kept the family on the move and financially challenged.

Eddie had an older brother Thomas (8 years senior) and an older sister Margaret (3 years his senior). Early in their lives when Eddie was age two, his father's health began to fail and he was refused life insurance. His mother Abby Dyer Allen enrolled in medical school in anticipation that she would become the primary breadwinner. With three children ages 2, 5, and 10 years of age, one can imagine the challenges even if finances were not an issue. Thus, much of Eddie's childhood was spent under the care of other members of the family. Grandfather Allen and Grandmother Dyer lived with the family as Eddie's mother worked to complete her medical school training. Aunts and uncles from both the Allen and Dyer sides of the family, helped throughout the children's lives. Upon her completion of medical school in 1900, Abby's health also declined as she suffered from recurring pneumonia. Edmund Sr. took Abby and Margaret to Arizona for 3 months while Abby recovered. Thomas and Edmund Jr. stayed with their Aunt Louise. While in Arizona Edmund Sr. was involved in some kind of "gold mining venture" which failed with the loss of significant family monies. The family returned to Chicago. Then in 1903, the family abruptly picked up and moved to DeLand, Florida, again apparently because of Abby's health. They struggled financially although the Florida childhood was remembered as a wonderful experience by Eddie's sister Margaret. The family lived for 5 years in Florida and then abruptly moved to Denver, Colorado, for a period (1907-1909). Edmund Sr. purchased a resort hotel "Strontium Springs" miles out of the city on the South Platte River in 1909. During this same time he also was involved with Mexican land speculation. With failing finances, Abby put her foot down and insisted they return to Chicago in 1909.

Struggling health and financial issues always seemed to be problems. Edmund Turney Allen Sr. eventually died in 1913 when Eddie was 17, Margaret was 20 and Thomas was 25. Thomas was in medical school and took a year off to work. Margaret was an undergraduate at Denison College (a small liberal arts school near Columbus, Ohio). Eddie had just graduated from high school in Chicago.

The first letters were written while Eddie was being cared for by his Aunt Jessie in Centerville, Illinois (near St. Louis, Missouri). Margaret was staying with their Aunt Louise in Morgan Park (Chicago). Eddie was fourteen.

                                          529 W State St
                                          (Oct 1909?)

Dear Father:

    I have been very busy this week but must drop you at least a line. This is Monday night. I have gotten my lessons. Mother left this morning. I got your letter and am so glad you will be here in a month. I hope this month will be short.

    Just after mother left this morning I went in and found that I have to pay tuition because my parents did not live here. I was notified to pay it at once - It is $2. a month $8. semester $16. a year. I wrote to Mother immediately and asked her what to do. Then I told the principal what I had done and he said it was all right and now I am waiting for an answer from mother. I do not know whether or not she wants me to use the $10. I have in the bank but now it is all right. Mother will tell me what to do.

    I am beginning to get lonesome already Oh! you dear Father I love you with all my heart and hope to see you soon.

                                                  Your son

                                                  Ned Allen

[The following letter to a 14 yr old ETA Jr., is typed, from ETA, Sr.'s downtown office in Chicago]

[Letterhead paper is from the year before in Denver. The speculative nature of Eddie's father is apparent as well as his strong religious beliefs.

{E. T. ALLEN, A.M., M. D., Ph.D. OFFICE: 1604 ARAPAHOE STREET
EYE, EAR, AND NERVE SURGEON PHONE Main 7216
Interested in Mexican Club Lands RESIDENCE: 3222 DOWNING AVE.
Better than U.S, Government Bonds PHONE York 3327
Safer than a Gold Mine MOTTO: "Better Safe Than Sorry"

References
C.E. Ford, Optician, Denver
J. I. Hansen, Jeweler, Denver
Perry O. Kennedy, Designer, Denver
Federal State and Savings Bank, Denver
M. W. Whitely, Credit Dept. of Hamilton National Bank, Chicago
Denver. Colo., U.S.A._ _ _ _ _ _ 1909}

92 State Street Chicago, Ill.
Feb. 19, 1910

My Dear Sonnie:-

Why have we not had a letter from you all this week? I wrote you a long letter on Monday, and we have been looking for word from you. I also wrote Aunt Jessie and sent her $12. to pay on your board and tuition and she has not acknowledged it. Please ask her if she did not receive it.

As an answer to your prayers and those of others God has completely healed my heart and it is cured. Now we must thank him. We have a wonderful God. He can raise the dead, and so of course he can cure the sick. Only it is the prayer of faith. A little girl had a severe tooth ache. She prayed and still the tooth continued to ache. Her mother told her to ask God for relief. She said, "I did, but God takes his own time to cure it." Instantly the pain ceased. You see God was testing her faith, and when she professed her belief in him in spite of the seeming unanswered prayer, then the answer came.

Margaret is quite well. Do you ever write her? or does she write to you? I can never get anything out of her.

Have you heard about the preparations that Germany is making for a great war? She has the best army in the world today. She is now building battle ships of the largest kind faster than even England is, and her fleet of gigantic war balloons[1] will within a year be able to transport an army as far as to Liverpool from within the confines of Germany in a single night. England is terribly alarmed at all this thinking that it is against her that all these preparations are being made. I do not think so for I study my Bible, and there I find that it will eventually be against Russia that the power of the anti-christ is to be hurled. And I am more and more inclined to think that the antichrist is the present Kiser of Germany. Perhaps not. I dare not be dogmatic on that point. But he is to be a Roman. And the Kiser claims to be the direct descendent of the Caesars. His title is Emperor of the Holy Roman Empire, or was until recent years, as you will see by reading history. Charlemagne was the emperor of the all western Europe about the year 800 A.D. His empire broke up but the German empire is the biggest fragment of it that is left. He claimed to b a Caesar, and his title has defended to the present Kiser.

Dan. 11:36 shows a king of grim determination like the present Kiser. He honors the god of force. (v38) In v.40 it shows that the king of the North (Russia) shall come against him, and the king of the South (probably England which owns Egypt (v.42) shall push at him, but in spite of that he shall enter the countries (Turkey) and overflow and pass over (the Bosphorus) and shall plant his tents in the "glorious holy mountain" (Mt Zion at Jerusalem). Yet "he shall come to his end and none shall help him" because he is to be destroyed by the brightness of the coming the Lord Jesus when he descends with all his hosts after the three and a half years in the air. Oh, what does it matter after all what we do or fail to do in this life! The great thing is to get ready for that life which is without end. We can however not be

---

[1] The Zeppelin "Blitz" of England from 1915 to 1917 was justifiably feared as 557 were killed and 1,358 injured just by these airships.

blind to the fulfillment of the wonderful prophesies which are taking place right before our eyes.

Now Sonnie please let me hear from you this next week without fail. It is very cold here. The ground is covered with ice, which resulted for the snow melting and freezing again. The same snow which fell early in December is still on the ground and the ice is piled up along the sidewalks in places three feet high, and frightfully dirty.

Give my regards to all there and do not neglect your prayer daily. With a heart full of love, your Father,
        signed   E. T. Allen

                            529 W State St
                            April 10, 1910

Dear Parents,
Auntie will not let us use her pen or ink and my materials at school and it takes away from my lessons if I write my letter there. That is why my last weeks letter was so late.

Only 7 more weeks and if I keep on as I have started I will get out of all my examinations and Auntie is going to try and get me off on the 20th that will be less than 6 weeks. My! won't I be glad.

In Commercial Geography class a few days ago we were studying China and Japan and a boy that sat next to me told me to ask Miss Portloch if the Chinese ate rats and mice. I laughed and told Russel to. He laughed, then we began studying again and I found where it said the Japanese seldom eat meat. I showed it to Russel and he said "that is Japan look in China so I turned over to China but could not find it. He said "O look in animal industries" so I looked and it said "The Chinese eat little meat". "Little meat, said Russel aren't mice little meat" I laughed so that Miss Portloch told me to come down and recite my lesson to her after school.

Yesterday I got a job across the street at Mr. Efaw's picking up shingles they just had their house re-shingled and the old shingles were all over the yard. I did not quite finish it. I will get fifty cents.

I am 4 feet and 99/100 + 2/200 feet over. The leaves are all out on the trees and the grass is green and all the fruit trees are white with blossoms.

The census taker came around last week on Aunties birthday and he was surprised to hear that she was 47 today. Auntie did not give my name. She thought you would give it. Has he come around to you.

I have heard the Halley's comet can be seen with a field glass before the sun is up. Won't it be very hot if the tail of Halley's comet envelopes the earth. I should think the forests would catch on fire,

Eddie Allen "Ned" Age 14

Today the sermon was by a G.H. Chapman. He stayed with us last night and ate breakfast with us this morning. He knows both Prices in Morgan Park. His home is in Chicago.

You knew that they had taken off $3000. from the $10,000 for the lots and the Parsonage to be sold and what do you think. Mr Hagleman the pastor bought the parsonage. Nobody knows where he got the money and now in return for what he has done, Auntie is

trying to raise $30 to send him to Chicago as a representative of the church to the congregation. She has got $13 already.
I must close.

>Your Son
>Edmund T. Allen

P.S. Dear Father   Would this stick do that is inclosed in the loft of the barn where they keep the pony, there are a whole lot of strips of bamboo just like this. It used to be a sunshade.

>529 W State St
>April 18. 1910

My dear parents:

Only 32 more days until I will start on my pilgrimage to Chicago. Aunties girl left her Friday and she has another now.

Yesterday it snowed all day. It also snowed Saturday. The trees are all in blossom and peoples gardens are up. I guess we will not have any fruit this summer.

What is "Partner Allen's" business. You said in your letter that he said he wished I were there. What does he want me to do.

I got 96% in my civics monthly test and 100% in Grammar test. I got 95 in Commercial Geog. We have started a review in arithmetic. Our first problem tomorrow is in multiplication and subtraction.

Do you remember the name of the man in Denver that lived right across the street from us when we lived in that 3 cornered lot (3600 Arapahoe St). He had two sons Roy and Bill and Gladys and Ruth were the girls. We used to play can-can. The name I think began with "E". I don't know for sure though. I want to write to Roy.

Miss Flord, our Civics teacher, read a poem to us the other day. It went like this. It was a poem on Longfellow's poem

"I shot a arrow into the air
And then I gave it no further care

But split some kindling and fed the hogs
And threw some bricks at the neighbors dogs

And did my chores with a joyous mind
And woe and trouble seemed far behind
That night a peeler (policeman) came to my bed
And broke his billy upon my head

And bore me off to a moldy cell
And there I sit on a stool and yell
And there its likely that I'll remain
My arrow ruined an airplane

It flew right into an airship's works
And made the rudder give mighty jerks
And knocked some cogs from the jingle jigs
And tore a hole in the thingumyjig

And as for that man that rode in style
He was knocked from his perch and fell a mile
And when he landed Alack! Alas!
He broke an acre of greenhouse glass

I'm charged with arson and larceny
And homicide in the steenth degree
And breach of promise and other crimes
And lawyers badger me for my dimes

I shot an arrow one evil day
And let it fly in my aleck way
There was wood to saw
There were chores begun

There were useful tasks that I should have done
But I fooled around like a useless clam
I shot my arrow and here I am"

It is a pretty good one, eh? We are reading "Titus A Comrade of the Cross".

Judge Lee a honorable judge in the city died Friday and today they had the funeral. All the stores in the city shut down. How are you mother? You must be busy as I haven't gotten a letter from you for a long time.
Never mind. How is Harold? He said in his letter that he was going to the country to work this summer. Isn't that too bad for me?
Well must close.

        Your son
        Edmund Allen

How are you Father. Are you getting any more business than formerly. ETA

P.S. Auntie just told me that she had made arrangements for me to stay over with Mrs Efaw.
Well I am going right now. Auntie read an article by Dr. Keene on vivisection. Then we read by Ella Wheeler Wilcox on antivivisection. Well I must close

       Your son Edmund T. Allen

  My precious Son - I am homesick for a letter from you. We have not heard from you since your Aunt Jessie wrote that you must find another place to stay. It seems strange too for of course it only takes about 12 hours for a letter to reach you. We thought we would get a letter back Wednesday, and it is now more than a week since we sent your letter. I suppose you do not realize that it is long waiting for letters. How would it do to carry a stamped envelope in your pocket *telling us just what occurred or just what plans you had made. If I only had the money I would come down to Centerville and help Aunt Jessie for a few weeks, but that is out of the question. I wish one of the doctors had a case or could work up several cases for your father.*

I am glad about the Church, so glad the money has been subscribed, and Aunt Jessie must be glad that she helped in it all. Last Sunday your father and Thomas and I all went out to Windsor Park to see an old School friend for Thomas and hear an old school friend preach. We took dinner at the house of this friend. They had six children, 3 of them in college. Three at home going to school. The youngest boy, 8 years old, lay on the floor reading Water Babies. I asked him if he read it all - he said no he skipped some of the hard places. I thought it pretty good for a boy 8.

I am so glad you are learning bible verses. They will be of use to you sometimes.

Mr. Allen[2] has come home from Mexico, and reports very favorably on your fathers land.[3]

Let us hear soon my Son   Most Lovingly   Mother

[Apparently, Eddie had been sent to live with his father's sister Jessie (Mrs. Edward Harris), because of the bad part of Chicago his parents were living in, or for financial reasons. Eddie did return to live with his parents in the Beulah Home Maternity Hospital subsequent to this letter. Aunt Jessie would die the following year (August 1912) while onboard a ship returning to Burma where she and her husband had worked as missionaries for many years. She was buried at sea.]

---

[2] Coincidentally, ETA senior had some partnership with a Mr. Allen, no apparent relationship.

[3] Other references indicate that ETA Sr. bought land in Mexico, either when they were in Arizona in 1901-02 and the gold mine made him temporarily wealthy, or when they were in Denver.-MRB

Dairy Farm 1913-1916

Eddie attended the R.A. Waller High School (now known as the Lincoln Park High School) in Chicago, Illinois, and graduated in 1913. His father died in August of that same year and with his father's death, Eddie was not focused on education as were his older sister and brother. The need to generate family income was also a factor. He took a job at the Natoma Farm (located about 20 miles west of Chicago) from 1913 to 1916 and with his separation from the family he found appreciation for the value of skill and achievement. In three years he progressed from a general handy man and mechanic in the power plant to some level of foreman in the dairy. Certainly, his letters demonstrate a unique interest in the technical aspects of his job. His mother and sister had to pressure him ("pry him away") to get him to enroll in the University of Illinois in September of 1916. He was willing to enroll in the school of agriculture but wasn't sure if Illinois was the best school for him, questioning that Wisconsin may be the better agricultural college.

The letters which follow were written when Eddie was about 18. They reflect many issues of a young man seeking a level of success in his chosen field. They also give view of the issues of the day (1914-1915), not unlike current issues of today.

Natoma Farm Hinsdale, Ill
August 8, 1914

Dear Sister,

I certainly did enjoy your letter. I always enjoy your letters. You write so easily and freely. I hope you will pardon the use of a pencil. It is easier for me to write with one.

The last two days I have been working in the dairy. Two fellows took a day off and I was put in to take their place. Dairy work is pretty nice for a change, it is fast work - something I like for some reason. Capping bottles just as fast as ones hands can move or icing cases of milk just as fast as possible or washing bottles or rinsing. It is real interesting work and Capps is such a queer fellow to work with. He is the head dairyman drawing a salary of about $85. per month. He is short and thin with whiskers all over his face. He used to be a captain in the U.S. Army and he has gotten some queer ideas about things. The principal subjects we disagree and debate on are Theodore Roosevelt, bacteria in milk, increasing the army and navy, Germany as a world power, women's suffrage, and the liquor question. He has a way of making a statement with his head bent forward and his eyes flashing as if that statement were law, and he were defying anyone to deny or question it and I deny every one of them and then he says, "Well, I'll just bet you my months salary against yours that etc."

    For about two weeks back I have been working with Mr. Craine the engineer. We have been putting in the steam heating system in the new bunkhouse. That was fine work. I do like pipefitting very much. Mr. Craine is a little fat man with a big "bay window". He gets up at 3:30 every morning and gets the engine started so the milkers (who begin work at 4:00 A.M.) will have light. This whole farm is electric lighted. The current is generated by a 125 Amp generator run by a 65 H.P. Westinghouse 3 cylinder vertical gas engine. They also have a Ball steam engine direct-connected to a 200 Amp. direct current generator. All the pumping of water (about 45,000 gal. a day) is done by a motor. The feed is ground by a motor. Of course there are other gas engines. There is a 6 H.P. horizontal engine that runs the dairy machinery (capping machine, separator, cooler pump, bottle washer, buttermilk starter and ice crusher), a 5 H.P. engine to cut ensilage and shred corn- a 2 H.P. engine to run an emery and grind stone and then two smaller engines to pump water over at Oak Brook, across the road. Of course you are not a bit interested in this but I got started and couldn't stop.

    Well, as I was about to say - Mr. Craine runs the engines from 4:00 to 6:30 A.M. goes home for breakfast and comes back about 7:30. By this time I have got started cutting pipes or something like that and he helps me or rather I help him until noon. He has to pump water once in the morning - outside of that time we are working together. He doesn't come down after dinner till about 4:00 o'clock but he tells me what to do for the afternoon so I'm not idle much. Then in the evening he "turns the lights" until 8:30
I call myself chief assistant engineer while I'm working with him. Mr. Craine is an alumnus of Cornell - mechanical and electrical

engineering - and has 30 years experience so you can see I am in a position to learn a lot and I'm learning a lot.

The superintendent of Natoma is going to leave (by request I think) and I have an idea Mr. Craine will be made superintendent. If so I'm going to fight for his job, but I have an opponent, one of my best friends out here, a man of about 30. I guess he doesn't think he is my opponent for he doesn't know the engineer's job will be open soon and also because he doesn't think I will try to get it. I know Mr. Craine will do all he can to help me. He has been awfully nice to me - showed me everything about the engine and boilers and he has asked for me every time needed help in his work. Perhaps I am a kind of smarty to think I can come out here and in a year get a job as engineer, take the place of a man earning $85. a month. "A little learning is a dangerous thing." I guess that is what I have got since I've been out here.

Much later: - events have developed rapidly since I wrote the preceding pages. Friday evening I went to Chicago and didn't come back till Sunday night. I had a fine visit with Thomas Friday evening and slept at the frat house that night. Next morning I saw Mother down at the office and we had lunch together. Then I went out to see one of Thomas' clinics in Gynecology. It was a horribly realistic lesson. I wish a good many fellows out here could see that - They have a good prof. who gives the minds of the patients something to think about as well as their bodies some medicine.

I then went out to Dr. Jones flat on Grand Blvd. and 47th, where I had been invited for dinner. After dinner we went to B. (Beulah?) Home to get some books and then met Thomas and went out into Lincoln Park to talk over things.

You know the office isn't paying for itself lately and Mother and Thomas have decided to give it up, and have Thomas devote his entire time to his studies. Mother is very unsettled. I felt so sorry for her. She seems lonely and almost overwhelmed with the financial burden she is carrying. You see Thomas must take a years internship when he gets thru and so he will be about 3 years before he will be making expenses. And Mother must bear the burden all that time - I tell you Margaret - we owe a mighty lot to our mother. Sometimes I think I can't realize it all yet - all I owe.

Well, by this time, I had found out that Mr. Benske would probably stay superintendent indefinitely - then Mr. Craine took a day off to see Mr. Butler in Chicago. In the meantime I took his place - kept up steam, cleaned the boiler-grate, ash pit, flues, and ran the steam engine to pump water and generate electricity for the lights. He came back about 7:30 and everything was in perfect order. He told me that Mr. Butler was intending to make some changes on the first of next month - Mr. Craine was to be master mechanic - I was to be chief engineer - etc. That is the most important part, and then Mr. Craine can be superintendent if he wants it. So your Bro. is chief engineer (to be) of Natoma Farm.
Hurray!

Then he told me Mr. Butler intended to put me in charge of the milking machines when they install them (Whaddaya think o' that). I'll have to close now or I'll be paying extra postage.

S'long
Edmund T. Allen

Natoma Farm
September 8, 1915

Dear Mother,

I have been wanting to write to you every day since Saturday, really wanting to write, but I have been short of paper and just got to this today.

Friday afternoon I went in to Chicago and out to the Pendray's for dinner. It was a doings in honor of Thomas's graduation. Saturday morning I made a decision. Really that is something unusual for me to do. I usually let circumstances and other people decide things for me. And this is my decision. That I will get out of the power plant work as soon as possible. I will go into the dairy at the first opportunity, Or if that is closed up I will go on to the farm. At any rate I will leave Mr. Craine.

And, a decision being such an unusual thing, you can't imagine how hard it was for me to stand by it. But I resolved to stick by the decision. If I shall have made a mistake it was when I decided. After once deciding, there was no turning back. What do you think of that. I doped that queer idea out myself.

The causes of the decision are many. Do you remember why I came out to the farm? Sometimes I think I have lost sight of that entirely in my race to get ahead. The reason I came out here was to build up physically. To make a-big-a-da-mus. So whatever happens I cannot regard my stay out here as a failure. You remember why I worked in the power plant? As you said I was disappointed in not getting field work. They absolutely squelched me there. And then this opportunity opened here.

Mr. Craine was so nice to me, I liked the work so well, and above all, I learned something every day. Every day for about a year. Imagine how much can be learned in a year. I got to be a kind of partner in here. Everyone regarded me as the engineer. Then Mr. Craine failed me. And then I see clearly I am where I can go no higher. I have almost reached the limit in this small plant. Of course I don;t know everything I can know, but than I am not learning something every day. Perhaps once a week I learn something. But the main thing, Mr. Craine has failed me. It surely is now time to make a change.

I cannot regard my stay in the power plant as a failure, but a longer stay undoubtedly would be a failure. It has contributed largely to my physical development. So having decided to make the change I went up to see Capps. I asked him if he would take me in the dairy. If he would I had previously decided to see one of the other boys about quitting. If he wouldn't quit I'd pay him for giving me his job. But all this was not necessary. Capps said, "On one condition, that you will have nothing to do with the power plant" "And that's the only condition on which I'll come in the dairy," I said.

Now here was luck. One thing that had troubled me when thinking of changing from power plant to dairy was that Mr. Craine would expect me to keep on taking care of the dairy engine–– something I would not want to do when in the dairy. Not only that, when He wanted to go away for the day, I would be the one he would want to run the plant for the day. He would want to take more of the dairy. And Capps could put the damper on that right at the start by speaking to Mr. Butler.

Now for some more luck. Capps told me that he was going to start another man in the dairy to help him—-Capps, i.e. to take his work and Capps is going away for a two-week vacation next Monday the 13th and this new man is to take his place. Capps had wanted me in there—-Mr. Amacher, the new general manager had spoken to him about giving me the job and Mr. Butler had told Capps that he was thinking of putting "Allen in the dairy". It seems like a strange coincidence, doesn't it, that I should have made that decision just at this time. Methinks it augers well.

Now if I can retain the friendship of Mr. Craine while making this change, I will be doing well. You see he might take it ill that I should leave him when he is so rushed with work. Secretly, I feel a great satisfaction in leaving him thus— — when he needs me. He has been so dissatisfied with my work at times that it has made me kind of sore. I am pretty sure that the next fellow he gets will make him appreciate me all the more, even if he won't admit it.

Now I am trying to bring this change about so that he will not be sore. And things are working that way finely. Mr. Capps has told Mr. Craine that he is going on a vacation and wants me in the dairy Friday morning so as to get used to the work before he leaves. Now I will go in the dairy and when Mr. Capps gets back—-Mr Amacher says I am to stay in the dairy and as I can get $5. per month more, I want to stay in the dairy. (Mr Craine will appreciate the last reason).

Then —-Mr. Capps says the dairymen are to have nothing to do with the machinery, engine, shafting, etc., as he won't have a fellow fool with dirty machinery and then come and handle milk. How's that. You see I've got everything doped out already.

It's too bad that I've made this letter all about myself when you are so worried over Margaret. I surely am an egotist. I am enclosing $5 for Margaret. More later.

Your son
Edmund T, Allen

September 8, 1915

Dear Brother,

I received your letter yesterday. Thank you very much for writing the way you did. You explained things I did not understand. If it was hard for me to get that point, how much harder would it be for a man who never had those ideals.

Saturday morning I heard some strange things. Purdy the foreman asked me if I was still here. He told me that when Mr. Craine got back that evening he was just raving. "That was a fine way to do——go off and leave Chester in charge of things". He didn't see why I had to take a day off just when he went away. You know when Mr. Craine told me he was going away on Friday I had given in ——said I wouldn't go—- Then he had told me it would be alright for me to go away about noon —-that Chester could take care of the boiler and he would be home about four o'clock.

So I went but it was about two o'clock before I could get away, When I heard that he was telling folks around here how unreliable I was, I decided to leave, to go in the dairy, That was Saturday morning. When I went after the cream, I took along "Decision of Character" and read it and I could trace myself every change of thought which the author speaks of in men to whom decision is unusual or foreign. There really are a good many reasons why I should not make the change and then as I am naturally over cautious about changing it was awfully hard to stick to my decision but I did ——and so that I could not back out of it I went up immediately to see Capps and told him I wanted to go into the dairy.

I found that there is a place in there that will receive me with open arms. Friday morning I make the change. Mr. Craine doesn't know about it yet and probably will not know for a couple weeks. Capps is going to take a vacation and he told Mr. Craine that he wanted me to take his place. When Capps comes back I will stay in the dairy. Mr Craine told me once that if I could get $5 a month more in the dairy to go ahead and take it.

I have told Charlotte that I cannot get Wednesday off because Capps will be away. And then one day a month —-but I will not feel like I am robbing my boss when I take it. I just finished a letter to Mother.

Your brother
Edmund T. Allen

Natoma Farm
September 25, 1915

Dear Brother,
  This is Saturday night and I surely am glad of it. We in the dairy have been working every afternoon for two weeks and we finished this afternoon. Mr. Amacher was well pleased with the work. He has not paid us yet but we are to get five dollars apiece for it. There are a few pipes yet to paint with aluminum but all the White Leading is done. Capps is also pleased. I like him very much.
I have suggested a few little changes in there and he (Capps) has taken them right up. Sometimes he lets me try out something new. But I am overshadowed by Emil Voss. Voss has a way about him ——a confidence in himself ——cocksureness, independence——beside which I look like a 1st grader. I like him and try to help him and profit by his suggestions but he hasn't much use for me. He says when I make a mistake or am not quite as clean as I should be about something, "Your a fine dairyman" or "you make some dairyman" or "Do you call that sanitary or "Nobody home" etc. In a way, that is a good thing because it makes me notice those things that otherwise I might let pass, and surely I never again do anything he speaks about.
  Capps thinks a good deal of Emil though he doesn't tell him about the orders and take him into his confidence as he does me. I am trying to do all of Capps work I can, but there is an awful lot of it and an awful lot of other work that wasn't done before I came in here.
  Mr. Craine is barely civil to me any more. But I won't take offense, and speak to him pleasantly every time I meet him. He has been trying to get a young fellow named Wesley Spaulding in to help him. Wesley has had a little experience around machinery and wants to go in there. He (Mr. Craine) told Wesley that if I came back to work

in the power plant he would tell me to go to Mr. Gensk for another job. Then he asked Mr Amacher to put me back in the engine room until he finished building the new sterilizer. Mr. Amacher said "Nothing doing" and telephoned to Chicago for a steamfitter's helper to put in the sterilizer with Mr. Craine.

There is one thing in the dairy that I have been trying to think out a solution for. You know the bottling room is at one end of the dairy and at the other end is the room where all the bottles are received, washed, sterilized and where the galv. iron cases are piled. Now all those cases and bottles have to be carried from the receiving room to the bottling room and all the wooden racks that the bottles have been taken out of have to be carried back——about 50 feet in a straight line —-75 feet all around——4 steps to mount or descend ——one narrow door to go thru—-several machines ——piles of cases and cans to dodge, etc. now you have never been there when they were carrying down anything but I think it would not be an exaggeration to say that one fourth of all the work in the dairy consists in carrying things from the receiving room to the pit. That means that if this was done away with we could get along with one less man easily. I have thought of a belt conveyor. This is used in large plants a good deal but not in small dairies. Why not? Is it too expensive to operate or is the investment to great? Now it is worth $35. per month to get those bottles and cases moved. Could it be done cheaper with a conveyor? Then these is the point of moving things around so as to have the bottling room right next to the receiving - washing - sterilizing room. This is, of course, out of the question at the present time. It would be very impractical to move the sterilizers at the present time when next spring the building is to be remodeled or rebuilt. Also if the bottling room were moved next to the receiving room the old bottling room could be used for nothing else. But when the building is changed it might be well to fix it so that the sterilizers would be between the receiving and bottling room - that the sterilizers have doors on both sides so that the bottles could be put in after washing and after sterilizing could be taken out right in the bottling room. The bottling room should be on the south with a south and east or south and west exposure full of windows - glass sides - new the refrigerator - cold storage - should join right on to the bottling room so that the cases could be put right into the cold storage immediately after bottling. That leaves just one side of the bottling room - on this side we will put the sterilizers - the washroom - now where are we going to put the pasteurizer so that we can turn the pasteurized milk right over the cooler into the bottler after being pasteurized - where shall we put the separator so that the cream may be cooled and bottled.

Suppose we close up the east side of the bottling room and put the pasteurizer - separator - dump tank - weigh tank - heating tank - on that side. You see the bottling room is the most important room - bottles - cases - and milk must come into there -0 all the bottles must come thru the sterilizer and all the milk thru the cooler. All the milk - in the bottles - in the cases - must go out into cold storage.

This isn't a thing that can be thought out in a day. There are several objections to having the pasteurizing room on the east side of the bottling room. In the first place it would be a great expense to build that room - next - It would not look well - then the farmers milk would have to be put into the dairy on the east side of the dairy - a thing that would not do at all - Mr. Butler would not like to have the farmers come into the yard - now they can go back of the dairy - then it would break up the grass plot in front of the dairy - and last it would cut off the sunshine from one side of the room which it is most important to keep sanitary and sweet smelling.

I guess I've written enough on that subject for to night. I got started on it even if I didn't get very far.

There is a boil starting on my left forearm on the little finger side near the wrist. It is very red and painful but not swollen very much - there is a little white head. When would it be convenient for me to come in to see you. I can come any evening - leaving the farm at 6 o clock or a quarter to, I may be able to reach Chicago before 8. Then I'd come back about 10.30 or 11 - Where and when shall I come -

Your bro.
Edmund T. Allen

[Please return to Mother. Have you read any of my letters? Esp. the one I sent to Canton?]

[This letter shows how Edmund puzzled over situations, trying to solve whatever problem he met; and his letters are often lengthy arguments with himself, first one side then another; trying to find the solution or the truth of a matter, rather than any convictions. I think "openmindedness" but constantly thinking things through was his chief mental characteristic from early childhood.] — MAR

[I think the above short note was written by MAR back at the time of receipt in 1915, perhaps at their mother's request, and the following note was written years later, when she shared these letters with, I assume, Thomas Collison or some other biographer-to-be much earlier than Richard Smith.] — MRB

Natoma Farm, Hinsdale, Ill., Oct., 9, 1915.

Dear Mother:

On Wednesday night I caught the 10:35 arriving in Hinsdale at 11:15. I ran almost all the way out to the farm and got here at 12 o'clock <u>warm</u> but tired and footsore - I washed my feet before going to bed for I had walked about 13 mi. that day.

My suit and shoes came to-day and I put them on immediately. I like the suit very much and the shoes fine. The lisle sox I got are kind of cold. I guess I'll wear two pair at once. My jersey is washed and my sweater & shoes sent down town - That letter is sent off to Uncle D.R.

I bought some pictures from one of the men here who took them. I'll send you one of part of our herd. It was taken from back of the barn - shows about half of the back of the cow barn - silos - hay barn - and in distance the house where we eat. To one side of the silos you can see the top of our new 65 ft. stack.

Friday morning Capps went away for a day and in the afternoon I was talking to Mr. Amacher. He said - "Say, Allen, you'd make a crackerjack dairyman if you went into that line." I said "Why". "Well," he said, "You've got the intelligence - and - then so handy with tools and machinery." I'd like to couple-up with Amacher. He would give me a lot of help and opportunity to make good. You know he is general manager of the farm but <u>especially</u> of the dairy. The other day he told Capps that if Mr. Butler would let him he would take him (Capps) right away from here and put him in charge of Natoma Dairy Co. at Oak Park. That is putting it pretty strong because Mr. Amacher's brother is the head of Nat. Dairy Co. now. if that ever happens there is going to be a vacancy on Nat. Farm. <u>Could</u> <u>I</u> <u>fill</u> <u>it</u>?

26 hrs. later. I have just returned from Church in Hinsdale. I went with a young fellow out here who is recently a product or rather - a recent product of Billy Sunday. He has evangelistic ambitions - has done some work among boys in Y.M.C.A. We arrived late for the Christian Endeavor but enjoyed the remainder of the service very much. They certainly have a live preacher there - He seems to have Christ's spirit - He is fatherly - tender yet firm and forceful - Something like I imagined the priest in Les Miserables. We were heartily welcomed both in Christian Endeavor and Church. The people are <u>so</u> cordial. Perhaps I'd better explain that it was a Presbyterian Church altho that doesn't make much difference. It <u>was</u> a <u>Christian</u> church. (doubly underlined.)

This young fellow's name is Wine. He has only had a grammar school education but can speak well - is not embarrassed. He is a member of Moody Church. He has some queer ideas - I should say convictions because they are certainly deeply rooted - that I cannot agree with at all. But I can admire him in spite of it - and I do admire him for talking to the fellows here the way he does. He is doing some personal work with two of the boys I know of and has got the whole bunch to talking of religion. Of course he enjoys it.

I am certainly going next Sunday night to hear that man and get acquainted in the church.

It's about eleven o'clock so I guess I'll go to bed.

Your Son
Edmund T. Allen.

Monday evening -

Shall I take Rochelle or Epsom salts or a mixture - how much of a dose and how often. - Is a little every day better than - more - not so often?

To-night we tried out some of our new machinery and piping - we tried it with water - The new <u>steam turbine</u> separator is up - the dump tank - heating tank - milk pump and cooler are all ready to be used for the first time to-morrow.

I have been trying to systemize (or systematize) my work, but it don't work very well. One of the boys in the dairy - the one that washed cans - took care of separator - pasteurizer etc. decided to quit the other day. We all thought he was an awfully good worker. - He worked hard and worked from 5 o'clock right up till 12 o'clock. So Capps didn't want to lose him. Voss, -- who washes bottles wanted to change work so Capps changed them and the fellow, Eric, stayed. Voss now has the milk handling job. The first day he had it he got thru about 11:30 - and now he gets thru about 10:00 o'clock - and then has about 15 min. work later on. He does the same work Eric did - does it better - washes things cleaner - and gets thru an hour and a half sooner -<u>system</u>.

Beginning to-morrow the work in our department will be changed all around - because of the new machinery. We will dump all the milk we use, into the new dump tank and pump it to wherever we want it - pasteurizer - clarifier - separator or cooler or heating tank.

I just received a letter from Maurice. He sent the pictures taken on that hike this spring.

Please send my list of high school studies & credits. Don't send the watch. - If I don't wear it, it would be useless to have it out here.

Natoma Farm, Hinsdale, Ill.,
Oct., 17, 1915.
Dear Mother,

I received your card and have been following instructions. The foot and mouth disease has come within five miles of Natoma Farm. At a farm just south of Hinsdale where they thrashed on Saturday, the foot and mouth disease broke out on Monday. Everyone who goes off of Natoma will have to be fumigated when he gets back.

Capps went away this noon and hasn't got back yet. I have done his work. The following seems queer: Before I started in the dairy the fellows were busy in there. Capps talked of putting another man in to do his work - then all he would do would be to wear a good clean suit, a collar & necktie and take charge of things. Now everyone is just as busy as ever and even more so. Capps works just as hard and I do all I can. I wish I could do more of Capps work but just when he is busiest I am busiest. I suppose Thomas will read this so I am going to ask him a question. How can I assure myself that Illinois is the best university or rather 'College of Agriculture'? Now that I have decided to go there I want to think it is the best in the U.S. I want to know it - I should say. Whenever I talk to anyone who thinks Ames or Madison will fit one better for a successful career I wonder if I am doing well to choose a second rate school.

I wrote to the registrar and he sent me a blank to be filled out by the principal of N.D. High School. I have sent this off. I'll have to take an entrance examination in one unit of English Composition - two units of English Literature - one unit of Algebra and three units of subjects to be designated by the university authorities. I'll have to do some studying. What subjects do you think the test will be - a science? Maybe two. A language? Latin??? That's the way I feel on that. History? Suppose I didn't pass the examination. There is a new rule there at Ill. that after Sept 1st, '15, absolutely no student will be admitted on condition. Say! I thought that at a state university there was no tuition for students of that state. Then why the scholarship?

I have been reading up on pasteurization. It is a very interesting subject. There are three things always to be considered in pasteurizing milk - Bacterial efficiency - Creaming efficiency and taste and odor. The idea is to get a high enough temperature with a long enough exposure to get a bacterial efficiency of over 99% - preferably 99.755 and at the same time get a good cream line and have no cooked odor or taste to the milk. There is an article in the Milk Trade Journal that deals with pasteurization in the final package. This, of course, is the ideal way for it prevents infection after pasteurization as the bottles are sealed before pasteurized. This article takes up the different temperatures and exposures - in minute detail. The bottles are dipped into a vat and held there for a given time and then taken out and cooled by a blast of cold air or some other method. The conclusion is that the best results are obtained with a vat temperature of 145° and an exposure of 30 minutes. There is a little trouble experienced in getting the center of the bottle as hot as the sides. I have been trying to think out a system of electrocuting in the bottle. If I could get a conductor thru the bottom of the bottle and then a metal cap in contact with the milk at the top I could send a current thru the milk. So far I have not been able to devise a practical method. There must be some solution to this problem. What is it?

I didn't go to church to-day. It was so wet and windy. Wine went to Sunday School this morning and thoroughly enjoyed it. Did you see the article by Henry M. Hyde in the Tribune this morning? I'm going to cut it out and send it to Uncle D.R.

Your Son
Edmund T. Allen

Dear Mother,

I have finished Pollyanna and The Fat of the Land and have started Pan-Germanism[?]. Will return all three in a few days. That "Fat of the Land" is great. I thoroughly enjoyed it.

We have been very busy in the dairy. Men have been away often and we had to work hard. Voss got a big boil on his neck and I've been taking care of it. I squeezed it out as much as possible but it was hard and I advised him to get some antiflogistine. I put this on and it drew out the core. The opening is about half inch diameter. He has named me Doctor Allen. He has talked of starting in farming for himself. I guess he wants me to go in with him, rent a farm and go to work on it. He says two fellows willing to work ought to make money on 160 acres with a capital of about $2000 to start on - I guess he has the capital.

I got a permit from Illinois U. to enter in Feb. - "conditioned ½ unit elective" also information to the effect that there are no more scholarships to assign this year and that I should not depend on my being able to earn any my expenses the first semester. Urbana and Champaign are small towns and employment is scarce. Voss took two years work at Wisconsin and he says he washed a doctors auto, took care of furnaces, waited on table etc. and made sometimes $12 a week. He said the work was plentiful.

I am awfully sleepy now so I'll close for the present.

Edmund T. Allen

NATOMA FARM
IMPORTERS AND BREEDERS OF GUERNSEY CATTLE
SANITARY MILK AND CREAM

Dec., 8, 1915
Dear Mother,

Somehow or other when I end my letter "more tomorrow" tomorrow never comes, or when it does come something very pressing is more important. The last "tomorrow" I wrote to Thomas instead and the next day I received a letter from Lida. She gave me the name of several men to write to about work, and also gave me a lot of taffy and encouragement. She's a good scout, all right. So I wrote and wrote, letter after letter and the funny thing about it is that each one became more boastful. Evidently I was encouraged by each letter to write more about myself. I wrote again to Dean Clark, to two men connected with the agricultural extension, to the dean of the Ag. College and to the employment secretary of the Y M C A.

Sunday evening is a red letter time for me. I look forward to it all week and then look back to it the next. The preacher is the finest man in Hinsdale and his wife the sweetest woman. The men and women of the church are fine. Last night I spoke in Christian Endeavor for the first time. I thought of something connected with the topic that had not been mentioned and so I got up and said it. The minute I got up I felt all the blood leave my face and my heart began to hit my ribs like it wanted to knock me down. It was quite a while till I got normal again. I didn't know I was so timid or susceptible to be stage struck. I think I'll have to say something every evening until I get used to it. Tuesday evening we heard a lecture on Christian Science by Mr. Sutcliff of Moody Bible Institute. He said that Christian Science, which was neither Christian nor Science, had been rightly called Satan's masterpiece. In another place he said he hated C. S. like he hated the things of hell. He showed all the way thru that Science and Health, with a key to the scriptures, (a key, which he said locked up the scriptures instead of unlocking them), and Mrs. Eddie's miscellaneous writings were contrary and opposed to the Bible. He believed just as truly that these books were inspired by the devil as that the Bible was inspired by God.

The fellows out here have contracted a mania for buying and selling. Every spare minute is devoted to an auction of sale of some kind. The trouble is that everything sells so cheap. For instance nobody will give over one dollar for my sheepskin coat, and the same for my overcoats. The trouble is I have everything to sell and don't need to buy anything.

I got answers to two of my letters. They simply state in answer to my account of my experience and plea for a job, "Your interesting letter…..received. You have certainly had quite a varied experience…It is up to the man himself to find employment, Registration day is the 8th of Feb." I am relying on my letter to the Y.M.C.A. emp. Sect. now.

This letter also states that a conservative estimate of freshmen Ag. Students' expenses is $401.25 and liberal about $475.

Shall I send all of Margarets letters to uncle D. R?

Edmund T. Allen

NATOMA FARM
IMPORTERS AND BREEDERS OF GUERNSEY CATTLE
SANITARY MILK AND CREAM

TELEPHONE 250　　　　　HINSDALE, ILL.　　Dec., 14, 1915.

Dear Mother,
　　I received your letter enclosing Sister's. I will be in Chi.

Dec.,14, 1915.
Dear Mother,
　　I received your letter enclosing Sister's. I will be in Chi. next Sunday soon after noon and can stay until late. If I don't hear from

you I'll come out to Morgan Park. Then I can have a day off in vacation week and can come in any evening, and if she wants Sister can come out to the farm. You see no one of us is entitled to more than his share of "days off".

Last night I went in to Hinsdale to a church social and had a fine time. Coming home I made a date to go coasting with a couple of girls Friday evening. Those folks certainly know how to have a good time decently. Mr. Houghton, the minister, is sick with the grip. We stopped in on the way down to see him.

Mr. Comerford, the bookkeeper, has given me an algebra [book] and I am reviewing it. I didn't know that I had forgotten so much.

I have a lot to talk over with you and Thomas about Illinois. I am not sure that it would be the wise thing to go in February. I can get no scholarship if I go then, in fact I can't get a scholarship if I enter conditioned at all.

Your son,
Edmund T. Allen

Eddie Allen Age 20

University of Illinois 1916-1917

Eddie's first year at the University of Illinois would appear to be a freshman's normal experience. The university is located about 135 miles south of Chicago in Champaign, Illinois. Initially overwhelmed by the social life of a young man who might have been sheltered in some aspects of life, he was 3 years older than his peers. That made a difference. He managed to balance his social experiences with his academics and seemed to enjoy success in both.

This year was also his first meeting with a progressive feminist professor, Allene Gregory, who would later become his wife in August of 1920, although not without controversy within his family. Their marriage would be short-lived. Also he meets Professor Carl Rahn, an associate of Allene Gregory, who would become a close friend of Eddie Allen although this relationship would also be one of some family controversy. Unlike Eddie's marriage to Allene, Eddie's commitment to Carl Rahn and to his sister, Hedwig Rahn would become lifelong.

At completion of his freshman year, Eddie's university life was interrupted by WW I and the need for young men to enlist in the U S Army. Academics became secondary and Eddie was attracted to the Army to fight for his country. So at the completion of his first academic year, Eddie volunteered to go fight the "Bosch". There were no thoughts of aviation when he enlisted.

As background, Woodrow Wilson was president from 1913 to 1921. He stridently kept the U.S. out of the war until 1917 when he asked Congress to declare war. Even then the country was split, but there was growing support for the war. Young Edmund had the patriotic fervor to save France. He referred to Lafayette as his "patron saint". He was 21 years old when he finished his first year of college.

Marshall, Ill
Jan. 31, 1917

Dear Mother,

I have had the best time you ever saw to-day and yesterday. Exams are over and I know I didn't flunk anything & so I feel great.

When I got Margaret's letter to be sure to come to Marshall, I made plans accordingly. Mrs. Price felt awfully bad because I have helped her out so much lately and she so needs my help now. She is inventorying in the library this week, and she has to take care of Mr. Price who is really not able to work, and cook the meals, and wash the dishes, and make the beds, (six of them) and clean the rooms and the whole house, and get to school from nine till twelve and one till five. She suggested several times that I don't go but I decided that I tied my own self down at Natoma and Wheaton and if I did it here it was my own fault. So I broke away.

Mrs. Price is awfully good. She is giving me my meals for taking care of the furnace and helping her with the dishes and the rooms. She said she wouldn't have anyone else eat there but I seem to cheer Mr. Price up when he gets the blues. During examination week when the ones on which we could cram were over we played "500" with Mr. and Mrs. Price every night. Mr. Price enjoys it as much as I do and that's an awful lot for I am learning to play real well and I have lots of fun.

Saturday night, we played till half past eleven, then I took a bath and got in bed about twelve-fifteen. At one-thirty one of the fellows from upstairs came down and dragged me out of bed to go up and take some flash light pictures. It was the last night for Jacques, one of the boys, and they were celebrating. They ate candy and took flash-lights till a little later than one-thirty.

I was supposed to get up at five-thirty to fix the furnace but I didn't wake up until seven-fifteen. It was a real warm morning so it didn't make much difference.

Monday morning at eight o-clock I had my Rhetoric examination. We had to give definitions and examples of a good many grammatical terms, correct and point out the errors of some sentences...

[This partial letter has holes eaten into it, and appears to be only half of the letter, the rest being lost.]

Freshmen at Illinois   (ETA at right)

1005 So. Fifth St., Champaign
February 27, 1917

Dear Sister,
I never was quite so busy in my life, I believe. During the last month I have besides taking 19 hrs and trying to get a continuation of my "above 90" average, besides working for my board, besides going to Sunday school and Church and organ Recitals and B.Y.P.U.[4], I have on Monday night gone out to the mission with the boys club, on Tuesday night gone out after fellows for the YMCA (visiting), on Wednesday night had a date to go to dinner (All the frats & clubs entertain Wednesday evening). Thursday night we have a Y.M.C.A. meeting of a bunch of fellows called the Inner Circle. Friday evening, a lecture by Lorado Taft[5] - (every Friday) and Saturday evening I am almost ready to go to bed early. My studies have gone to the blazes. This week I told the fellows I couldn't be out at the mission, I resigned from the Y.M.C.A. I swore off on dates (for a while only) and I am sticking close to home every night - studying like a good boy. This evening I have been over in the library looking up material for a debate we are to have in a couple of weeks. I got into the congressional records, reference books, gov. documents, etc. and soon felt like I was in the middle of a Christmas pudding trying to eat my way out. I guess I told you I got C+ on my first theme. I got B+ on my second. I just handed in the third. In order to keep up the speed I'll have to get A+, won't I?

But this isn't telling you about the Military Ball. And that is really what I wrote this letter for. I had an awful time getting a cab in the first place. They were almost all gone, and I ordered it over a week in advance. Then I tried to get dances traded but no one I knew was going. Some who were going said they intended dancing a straight program with their partners. I was terribly afraid Mary Jane would not enjoy anything like that so I continued to try to get "trades". I got several leads but at the last minute they failed me and so I went on the floor scheduled for eighteen dances with Mary Jane. (Very sweet of her)

---

[4] Baptist Young People's Union

[5] Lorado Taft - a celebrated sculptor of the time

she said she would rather have it that way. I got on famously, enjoyed myself immensely. She seemed to be enjoying herself. She said she did anyway. There was only one trouble - Mary Jane and I are crazy about waltzes and there were only two on the program, the fifth and the twelfth. We had dinner during the twelfth and thirteenth dances and so we missed that. And then the Peoria orchestra played the waltz for the fifth dance and they played it so fast we couldn't waltz to it. We started out to do it but couldn't get in time to the music. Finally we got out into the middle of the floor and just stopped and tried to get started again but all to no purpose. We made the best of it and one-stepped the rest of the time. Of course, among two hundred and fifty other couples we weren't noticed but we felt awfully embarrassed. I surely do love to dance.

By the tenth dance I got kind of limbered up mentally (and physically). Before that I had felt kind of held down by the formality of it all. Once we went up in the gallery to look down on the dancers and once they put out all the lights except four search lights, which shed on orange, red, yellow, purple and blue and changing light over the decorations and dresses and dancers. I believe it was the prettiest sight I ever saw.

We had the most fun after dinner. Dinner was at midnight. After it everyone was more informal. We had seven dances after dinner, every one of which I hated to see go by for it meant one less. This sounds awfully mushy, don't it? I'll cut it. We got home about two-thirty.

Saturday (next day), I worked all day for Mrs. Price with the result that I was missed at Church Sunday morning. Sunday evening, the two Mrs. Weir were at church. They dropped in to surprise their daughters.

I almost got a job in the library this afternoon. I had left my name, so when they needed someone they got hold of me. Of the ten hours they needed filled, I could only fill seven each and so could not take the job.

Your brother,

Edmund T. Allen

Eddie Allen   Age 22

Army Basic Training/ Pre-Flight Training 1917

Woodrow Wilson's plea to Congress for support of the war in France was met with enthusiastic support by many American youth. Support of the war required enlistment of 400,000 young men in the first year. The support of the French and the hatred of the German actions in WW I diverted many young people to enlist in the military with the sacrifice of their college attendance. Edmund had the fever to join in to fight with the French and since the University of Illinois was a land grant college, the ROTC provided an easy entry. Eddie thought that a commission as an "Army Captain" might be the best fit for him. But first he had to go through the Army's Basic Training program.

The Army Basic Training was held at Ft. Sheridan, Illinois, a few miles north of Chicago. The basic training was an 8 week program which was probably similar to the basic training of today in that it gets the recruits in physical and mental condition and impressed with the basic military skills. Eddie's description of the battle field tactics to be followed by the infantry are amazingly detailed. Following this training, recruits go into specialty training. Eddie found that because of his height and weight he was a candidate for the Army's aviation course. In his letters he is passive in his description of the aviation alternative. He seemed more interested in the potential for a commission and the salary of the different ranks.

His pre-flight training is conducted in classes back at the University of Illinois in Champaign. While he is billeted at the YMCA facility, he finds time for many of his friends from his freshman year at school (including Miss Gregory).

At the conclusion of his pre-flight classes and exams, some candidates are sent on to France (for flight instruction) and some are sent to Scott Field in Belleville, Illinois, for their flight training. While initially disappointed at a perceived failure to get to France, Eddie seems to rationalize that he may achieve greater opportunity to fly at Belleville.

May 27, 1917
Co 6 R.O.T.C.
Ft. Sheridan

Dear Mother,
Today is Sunday and cold as can be. It has been cold, cold, cold for the last two weeks except for a few days. I have only light underwear out here so I am hoping every day it will get warm.

Monday. I didn't get much written yesterday. To-day we have had an awful workout. Everybody is tired out I am just getting over the effects of my third and last typhoid vaccination. It was the worst of the bunch. I had fever; my arm was awfully sore and I also had headache and stomach ache. But it only lasted two days so I am all right again and immune from typhoid for three years. Small pox next Saturday. The worst is yet to come.

As for the effect of the drill it is certainly hardening us up. Bayonet exercise is the most exhausting work I know of. We started out drilling for about one minute at a time. Now we can stand about five minutes. We will be drilling about fifteen minutes at a time before we get thru. To-day we had skirmishing. We were all in skirmish line - about 150 yards of front. The captain gave us the command to lie down. We all dropped as close to the ground as we could lie, in order to escape fire of the imaginary enemy. The captain then told us where the enemy trenches were (about 750 yards distant) and gave us the range so we could set our sights. Then we got the signal "commence firing" after we had loaded our pieces. After we had been firing some time (we supposedly got the fire superiority over our enemy) we got the signal "advance by rushes" and the signal for the right squad (8 men) to rush first. The right squad ceased firing, loaded and locked their pieces, laid their sights, and at a signal from the corporal all jumped to their feet, ran at top speed for 25 yards and dropped to the ground again, resuming firing immediately. As soon as they were on the new line, the second squad rushed and dropped on the new line. All the men not rushing were firing so as not to allow the enemy to get out of their trenches. When all 16 squads had gotten up on the new line, the captain gave the signal for platoons rush. So the first platoon (4 squads - 32 men) ceased firing and rushed another 25 yards. They were followed by the second platoon and the third and fourth. Then the captain gave us all the signal for cease firing and we knew the whole company was going to rush. He broke through the company a second later and led the company for a rush of 50 yards. So we kept on by squads and platoons and company till we got within 50 yards of the enemy trenches. Then we got the signal - fix bayonets. Almost without stopping firing everyone fixed his bayonet. The captain led the charge that followed and we bayoneted every man left who had escaped our bullets.

This afternoon we had formal guard mounting. It is a lengthy and impressive ceremony. We have been studying it in our Manual of Interior Guard Duty for the last week. To-night we study Battalion drill and Field Service Regulation. Everyone here has had Military training or we would never be able to go as fast as we do. Some of the things on our program are: Position and aiming exercise, signaling, semaphore and wigwag - gallery practice, mounting guard, infantry problems - range practice - pistol practice, sketching Army regulations - estimating distances and use of range finder, company in defense and attack preparation of field fortifications, battalion drill Field work in scouts and patrols, battalion in advance guard and attack, Studies in minor tactics Battle fire training - Determination of ranges, target designation, fire distribution, Manual of Courts-Martial, Bayonet fighting, tent pitching. Battalion camp overnight with outpost. Battalion in defense, siting, construction of trenches, communication, signals, transmission of fire data, fire discipline, application of fire, Court-Martial, machine guns, combat firing, rules of Land Warfare, Personal and military hygiene, camp sanitation, sanitary service in camp and combat. Trench warfare - <u>grenades</u>; use of and defense against gas; attack and defense, Billeting - First Aid - Wounds asphyxia etc.

Aug 8 - 11 inclusive will be devoted to maneuvers. After the first month the cavalry and artillery units will leave for other camps. About half the fellows in our company are out for cavalry or field artillery. They will all leave for Ft. Sill or Newport News. From those who are left will be chosen 49 officers 3 majors 15 captains, 16 1st lieuts. 15 2nd

lieuts. I am hoping for a captains job. The company is the real unit of the army and the captain is the most important man in the army. I think my usefulness will be much greater as a captain than as a 2nd lieut. The question is - can I make it? I am afraid that I have little chance of being one of the 18 most able men from 150 mature men all older than I. I am in for it all I am worth. As for going back to school next year, as I see it, the reason I came here to camp for the summer would make me stay till the end of the war. I cannot relieve myself of the responsibility that I must bear in this war, and I am in it till it is over.

    I got into an awful argument with our corporal, a man of 35 years of age. He bawled me out for getting ahead of him in the rushes - He said he was the leader of the squad and I was supposed to follow him. I said we were supposed to run as fast as we could; the corporal was supposed to be the fastest man in the squad and if he wasn't, that was _his_ hard luck. He said if I was in his squad I was going to keep behind him. I said if we were under fire he could bet I wouldn't stand up and be shot waiting for him to catch up - and so it went -

    You won't know me at the end of three months - I'll be like iron and stand like a soldier -

    Your son
    Edmund T. Allen

Co 3 ROTC   Ft Sheridan, Illinois
June 15, 1917

Dear Mother,
    Just a word to let you know my correct address until further notice. I have been transferred to Co. 3. Our company has been broken up, in fact, the whole regiment has been broken up. Half of our company goes to Co. 1, about half goes to Co. 2, and a chosen few to Co. 3. Our company is to be made into the first battery of field artillery. Co. 3 is mostly composed of those going out for aviation. The infantry -.....over in 1 to 5 inclusive.

We get paid tomorrow $19. to the first of June. 3 bucks a day I am earning now besides my room and board. (Better than Mr. Pierce ever offered me).

I have not written to Miss Gregory. Somehow I don't know how to write her. I have started a letter several times and torn it up each time.

I got a letter from Dean Clark the other day.

We have lost our Captain. He was transferred to a Wisconsin company. The fellows liked him so well they are giving him a pair of $75. or $80. field glasses. We took up a collection and the money flowed in.

I heard Harry Pratt Judson last…..here at the Y.M.C.A. I am now ….a French class. We are learning….French and are speeding along … rate.

Studying minor tactics and Army Regulations is our major occupation besides drill now a days.

         Your Son,
         Edmund T. Allen

         Co. 3 - 11th Reg R.O.T.C.
         Ft Sheridan Ill.
         July 2, 1917

Dear Mother,

I don't know what has happened to my mail. I haven't gotten but about four or five letters for a week or so. Be sure to put 11th regiment on them for some of my letters have gone first to 10th Reg.

Lida obeyed my order and came to the dance - she said it wasn't a bid - it was an order. Anyway she accepted and I went to Chicago on the 12:30 and came back on the 1:50. I had a good time. I am afraid I bored Lida to death, however there was enough diversion other than myself to keep her alive.

On the floor I met Della Frazier. I was more surprised than she. She seemed awfully glad to see me and I was glad to see her. Isn't it funny how our wit and spirits rise in direct proportion to the companion. Whenever I am with Della I have lots of fun. As soon as I met her we started a line of wit I didn't think I, at least, was capable. She is much the same that way as is Mary Jane.

Lida stayed for dinner at 5:30 and met some of the boys. We then went down on the beach. We went home on the electric arriving about 10:30. I stayed all night and met Charlotte next morning. Then I left for Drysdale's. I had dinner and supper there and went to church in the evening. I got perfectly disgusted with the Christian Endeavor meeting they had. In fact I believe I would be disgusted with almost any of the meetings I used to enjoy. They don't appeal to me at all. The kind of religion that appeals to me is the kind that is not religious on Sunday. It is not religious all the time either but is never religious. In other words, instead of acting as if every day were Sunday why not act as if there were no Sunday. Why can't we have our religion so much a part of us that Sunday makes no difference with us as to our thoughts and actions. I don't believe I have made myself very clear - Anyway, you never could have told me we were at war or there was a war from all you heard at that meeting. It seemed so atrophied to me to whom the war means everything. I am living, healthy, thinking, the war until it is a part of me- and then to find these folks for whom we intend to fight and die acting absolutely unconscious and unworried

Well Goodbye

Edmund T. Allen

Co. 3 - 11th Reg R.O.T.C.
Ft Sheridan Ill.
July 7, 1917

Dear Mother,

I did get "Mr. Brittling" and also the picture. I find Mr. Brittling very hard to read. I have but few moments during the day to glance at it and then do it at the expense of my other work. However, I have started it. The work here is making greater and greater demands on our time.

Eight weeks have passed. Five more are to come. I don't know what the future holds in store for me. I understand that twenty of the twenty eight who applied for aviation from our company were recommended to be allowed to take the examination. I don't know yet whether or not I was recommended. The insurance man informed me when I took that examination that I had high blood pressure. I am thinking that would be against one as a aviator.

If aviation fails for me I have the infantry. Infantry is really <u>the</u> branch of the service. I would be greatly elated to get a commission there.

We have been digging trenches this week and last. I am enclosing some of the prints. I am also sending some post card views of the camp. (#1) In the birds eye view you see just half of Ft. Sheridan - the Illinois half. Wisconsin and Michigan are in the brick barracks form the top of which this picture was taken. As you see there are four buildings to each company. The lake is off the bluff to the extreme left. (#2) shows our regimental street. The company streets branch off between every building so our company street is where I marked X. The first barracks is unoccupied & Co, 1 is the second. (#3) is one end of one end of the barracks of one company. There are 87 men in our barracks - 160 in our company. The two center buildings of the four are barracks. The picture also shows but half of the mess hall.

(#5) shows one of the company at semaphore signaling. In our company every man signaled. We paired off and each pair communicated with each other. In #6 we are shown starting our trenches. We now have the most efficient system of trenches in the

United States. The engineers lay out the trenches using the white tape shown in the picture. We then take off the sod and dig to a depth of five and a half feet - strengthen the wall with sand bags and cover the whole with sod. We have about one hundred and sixty acres entrenched as it actually is in France with first, second, third, and fourth line trenches. Everything is connected up with communicating trenches, bomb proofs - first aid stations, snipers' positions, listening trenches etc. The whole is so hidden with sod, brush etc that it cannot be seen from an airplane.

We don't have much calisthenics anymore. Trench digging and skirmish work is replacing it, however No 7 shows us as we did it last month.

Several times a week we spend a half day on the target range. While waiting for our turn at the shooting some sleep - some read - etc. Several of us have been spending our time doing battalion and drill in miniature. One of us acts as major and the rest as captains and we actually go through all the movements in the drill regulations. Then we practice signaling or study tactics. In No 8 most of the fellows seem mainly concerned with getting their pictures taken.

No 9 must be medical corps work. I never saw it done in our company.

10 speaks for itself. Col. Nicholson is a wonderful man.

I went to the Winston's three times. The first time was to Sunday dinner. The Colonel did not get there until late in the evening. I went walking with Mrs. Winston and the kiddies. Two weeks from then I went again for Sunday dinner. Thomas came also. We spent the afternoon on the beach. We left soon after dinner at night. Thomas came back to the Fort with me. We walked back. It is only two miles. Then 4th of July Mrs. Winston invited me down to a picnic. The people of Highland Park had invited all who would to parade in the morning. So I went down with the parade and then went home with Tom Winston. The guests arrived one at a time. There were three automobile loads. We went down to the beach to have dinner but they wouldn't let us park the auto there so we went back and ate in the Winston's yard. We had a real good time playing old-fashioned games etc. One of the guests was a Miss Ingram, one of the artists at Ravinia

Park this season. She is one of the best contralto's in the United States. I was much interested in her and she was quite interested in me. She talks French almost as well as English. I had never met anyone like her before and so tried to find out what she was like. Evidently she had not seen many folks just like I was. I don't know just how to describe

Digging Trenches at Fort Sheridan

her. She seemed sort of out of place at the Winston's. Yet she enjoyed it quite well.

Last Saturday the dance was out here at the Post Gymnasium. Lida did not have a very good time dancing but evidently enjoyed all the rest of it. Sunday morning I went to see Drysdale. He lives in Ravenswood - Sunnyside Ave. The folks are on somewhat the same order as the Barr's.

This is Saturday night. I am staying here this weekend. Charlotte and Lida and Ruth are coming out tomorrow in Ruth's car. Tom is coming with them. To-night we have been entertained in the Y.M.C.A. with movies and a talk by Richard Henry Little of the Chicago Herald. I have decided to stay at the Fort weekends after this. I need the rest and can not afford to disperse my energies - mental etc - all over. This is too serious a business here.

We had a lecture this morning on infantry. It took up the history of warfare and the theory and practice of modern tactics. The lecture was fascinating as it sort of connects up all this stuff we have been learning so far.

I now understand that the training period at Rockford where we will go to train the new army will be but three months or less. Then we leave for France where we will be trained behind trenches.

I hope this letter is more satisfactory than last Saturday's.

Your son,
Edmund T. Allen

P.S I got a wonderful letter from Miss Gregory last week. Want you to read it sometime.

Co. 3 11th P.T.R. Ft. Sheridan
July 28, 1917

Dear Mother,

Saturday afternoons are awfully hot at Ft. Sheridan. I'm going in swimming this afternoon or tomorrow morning. We have a fine beach here at the Fort. This morning we had inspection and then a written examination. I decided not to leave this week end. I have already written five letters - one to Lida to Charlotte to Dean Clark, to Mr. Bryant of the Champ. Baptist church and to Bob Drysdale, my roommate last yr. Last night I wrote to Mary Jane and Della Frazer. I kind of think I have done my duty for one day. However I must write to Alice Frazer yet.

I wish you could read some of the letters I am getting these days. They would cheer the old heart of sphinx. Mary Jane's are certainly wonderful. I kind of think our friendship has passed the preliminary stage and we are honest to goodness friends now. Same with Della Frazer. Dean Clark writes so encouragingly and Miss Gregory's letters carry me off for an hour or more into a different world. Friends are wonderful things, aren't they?

This week in fact yesterday the regimental board visited our company - a most auspicious visit. Its purpose was to ratify or reject the captains recommendations for commissions. No one knows what he is to get or if he is to get anything. Our captain has been closeted for the last week going over our records and deliberating on each of his 178 children as to what kind of a piece of pie each will get and who will get the axe. He described it as the hardest task he was ever called upon to perform. About 68 are to be dropped, for there are not enough commissions for everybody. I did not get called before the board. At night the captain sent for me and told me I was designated to go to Champaign next Saturday - Aug 4th for the course in aviation. I will live in the new woman's dormitory which has been delivered over to the aviation candidates. I am to go as a first class private in the enlisted reserve corps - aviation section of signal corps. I get $100 a month till the camp is over, then $36 a month (I think) until I pass the flying examination when I get a commission in the aviation corps as 1st. lieutenant at $180 a month (I think).

My next letter will probably be from Champaign.

Your son
Edmund T. Allen

THE ARMY AND NAVY
YOUNG MEN'S CHRISTIAN ASSOCIATION
HEADQUARTERS:
124 EAST 28th ST., NEW YORK

The Association Follows the Flag on Land and Sea

Public Correspondence Table at Co 3 - 11th Reg. R.O.T.C.

August 12, 1917
Y.M.C.A. Building
[Champaign, Ill.]

Dear Margaret,

This is a birthday letter, Sister, from your aviator brother. This is your second dozen - the last year of your second dozen, isn't it? What does it all look like, when you look back on 'em all? And isn't it great that the best of it all is yet to come. You know, the intenser - the more intensely we live, the more we get out of life. And opportunity is surely coming for us to live intensely. How I hate a vegetable existence - the smug small town life of petty passion, clothes and small talk. Not that a person in a small town cannot live intensely - on the contrary, I think such a person often lives more intensely than one in a big city or in the trenches but the general small town attitude, the society folks or even the stay-at-homes, is typical of what I mean by leading a vegetable existence. A vegetable mind more often goes with a vegetable body than the contrary.

Eleven of us from Ft. Sheridan arrived here last Saturday night. We had a special car on the Panama limited from Chicago. We came right out to the Y.M.C.A. building and were given cots, mattresses, pillows and blankets by the quartermaster. I am sleeping in the big library room on the main floor. Even the lobby is full of cots. The Y.M.C.A. has moved over to University Hall and turned everything over to the U.S. "School of Military Aeronautics at the University of Illinois."

We are still members of the Ft. Sheridan training camp on detached service. After the 15th we will be enlisted as 1st class privates in the A.S.S.E.R.C. - Aviation Section of the Signal Enlisted Reserve Corps and our pay will be $33 per month. There is a bill before Congress to raise it to $100 but no one knows if it will be passed. We were put in "C"

Squadron, skipping "A" & "B". Each week they start a new "A" and move everybody that passes the weeks work up one squadron. This week we will be in 'D' - WE PASS. About half of our bunch didn't pass and will have to take "C" over again. I got 92 in Machine guns - very good in Wireless and passing in Military law and passing in drill so I guess I go on to "D". We didn't have any examination on Theory of Flight.

Our work is very interesting. We are marched to every class and dismissed after the squadron commander reports to the instructor. Everyone makes a rush for the front seats and they are on the edge of their seats during the whole class. At night we study as hard as we ever did at Ft. Sheridan. I can now send and receive about six words a minute by wireless. I can take a Lewis machine gun apart, replace any broken part, fix a jam, explain its workings, in record time. We have learned a lot in a week.

And they don't neglect our physical development either. We have 4 ½ hours of drill every day. This is worse than we ever had at Ft. Sheridan. The drill is very exacting. We march at about 135 steps per minute and if you get out of step it's almost ground enough to put you back a week not to mention throwing you out of school. If an officer passes you three blocks away and you don't salute - either from not seeing him or from not recognizing him to be an officer, you stand a good chance of being kicked out. The close order drill here is very strict. They pay a great deal of attention to the minute details. The drill masters crab all the time about putting more pep into the drill and they do it in just the right way to take all the pep out of us. I have felt an almost irresistible desire several times to swing my rifle around my head and let it fly at him and I know I have muttered under my breath some language that would have shocked even the man next to me if they had not been muttering the same thing. But that's all in the game and I'm not kicking a bit.

After we finish here we go to either Rantoul or Dayton, Ohio, or Belleville, Ill. Or straight to France for our flying training. They are sending the men with the highest grades straight to France. That gives us an added incentive to work, for every man here is impatient to get on the front.

Some of the boys who went from here last May in the Ambulance unit have written about their experiences. Experiences like those bring out everything there is in a man. It makes character. It makes men.

Yesterday afternoon I spent on Crystal Lake canoeing with Miss Gregory. We had a wonderful time. She has commanded me to use her canoe whenever I want to, and to take her along as often as I care to. She is really the most wonderful girl I ever knew in my life. She is an example of one - living, in a small town, very, very intensely. Yesterday was the first time I was ever out with her and I was very inarticulate I guess. Somehow, chatter seems out of place and I am so used to chatter that I didn't know what to say - In consequence I was a very un-companionable companion whose conversation consisted mainly in friendly silences.

Monday - We are in "D" Squadron now - about five of Squadron "C" went on to "D". The rest of "D" is composed of those who failed "D" last week.

We started on engines this morning - learning the theory of gas engine operation and wherein airplane [he had overwritten "aeroplane"] engines are different from others. We still have Machine Guns and Wireless with us but have in addition Rigging - Tools - Ty Machinery and Insts. Soon we will get over to artillery direction from airplanes and photography.

I received Mother's letter this morning. I have a very queer feeling about Thomas going to war. It seems that he could be of so much more value at home and to sacrifice such a life seems a terrible thing. With us here it's different. Most of us aren't worth much to civilization anyway and then it seems so natural for us to go. If something should happen so that I couldn't go, I believe I would - well something awful anyway - I might die of a broken heart. Our place is in France. It is almost our chief end of life - and if that was frustrated - it would be worse than death.

Isn't it interesting to watch the Psychology of war working in our minds? To see how thoughts of life we had but six months ago are all wrong to us now, - To see what a hold the military life has on us - on our desires and ambitions - on our outlook on life in general.

Yesterday I called on Dean Clark and met Mrs. Clark. I enjoyed the call very much. They are interested in everything. I have seen a good deal of the Prices lately. They are very cordial and welcome me as much as ever.

      Best of good luck for your birthday,
         Edmund T. Allen

      U.S School of Mil. Aeron.
      Champaign
      Aug.13, 1917

Dear Sister,

Things have turned out so that I can get a leave and go to Hamilton Saturday. Thomas is going to Hamilton. We want you too. Perhaps you could fix it to get on our train. It leaves Chicago at noon Saturday. I'll get on here. Where could you meet it - or couldn't meet it? Will telegraph anyway just how our plans come out. Exams are awful just now. I headed the class in two subjects, tied for first in another and got 90 - an average - in another. 5 more to go.
         Edmund T. Allen

United States School of Military Aeronautics
Champaign, Illinois
Aug. 19 - 17 -

Dear Mother,

It seems to me that Margaret is one of the last persons in the world who would lead a vegetable existence. Her nature is just naturally opposed to it and her habits of thought make it impossible. When I visited her in Marshall I could easily see that she was living intensely. For instance, she was constantly, perhaps unconsciously, planning how she could improve those youngsters minds, how she could make them think and understand and learn, and how she could make what she was teaching of practical value to them; and all the time she was *setting them an example that tended and did uplift the community. You understand what I mean by "uplift"; not the usual kind, but the kind that gets people to thinking and working out their own problems, especially their problem, the problem of life, and the relation of their individual lives to the larger life, the community life and the State's life - the problem we call Democracy.*

And as for her summer, I'll vouch for it, it hasn't been vegetative. Even Edisons need rest, you know. I can easily imagine the laziest vacation, in which the mind was the farthest from being vegetative. Imagine the kind of summer vacation in which the young lady goes to a fashionable resort, flurries around to get clothes which will outdo the Million buck girl, goes motoring with all the dancing partners, etc. etc. Margaret vegetative? Never!

[written sideways in the margin:] *Hooray for Thomas!!!* [Who wrote that?]

This is some place! I've lost five pounds since I arrived - and I had gained over ten at Ft. Sheridan.

Thank goodness I'm out of "C" squadron where we did 4 ½ hrs drill a day. In "D" we have but one hour a day and I don't find it enough exercise. I have been going over to the Gymnasium in the afternoon. We have lots more studying to do; but it really isn't at all hard. If one pays attention in class, he needs to do very little studying. I hope I don't get put back any. I already have three demerits and five puts you out of school. I got the first for having a dusty room, the second for leaving my book outside when I went to drill and the third because someone on our floor left the water running in the bathroom and flooded it.

The fellows that come back from the flying school on visits are just crazy about it. It has a most wonderful fascination about it. When going to class the other day a machine flew right over us and looped the loop, about 800 feet high. He then did a spiral dive. It is the most beautiful sight you can imagine. And safe! Why the Curtiss training machine which is used exclusively in the camps, is the safest machine made. If the pilot looses control of the machine all he has to do is to take his hands and feet entirely off the controls and fold them behind his back and the machine will right itself. It has stability inherently built into it. On account of this they will not use it on the western front in France. You see if it is hard for the machine to get out of equilibrium itself, it is also hard for the pilot to get it out, and over in Europe they want a machine in which it is easy to make nose dives or tail dives or roll or flutter down like a bird that has been shot on the wing. Maneuvering and fooling the enemy are of primary importance. Of course, only experienced pilots run these machines and in their hands they are as safe as the Curtiss is in the hands of an amateur. You will notice that every one of the accidents that have occurred here have been due to some foolish move on the part of a student. One was caused by a student who attempted to start his own engine by spinning the propeller. He had no one holding the machine so of course when the thing started, it ran over him and the propeller beat him to death. Anyone would have sense enough to know that was a fool thing to do. The others were cases of the young pupil getting crazy up in the air and going out of his head. He gripped the steering wheel and before the instructor could knock him over the head so as to make him unconscious (a common occurrence like that of a man rescuing a crazy drowning man) they had crashed to the ground.

There are now thousands and thousands of men in training at the present time in flying schools thru-out the country and I have heard of but five accidents, a rather small percentage. Over on the western front, the percentages of deaths are: aviation, 5th. Before it are Infantry, Engineers, Artillery, etc. and the infantryman has 14 chances out of 15. (I'm glad I didn't know this until after I had applied for aviation.) We're in the war to sacrifice to get out of sacrifice.

The major asked for volunteers to go to Italy for training and ¾ of the fellows eligible applied.

The bill has passed giving us $100.00 a month while at the flying school. So I only have 5 weeks at $33.00 a mo.

Our work in airplane construction and design and Theory of flight is very interesting. I can't understand it all however. For instance, there seems to be a maximum speed above which they cannot go. A propeller is not efficient above 1400 revolutions per minute and if the engine goes 2500, they have to gear it down to the propeller. The maximum speed on a straight flight is about 150 miles per hour (seldom obtained). The Germans have a Fokker which, they say, will fly rings around any of our machines. They go up to great height and dive directly at you firing with a machine gun all the time. It doesn't seem reasonable that we have reached the maximum. If we could just perfect a machine capable of 200 miles per hour, we would soon have supremacy of the air. Of course sooner or later Germany would get hold of one and copy it and then she could regain supremacy again.

Supremacy of the air means supremacy on the ground. You remember I told you about Ruth, who goes with Charlotte? Her brother has invented an apparatus for mixing Pyrene with the Gasoline [in the] tank automatically whenever the latter is pierced by a bullet, thereby removing one of the worst horrors of the aviators life - being burned to death in a flaming machine.

Our major here is very much exercised least we get to feel that we are officers. He tries his best to make us realize we are the scum of the earth.

I'm enlisted now - for 4 years - (only 4 wks more here)

Edmund T. Allen

U.S. School of Military Aeronautics
Champaign, Ill
Aug 25, 1917

Dear Mother

I am certainly having a wonderful time here. I am renewing lots of my old friends and making new ones and spending two or three evenings a week on Crystal lake with Miss Gregory (Aunt Allene, as she has said I might call her).

She has certainly taken her responsibility seriously for she is a most perfect aunt and I am a most awful bore as a conversationalist. I really believe that just because she promised you she'd take care of me

Eddie and Miss Gregory

She is putting herself out to be good to me. You see, she has told me to call up whenever I have any spare time and we'll go out canoeing. So I call up quite often and then when I think I am bothering her to death and am ashamed to call up again, she calls me up. I spent yesterday afternoon out on the lake and had a really wonderful time. We got out on one of the tiny islands and got to talking and all of a sudden I realized it was 5:45. Well! We rowed to shore and I just left everything and beat it - hurdled three fences and finally struck a road and hailed a passing Ford which got me to the YM. Just as they were playing retreat - two minutes late. Well, I have one demerit left. I'll have to make that do three more weeks.

I was wondering what I would do with to-day (Sunday) and wishing I could go out canoeing again but really ashamed to ask again when Miss Gregory called up and asked what I was going to do Sunday afternoon. Of course, I was delighted to go canoeing again. And we are going. Did I tell you that when the war is over and John and I are back, we are going on a long camping canoe trip. You know, that is the first thing that has made me want to come back from Europe.

I call on the Prices real often. They are going to move next week to a new cottage over on 1st St. They will not have roomers next year. Mrs. Price is going to keep house and Mr. Price is going back to work. He looks real well and feels much better than he has for a long time. He has a good job in the library - not too hard at first. Mrs. Price has a job of playing the pipe organ in one of the Urbana churches. They are much happier than I have ever seen them before. Mrs. Price is certainly a brave, courageous woman and Mr. Price -- . He told me last night that for the last two years he has been ashamed to meet his fellow men on the street. It must have been a hard nervous drain on him to have his wife supporting him by hard manual labor. He says now that he feels like a human being again that he is going back to work. Work is after all the best thing for all of us, isn't it?

Pearl Weir came through Urbana last Monday but I saw her for only five minutes Tuesday morning.

I called on the Gordons in Urbana last Sunday and they seemed very glad to see me. They inquired about you and Sister and Brother. Mrs. Gordon read me one of Bob's letters. He is in Jacksonville now. He tried to get into the army - the officers reserve corps - the chemistry department and everything that he might be worth while in, but, on account of his eyes he was rejected. Poor Bob! How he does want to kill some of those Germans!

There was an aviation dance here last night. I am rather sorry I didn't go. You see, there were no girls that would go with me and so I didn't try very hard to look one up.

Is Sister coming up here before she starts school? I wish she would. I want to talk to her. She could stay at the Y.W. and I could get excused so as to be with her more than I could ordinarily.

I think probably I will be sent to Dayton. Unless I can get very high grades and be sent to France, Dayton is the most likely. At Dayton we will be off from Sat. noon till Monday a.m. and off every night so I can see you lots.

We all got paratyphoid prophylaxis vaccination last week. Everyone has sore arms. They are certainly immunning [sic] us from almost everything.

So long, Mother. I got to eat.

U.S. School of Military Aeronautics
Champaign, Ill.
Sept. 2, 1917

Dear Mother,

Most of this weeks graduates go to Fairfield, and so I expect I will go there too. Two more weeks and I'll know. We start in "G" tomorrow and "H" the next Monday. We have had one final - sending in wireless. I flunked. Then I got to work and worked on the wireless key for about three hours and got 90. I sent eight words in about 52 seconds.

Next Tuesday we have to receive eight words a minute. Then we have an exam in theory of wireless - one in machine guns, one in Airplane Rigging - Photography, Astronomy, Miniature Range, Engines, Motor Transportation, map reading and Reconnaissance, Meteorology, and Artillery Observation.

The finals are an awful nerve strain on the "H" fellows. They are always worried to death over whether or not they passed them.

Starting to-day we get $88.00 per mo. We have to pay $12.00 a month for board. The government furnishes $100.00 a mo and $.60 a day for mess and we have to pay the other $.40, for your mess costs $1.00 a day.

We are still in the Y.M.C.A. building. The new Women's dormitory is not yet completed. When it is done probably this building will be used by the Juniors and the other by the Seniors. They are planning to increase the capacity of this place to 3500 a year - about 75 graduate a week or 500 - 550 here at any time.

Last night I went to a movie with Mr. Price. He is certainly improving fast.

Mary Jane is coming early this year on account of their trouble with the new sorority house. I expect she will be here about four days before I leave but I will be awfully busy with my finals during that time.

My whole life seems awfully self-centered now. I can't think of anything but how soon I can get to France. Hope you'll forgive me.
     Son, Edmund.

U.S. School of Mil. Aer.
Champ. Ill,
Sept. 5, 1917

Dear Sister,
My! You must think discipline is lax here. If I ever left here for a day, the major would write to the adjutant general of the Army about it, not to speak of Court Martial. Desertion in time of war - (death), absent without leave, etc etc

As for getting a leave - absolutely impossible - unless I wrote a special telegram to Prexy Wilson telling him I was deathly homesick and had a lovely sister pining away her heart for her brother's presence.

You see, passes are a rare article to be obtained only by long suffering and extreme perseverance and then only for a few hours - i.e. to be absent for a meal or for one night, etc.

In the meantime all my final exams are clamoring for attention.

I just passed the final in wireless sending by the ------- after I had flunked it twice.

I told Miss Gregory I had flunked and she just laughed - no sympathy at all. She said it just struck her as funny - my flunking.

I just made a new friend - May Ralf. I believe I am going to like her very much. The only trouble with friends just now is that they take up too much time. I am going to cut 'em all soon when the final in Engines comes up.

Less than two weeks more!!! I don't believe you can imagine how crazy I am to get away from here.

Well - so long

See you here soon -

Edmund T. Allen

Champaign, Ill.
Sep 11 - 17
(postmark)
Postcard addressed to:
Miss Margaret M. Allen
Marshall, Ill.
Dear Sister,

I think I may be given a leave of absence Sat. noon. Last week's class was the first one given the leave and they were not notified until Friday noon. If I fail in one little subject I'll have to stay over next week. So far I've passed 8 finals - have seven to hear from. There awful.

Will telegraph Friday to both Tom & you whether we shall all meet in Champaign or Marshall on Saturday.   E.T.A.

Eddie, Sister Margaret, Mother Abby, Brother Thomas

Flight Training and Flight Instruction 1917-1918

Eddie completes pre-flight training on the campus of the University of Illinois and is sent to Scott Field at Belleville, Illinois, for flight training. Scott Field is in western Illinois near St. Louis, Missouri. He is disappointed as his perception was that some of his compatriots were more likely to fly in France before he would have the opportunity. However, he becomes enthused when he sees the "new machines" and facilities at Belleville and concludes he may actually get more flying time at Scott Field.

Eddie finds that "learning to fly" the Curtiss JN4D "Jenny" does not come easily for him. His letters laboriously reflect the trials of a student who analyzes and questions every step. He details each flight and his instructor's criticisms of his performance. He requires 32 flights and more than 8 hours of flight time before he satisfies Mr. Jones, his instructor. He is one of the last to earn the right to "solo" but when he finally achieves his goal, he is recognized by his instructors as one of the best pilots. His letters reflect his frustration but also his persistence, his attention to detail, and his understanding of the mechanics of flight. At the end of his flight training he is selected to be one of the small number of flyers to be designated for the role as "instructor". Obviously, the Army is short-handed when they would designate their best newly minted flyers as "instructor pilots" rather than send them on to fight in France.

Cadet Squadron Barracks
Belleville, Ill.
Sunday night.

Dear Mother,
Arrived here this morning after 48 hours without a wink and went to bed immediately. Slept for 6 hours. Meals are wonderful. Barracks great! We are almost charter members here now and we have 25 machines. Probably won't fly for a week or more. Everything is brand new.
Please take the card and verse out of "The Turmoil" before you return it. "The Turmoil" is fine but I don't quite understand how it applies.
I'll return the suitcase by mail immediately.

I'm all out of pep and so will go to bed and try to rest up. I read all night Friday night and arrived in Champaign Saturday noon. They had our orders right there. We left at 12:45 A.M. and arrived here at 6:30 A.M. No sleep. Sat. night I went to a reception with Miss Gregory and then went home with her and had a good talk. I saw Mary Jane Saturday afternoon.

Goodbye Mother. I hear there's a lot of study ahead of us.
Will write in full later.

Edmund T. Allen

[This letter was out of order and undated. The next letter to his brother was also undated. The next dated letter was to his sister Sept. 24, ruling out Sept. 25 as the Sunday of this letter. Therefore, I've dated it Sept. 18. – MRB]

[scratched out: *"Headquarters Detachment Scott Field, Belleville, Ill]*
A.S.S.C. 1ST Cadet Squadron
Dear Bro,
Arrived yesterday at 6:30 a.m. and after getting a cot and blankets at the Q.M. went to bed and slept all day - Friday night I spent coming in from Hamilton - no sleep - I read all night - Saturday night I went to a reception with Miss Gregory and then left at 12:45 am for Belleville. The train was packed and the bunch didn't sleep a wink, so that when I arrived I had 48 hours of wake to my credit or maybe to my debt for I began paying up immediately.

Everything is so new here it shines. There are only 54 fliers (with us) and 6 instructors. But there are about 25 machines and more coming tomorrow.

Mess is perfectly great after Champaign. Barracks are all right.

I went on K.P. (kitchen police) immediately and so am engaged in washing dishes. But this is for only a week and I will be off when we begin to fly.

There is an order posted to the effect that men must be equipped with leather coat-helmet and triplex goggles or celluloid goggles. I bought a leather coat from one of the men for $18.00 this a.m. and am well satisfied with it. I can probably borrow a helmet but must have goggles. All the men want triplex as celluloid get spoiled so soon. Can you get some for me. I'll pay up to $10.00.

If you can get 3 or 4 pairs - the fellows here are crazy to get them and asked me to try and get them for them.

If you go to Spauldings you might look at helmets and see what they have. I'll have to have one. Some of the fellows have football helmets and like them better than the regular aviation ones -

      Your Brother
      Edmund T. Allen

Cadette Squadron A.S.S.C.
Belleville, Ill.
Sept. 24, 1917
Dear Sister,

My vacation was certainly a decided success. From the time when, burning up with mortification and anger I left the commandants office to catch a train for Marshall, to my return Saturday noon, I was very happy. First of all, I had said goodbye to several folks in Champaign; and you know, goodbyes can be very enjoyable. Then you were so happy when I met you, - and all the time, - that it just scared all the glooms away. Our ride Sunday afternoon was enjoyable if not very comprehensive. I had wanted to talk to you but as it is, we will have to commune telepathically, - or else write letters. The latter is rather uncertain.

The Ralstons are good sports all right. They are most satisfactory people to visit with. I might fly up to Marshall some day and take you and Lucille for a ride - How'd you like that?

In spite of my forgetting to wake up when the Big Ben called and the worry and bustle incident thereto, our leave taking was quite satisfactory for it was short and quick. We arrived in Hamilton and *not seeing any committee of leading citizens to welcome us, we went in search of same (both welcome and leading citizens) and found both at 132 Ludlow St. The cold in Hamilton is speedy as can be, but it never gets away, for Margaret Byard never lets any grass grow under her feet and she grabs one every chance she gets. By Friday, however, she had given most of it away and shed [?] the rest of it.*

We took a trip to Dayton and visited the largest flying school in the world at Fairfield. This field is a very secret affair, no one being admitted without a pass and no passes being issued. Difficulties only exist to be overcome, so we headed for a notary public, got a marriage license to show in case we were stopped and, with my officers cap, uniform, puttees, and mustache, and with my wife hanging on my arm with a bored expression on her face we sallied forth to brave the guards. The first one we came to didn't know whether to salute - present arms or not. I looked at him - ready to return his salute but, as he hesitated, cast a disgusted glance at him and walked on, and, while he was wondering whether or not he had committed a military offense, we got in and it was too late. Margaret carried her part along perfectly and looked as if she wished the airplanes would come down and quit their infernal noise.

While we were there about 25 machines were being used. They were practicing starting and landing. One machine landed almost every minute and there were 8 to 11 machines in the air every minute of the day.

We ate dinner at a hotel that resembled a country farmhouse more than anything close. We washed in a basin after dipping our water from the rain barrel. On the table were two kinds of meat - three of potatoes - three of pie - 5 of preserves etc etc.

We got acquainted with some of the civilian instructors but were unable to get a flight.

I learned how to drive a car while in Hamilton. The only trouble was I couldn't go fast enough. Miss Beardsley owned the car and she isn't a daredevil. So we went along at 30 - 35 miles per hour and had to be content.

Uncle and I had lots of fun giving military orders to the civilian authorities. These last named parties however, did not take kindly to our tutelage and so we had to hold some court martial and executions.

On Wednesday the selected men went away and all Hamilton turned out to the funeral. There was wailing and gnashing of teeth in anticipation of the place they are sure they are going. I went down with Margaret and some of her friends. The girls didn't have any relatives or close friends going so they couldn't weep on anyone's neck but they started to cry anyway - on general principles. - All except Margaret. She didn't cry.

Nice pleasant feeling they send away their warriors with, isn't it? It isn't easy to smile - I saw that when Mother went down with me to the station. I was quite expecting to leave for France and they were going to Chilocothe yet Mother sent me with a smile.

As soon as I got back things began to go wrong. But to digress - Before I left I asked the tactical instructor if it would make any difference on going to France - if we took a furlough. He asked the Commandant and told me he had said, "No." And now I find that those who did not take a furlough were given an opportunity to apply for France and most of them were there because they had flunked and had to stay over and take exams again. So the flunkers were sent to France and we came to Belleville. But I am becoming reconciled to it now, for this place is so fine and new and well equipped that I believe we will get more flying than those who go to France.

When we got here 6:30 A.M. Sunday after 48 hours of no-sleep traveling we got cots and blankets and went to bed. The meals here are as good as at Ft. Sheridan mess. We certainly appreciate it. The only thing we don't like is K.P. (kitchen police) which each one has about every other day. However we get out of drill and have a banquet so it isn't bad.

The flying here is done in the morning. They are short of instructors and so we will have to wait a little while before getting on the flying list. There are 54 cadets here and only 5 instructors. More instructors are coming - we got one yesterday. The commandant wanted to try him out so he took him up this afternoon intending to get into all kinds of sideslips, tail spins, loops and nose dives and make the new man get him out. They had just risen off the field and were making their first turn when they got into an unexpected side slip at about 100 feet and struck the ground before they could get out of it.. They broke off both wings and crumpled the landing gear. They were both mad as could be as they crawled out but it was nobody's fault so they made up and charged up the $5,000 to the government.

This is the first busted machine. I hope our good luck continues so that all the machines won't be busted up before we get on the flying list.

We have infantry drill, telegraphy, and Officers correspondence (military table work). We used to think eight words a minute was pretty hard, but here we must send and receive twenty before we leave here.

In the evening we read poetry. Honestly the bunch read Service and Allan Seger. The fellows certainly appreciate it and we pass the book around.

All the men fly in the morning. The instructors only fly half a day as it is a pretty hard strain on them.

In the afternoon the officers fly. They just go up to keep in practice. They loop the loop all the time. This afternoon they have been stalling. They turn up [dotted line with arrows showing level flight turning up to near perpendicular] till the machine stops. [Dotted line slipping back down - see scan last page.] Stalling. It then slides back and falls forward to a dive of about 500 feet. Then they come out of the dive.

Sometimes they do spiral nose dives and once in a while a tail spin. It is great to watch them.

I hope to go up next week. The first flight is a joy ride. The second the men "follow" the instructor with their hands on the controls. Then gradually they take hold.

The field here is a mile square. Hangars are all along one side of it with barracks just across the road. The cadets and enlisted men are kept separate and are not expected to mingle although some of the enlisted men are college graduates. We are candidates for commissions and so the aristocracy of the army is felt among us.

       Your bro.
       Edmund T. Allen

First Flight - Truly a Joy Ride   October 2, 1917
Start: 9:38 AM
Finish 9:52
Time - 14 minutes
Wind - South- very strong -puffy
Altitude - 1600 feet
Pilot   Mr. Jones
Machine - JN4D

    I experienced a queer detached sort of sensation. I could hardly think that there was another human being in the back seat. The world got smaller and smaller and got farther and farther away. It all seemed wonderfully orderly down below. It was like a large checkerboard. Every field was straight and regular as if it had been planned by a landscape gardener in an airplane. I could see about five miles in every direction. Beyond that the mist hid everything. It looked just like that map of Belgium we used in artillery observation at ground school. There were regular woods and square plowed fields of brown and rectangular pastures and perfect orchards.

    As we got higher we got more and more detached. The motor and plane were the only real things in the world. One imagined the regular motion of the rocker arms on the motor to be the breathing of some great being on whose spacious back we were being carried to — nowhere from — nowhere. The world was far away and had little to do with us. We could imagine that if anything happened to our host, we should not fall but keep floating about forever up there. Even if we should fall we might miss the earth entirely.

    One becomes extremely egotistical and thinks of himself as the only being in existence. The world is far away and matters very little. I thought once of fighting in the air. How useless it all seemed: Why should one man shoot down another? Was there not room enough up here for all? And yet we are on the defensive there for it is the Teuton who seeks to impose his rule, his ideals upon the world.

    And to think that this ride must end. In a few minutes I must go down and walk on that earth far below. Compare this to riding in an automobile! Truly this is the sport royal of the world. Now we dive and I think with regret that my joy ride is over. Realities were not pressed home, however, until we struck the ground and went bumping along on the surface. There are no bumps in the air. The so-called bumps are billows over which we rise and fall as a board rides over a giant wave. Imagine getting off that board on to a lumber wagon. It is the same sensation.

How all the boys envy us. And how mad are those whose names begin with "W". Arnold and I are lucky indeed.

1st Cadet Squadron
Signal Corps Aviation School
Belleville, Ill.
October 2, 1917

Dear Mother,
I have received one letter since I've been here and that was from you. It was appreciated fully.

Please don't worry about the instructors or the machines. The government doesn't want to lose us from even a mercenary point of view. They have spent lots of money on each of us and they don't propose to lose us. Then the Commandant is a man of real human kindness. He sees to it that the men are properly equipped and the machines properly inspected. One man went up the other day without a leather helmet. He was immediately taken off the flying list and given a week of kitchen police. Each machine as it comes down comes to the "circle". A squad of men - "enlisted men" run out and go over the whole machine. They feel of every strut. They sound every wire. They snap each wing. They examine the landing gear and tail skid. This is done after each flight - perhaps thirty times during the morning.

And flying ————— I wish I could take you up. You wouldn't be the least bit scared. Its like riding on the magic carpet. When you come down you wouldn't worry any more about me. I'll take you up sometime, when I know how to fly real well.

I've started a diary. I'll send it to you instead of keeping it and you can send it around if the others would be interested to read 20 years from now.

I can hardly believe I'm really on the flying list. But it's true.

Edmund T. Allen

Did you get the suitcase and book?

Second Flight　　　　　　　　　　　　　　Oct., 3, 1917
Start　　　8:45 AM
Finish　　 9:01
Time　　　16 minutes
Wind　　　SW very strong
Instructor　　Mr. Jones
Machine JN4D Curtiss - No 1216

Mr. Jones told me to take the controls at his head nod. When he nodded I took hold but imagine he also had hold. The ship tipped for no apparent reason and I had an awful time righting it. Twice I almost put her into a stall by pulling back unconsciously. Mr. Jones told me to turn with the aileron following with the rudder. "Turn the aileron till you get the right amount of bank. Then push the rudder slightly. Then pull the aileron back until it tends to come out of the bank. Hold it there. When you get around straighten out the rudder and work the aileron very slightly to straighten yourself. Keep a constant pressure on the right rudder bar. To make left turns release the pressure. To make right turns press harder. Be careful about keeping level. Pick out some point on the engine that is in line with the horizon. [I found it to be the rocker arms]. Hold the machine there". I foolishly watched the aileron most of this time.

Oct., 3, 1917
1st Cadet Barracks
Signal Corps Aviation School
Belleville, Ill

Dear Bro.

I received the letter this morning and the goggles last week. At first I was awfully disappointed because I thought they were plain glass. Then I tried them with a knife and found I could stick the point of the knife into the edge of the lens. Of course it was all right then. And I noticed how clear they are. I became more and more pleased until Major Wheeler, our commanding officer noticed them and came over and asked me about them. He thought they were plain glass at first. (If they had been I would have been put off the flying list and given a week of kitchen police.) He asked me all about them - where I got them, etc., and said he would like to get a lot like them.

Is the glass triplex? I thought triplex was two pieces of glass with celluloid between. These are all celluloid aren't they? The only trouble is that they will scratch easily. Before I realized that I cleaned them with a piece of paper and found I had made a whole lot of very fine scratches across them. It interferes with their clearness a little. Can they be polished when they get all scratched up?

I was put on the flying list yesterday and went on my first "joy ride". I was a most wonderful joy ride. I wrote Mother about it. She will send you the letter.

Do you remember the picture in our old Dante's "Inferno", of Dante' and Beatrice being carried on the back of that big black creature with bat-like wings? My joy ride was like that.

This morning Mr. Jones told me that as soon as we got up he would nod his head and I was to take the controls. (I sat in the rear seat this morning.) Well, I didn't have any time to look around and enjoy the scenery. And when he nodded his head I felt about as confident as an old hen in a lake. And immediately my confidence began increasing -- in a negative direction. Believe me, running an auto is like wheeling a baby carriage to running an airplane.

They used to tell us the Curtiss Biplane would run itself but I was mighty glad Mr. Jones was in the front seat this morning. The only things that bothered me were keeping lateral, longitudinal and directional stability. The old thing rocked like a boat without the slightest provocation and then perhaps the worst thing was pitching and tossing. Before I would know it the nose would be up in the air and almost ready to stall! Then I would shove the control forward and she'd dive. While I would be doing this she would tip over sideways and veer around in another direction. Then when I would want to turn, I couldn't. I would never remember to use the rudder and aileron at the same time and to nose down while doing that.

Then Mr. Jones shook the controls and I let loose and he brought us down. I was expecting to get bawled out very properly but all he said was, "You grip the controls too tightly. I'll get out early tomorrow morning and I want to take you up first when it's calm." He doesn't want to take any more chances with such a pilot when the wind is blowing so strongly. I went up twice this A.M.

The major said it was good for us to get away from here Sundays so I went to St. Louis. I didn't know a soul, so I got on a street car and went to the end of the line. Found myself in Creve Cour, a beautiful park about 16 miles west of St. L. with a large lake and hills and trees, etc. I couldn't get the nerve to try to pick up a girl ("jeune fille") and I couldn't disgrace my uniform by trying the other kind in the cheap dance hall there so I roamed about alone.

Eats are wonderful even if we do have K. P. And the barracks are a whole lot better than those at Ft. Sheridan. Everything is new. We are the third class here. The others are two weeks ahead of us. More later.
ETA

Curtiss JN4d Jenny - 1915

They are flying the Curtiss JN4D, a 2 place bi-plane nick-named "the Jenny" and really the preferred airplane over the alternative Standard J1. The Jenny has a maximum speed of 75 miles per hour and a cruising speed of 60-65 mile per hour. The Jenny is the most widely used training aircraft in WWI and is considered a "forgiving" reliable trainer. More than 6800 were built by 1918. After the war, new JN4D's in the box could be purchased for $350. It became the standard for the "barnstorming" flyer after the war.

The "Jenny" is not considered to be competitive with aircraft being used in the war so it never achieves more than status as a trainer.

Third Flight   *October 3, 1917*
Start   11:15 AM
Finish   11:26
Time   11 minutes
Wind   SW <u>very</u> strong
instructor   Mr. Jones
Machine Curtiss JN4D, No. 1216

Mr. Jones said he would takeoff and then turn the controls over to me. I had a bad time due, I think, in part, to the wind. I had to grip the controls awfully hard to make one turn. I learned later that Mr. Jones had been pulling against me.

I came down with the most discouraged feeling I ever had. I thought I would never make a flier. I couldn't understand how to make my turns. I didn't nose up any more, anyway.

Mr. Jones told me I gripped the controls too tightly. He said he wanted to take me up first tomorrow morning when it would be calmer. Evidently he doesn't want to take any more risks with me in a wind.

Fourth Flight   Oct. 4, 1917
Start   9:10 AM
Finish   9:26
Time   16 minutes
Wind   SW light - misty
Instructor   Mr. Jones
Machine Curtiss JN4D No. 1216

Mr Jones told me not to grip the controls — to handle them lightly, easily. I took the controls when we got up about 1000 feet. I remembered to turn the controls back immediately as soon as I got into the bank. Several times I saw Mr. Jones hands on the cowl but I did not know whether or not he had his other hand on the wheel. Several times he kicked the rudder. When we came down he said, "When making a bank, turn the wheel till you have the right amount of bank. Follow immediately with the rudder and turn the wheel back to neutral. Then when you want to come out of it, you only have to turn the wheel a fraction of an inch"

I am awfully envious of Arnold who started at the same time I did. I constantly question him to find out if he is making better progress than I. And his answers give me ground for envy, for he is evidently making all his banks alone and he did a figure 8 alone this morning.

Oct 4 - '17   Fifth Flight
Start   11:07 A.M.
Finish   11:24
Time   17 minutes
Wind   S.W. strong - puffy.
Instructor - Mr. Jones
Machine - Curtiss JN4D no. 1216

I took the controls and made a turn without any assistance. We struck several vertical air currents and I would always nose up or down to keep her level. Once we dropped straight down and I felt very queer inside. Really I was scared stiff and I didn't know what to do. Mr. Jones suspected it and looked around to enjoy my expression. He was grinning from ear to ear. He turned back, shook the controls for me to drop them, and turned around in a steep bank. Then he gave me control again and I took her around again.

When we got down I asked him if he had his hands on the controls all the time. He replied, "No you were running it all morning. I had both hands on the cowl."

Oh! Boy -

I'm getting crazier than ever about it.

Another caution of Mr. Jones, - "Don't mind the air pockets. Fly right through them. If you drop do not nose up. It may be because of lack of speed so nose down to get up speed."

Oct., 6, '17
Sixth Flight
Start -   8:20 A.M.
Finish - 8:38
Time -   18 minutes
Wind -   S.S.E. light - very light
Instructor - Mr. Jones
Machine - same - Curtiss JN4D

I followed through on several take-offs and landings and then did some figure 8's, alone. The air was smooth and I just sat still and didn't move any control at all - clear across the field. On the left hand turns all I did was release the pressure on the right rudder bar and the machine went up in a bank itself and turned around easily to the left. The right hand turns are harder.

Landings are a puzzle to me. I cannot understand what is so hard about them and yet even Mr. Jones sometimes "pancakes." You cannot tell whether a machine is going to make a good or bad landing until it is over.

Oct., 6, '17
Seventh Flight
A brisk breeze had sprung up.
Start   10:30 [11:30?]
Finish   11:49
Time   19 minutes
Wind   S.E.
Instructor - Mr. Jones
Machine - Same

On my right-hand turns I thought I was banking steeply but Mr. Jones shook the controls and turned her up twice as steep. When we landed he told me I did right hand turns poorly - To use more aileron and not try and do it all with my rudder.

If I could only get that right hand turn better I would be real confident of air work. Then comes landings and taking-off. They're hard.

Headquarters Detachment
A.S.S.C. Signal Corps Aviation School
Belleville, Ill., Oct., 8, '17

Dear Mother,
First to answer your questions - My trunk had not left Champaign so I took it with me. I now have it here at the foot of my cot. My trunk is larger than any of the others here. I may have to get a "regulation" trunk when I leave but I will wait and find out. My red sweater is in

my trunk. Perhaps I had better send it to you. If I get another light one I will not need it. I could not wear it except in the barracks anyway. We were issued winter underwear socks, gloves, breeches, hats, blouses, shirts and an overcoat the other day. We are to get leather coats and helmets but should have a personal coat as these must be left in the machine. I brought a cheap leather coat for $18.00 I had to have it and knew nothing of the proposed issue. I also ordered a $10 helmet and a $7.50 pair of goggles. The issue helmet and the goggles are $12.00 goggles so I do not think I was extravagant. Most of the fellows bought about what I did. They have to have them or borrow them.

Did Uncle David get that suitcase of his that I sent from here by parcel post? I sent it a long time ago.

Flying is going quite slowly. One day it was too windy to fly and one day it rained.

I have gotten some wonderful letters lately. One from Miss Gregory came this morning and I have been spending the day enjoying it. Then I got one from Mary Rolfe two days ago. One from Tom and yours came together and I enjoyed them wonderfully.

Last night I went to a dance at the Belleville country club. The chaperones all wore dress suits and the girls low neck dresses. I learned that you are supposed to ask the girl for a dance when you exchange dances with some fellow. I quite love to dance now. I am getting a little more confident than I was.

Affectionately,
Your youngest.

[along margin] Please send me Mrs. Gordon's initials and address in Champaign.

I have a letter all written to her ready to go. You have her letter.

[Uncle David must be Uncle D. R. Byard in Hamilton, Ohio.]

Oct. 8, 1917
Eighth Flight
Start   8:59
Finish   9:14
Time   15 minutes
Wind   N.W. brisk, cold
Instructor Mr. Jones
Machine Curtiss JN4D, #2016 [since he makes no comment about its being a different plane, I think he meant #1216 as previously used.]

My progress seems rather slow. I made a good right hand turn this morning. The air was very bumpy but Mr. Jones let me handle the machine alone and I kept her level easily. When she would pitch and tip I would sit tight and let her right herself. I don't quite understand Mr. Jones signals. Sometimes he wants me to turn around 180° and I only turn 90°. Then he will jerk the wheel for me to come around some more. Then sometimes he straightens me up at 90° when I want to go clear around. He says very little.

Oct., 8, '17
Ninth Flight
Start   10:35
Finish   10:48
Time   13 minutes
Wind   N.N.W. puffy - strong - cold
Instructor and Machine - same

I made the mistake of nosing down too much on my right hand turns.

I really knew I was nosing down an awful lot but I didn't expect a reprimand for it. My left hand turns were all right!

Once we passed over a dark colored field and got into a strong descending current of air. The machine shot down about fifty feet (not nose down but straight drop). It went down so fast that I was lifted right out of my seat. I felt my safety belt tighten across my stomach; and my feet left the rudder and shot up to the front seat. Then we stopped and I hit my seat again from about two inches above it. It felt awfully queer inside but I sat tight and "didn't lose my calm". Of course fifty feet doesn't make any difference when we're up 1000.

Remember - level up on those right hand turns.

Oct., 9, '17
Tenth Flight
Start 9:29
Finish   9:42
Time   13 minutes
Wind   South - smooth light
Instructor and machine - same

Mr. Jones said, "I will make the first landing. You follow me and then - " the rest of his remark was drowned by the roar of the motor. I followed him through his first and tried to do the others but he kept his hands on the controls and didn't let me have them. I went up to do air work and was determined not to nose down too much on my turns so I watched them carefully especially the right hand ones. When we came down Mr. Jones said, "You must be careful not to climb on your turns. Watch that."

It's always something. I can't seem to watch everything at once.

Oct. 9 - '17
Eleventh Flight
Start   10:54
Finish   11:05
Time   11 minutes
Wind   S. puffy
Instructor and machine   Same

We did air work only. I watched my turns and kept the nose right on the horizon all the time. When we got down - "You bank too steeply. Be careful about that." See? If it isn't my rudder it's my elevator and if it isn't my elevator it's my aileron. I was plumb discouraged and decided to speak to Mr. Jones about it. So I waited till he was through the mornings work and then said, "I seem to be making awfully slow progress, Mr. Jones. What is the matter? Am I inattentive or just naturally slow to learn?"

"O, It will all come to you in time. Just remember your mistakes and think about them and how you would correct them and then watch them when you go up next time."

As if I hadn't been doing just that.

Oct., 10, '17
Twelfth Flight
Start -   8:44
Finish -   9:02
Time   18 minutes
Wind   N.W. very strong
Instructor and machine same

I went up determined to concentrate all my attention when banking to get just the right amount of bank, to keep her level, and to use just enough rudder. The wind was awful. On the right hand turns the blast would almost take the skin off my face.

When we got down Mr. Jones said, "Don't forget you've got a rudder. You didn't use a bit of rudder on some of your turns. That's what the rudder is for. Use it."

It's getting almost funny. I never get everything right.

The wind got stronger so they stopped for the day. It was very cold this morning. But I wonder what we will do when it gets below zero.

Headquarters Det. ASSC
S.C.A.S. Bellville, Ill.
10/11/17
Dear Mother,

I wonder whats the matter with all my correspondents. I write about two letters a day and get one or two a week. You are the only one I can depend on.

Almost everyone in our barracks is flying now and its lots of fun to listen to them all tell of their experiences. We've had all kinds of accidents. One this morning was caused by "pancaking" -landing by just dropping down and bouncing up then dropping again. The tail skid broke off. The fellows in the machine didn't know what was the matter. They tried to taxi i.e. run around on the ground, and it wouldn't work; so one of them got out and found the tail skid wiped off. You see what kinds of things they call accidents. The worst accident since I came here happened yesterday. One of the solo fliers lost his flying speed when about twenty feet above ground. He nosed down to try and get up speed but it was too late and he hit the ground, and, as he was nosing down, he nosed over and turned on the back of his machine. It smashed one of his wings and the propeller. He wasn't even nervous, except for a little fear that he would have to pay for the machine $7000.

The most dangerous place in the air is within 50 feet of the ground for within that distance you hit the ground before you can right yourself. And who would get hurt dropping 50 feet in a Curtiss. There is no position the machine could fall that could hurt you for you sit between the two wings and between the engine and the tail.

I am getting along very slowly. I have had about 200 minutes in the air and have to be bawled out for not using any initiative. Believe me, I'll show 'em tomorrow.

            Your son, Edmund

Please send me Mrs. Gordon's address & initials when you write. Is it G.O.? and Stoughton?

 Thirteenth Flight   Oct. 11, 17
 Start 8:44
 Finish 9:00
 Time 16 minutes
 Wind S. Strong
 Instructor and Machine same

This was my lucky flight. I practiced landings and take-offs. I made two perfect take-offs except for the throttle. Mr. Jones still handles that. I followed on the landings and believe I got the idea of them. Mr. Jones was pleased. He cautioned me not to make too steep banks. Arnold is making landings now. He will be soloing long before I. But I am real confident now.

One of our men, McAteer, started soloing and absolutely failed. He was taken back on the list. I never saw such a discouraged fellow. He was utterly disgusted with himself.

Fourteenth Flight   Oct. 11, 17
Start   10:25
Finish   10:42
Time   17 minutes
Wind   S   *strong*
Instructor and machine   same

It was so windy Mr Jones didn't want to do landings so we did air work. I tried my best to do everything exactly right on the turns - to bank not too steeply, - to use the rudder and to nose down a little. Once Mr. Jones shook the controls and turned up into a very steep bank and gave it all the rudder there was. When we got down he was mad. He said, "You're up there to *fly* the machine. *Fly it.* Don't monkey around with every little puff of wind. When you make your turns - pull the wheel over and shoot the rudder into her. Don't fool around".

Now what do you know about that? Evidently I don't show enough initiative. And yet he doesn't want me to bank too steeply!

Oct., 12, '17
Fifteenth Flight
Start   8:39
Finish   8:56
Time   17 minutes
Wind   west - strong
Instructor Mr. Jones
Machine JN4D Curtiss no. 1221

21 is a little slow on the controls especially the ailerons. I went up determined to show Mr. Jones that I was flying the machine and not letting it fly me. When he would signal a left figure 8, I would instantly pull, almost jerk, the controls over to the left, nose down quite

perceptibly and push on my left rudder real hard. Same on my right turns. Sometimes the wind got under a wing and whipped me around and I would get mad and jerk the controls back. When we landed Mr. Jones said, "That was very good. When you come out of your turns and level off, don't forget to release your rudder because you keep on turning".

Well, at last I got a "very good"

Oct. 12, '17
Sixteenth Flight
Start   10:23
Finish   10:37
Time   17 minutes
Wind   N.W. very strong
Instructor - Mr. Jones
Machine JN4D no 1217

17 is a thoroughbred. 21 stopped and 16 is being overhauled. 17 answers the slightest touch of the wheel. She is as dainty a little machine as you want.

I got up real high and did figure 8's. I banked up real steeply and pulled back on my elevator just a little. She went around corners like a whiz. I felt Mr. Jones hit the controls several times and thought I was in for a bawling out for banking too steeply. Instead of that I had been climbing on my turns. I thought I was nosing on all my turns but that little stunt of pulling back on the elevator evidently don't work. It makes me climb. Anyway I've shown Mr. Jones I'm not afraid of the old buss and I have a little initiative.

Arnold says he is going to solo tomorrow. And he has had only 15 minutes more than I! Slow is me!

Oct, 13, 17
Seventeenth Flight
Start   8:40
Finish   8:56
Time   16 minutes
Wind   South - very mild. Air - smooth
Instructor - Mr. Jones
Machine - Curtiss JN4D No. 1217

We did take-offs and landings. I took off four times with the fault of not climbing steeply enough. On one of my first turns near the ground I banked a little too steeply. The others I did well. On landings I did rather poorly. I leveled off too soon. That seems to be the commonest fault of all. The earth looks like it is about one foot from your landing gear when really it is about fifteen feet off. And fifteen feet is too high to level off. I taxied and used the gun on all but the take-offs. It's lots of fun to cut off the throttle and nose the machine into a steep dive. I have trouble knowing how fast I am going. One reason I leveled off so soon was to slow me up. I thought I was making too fast a landing.

Oct., 15, '17
Eighteenth Flight
Take-off - 8:56
Landing - 9:16
Time in air - 20 minutes
Wind - very little S.W.
Pilot - myself
Instructor - Mr. Jones
Chances for soloing soon - good - about Thursday
Machine - Curtiss JN4D. No. 1221

We got old 21 back. She is the slow old boat. We did landings. I begin to realize why McAteer failed on his first solo take-off. My first one without Mr. Jones feet on the rudder was made this morning, and it surely was a curved one. It went like this

My landings were fair to middling Mr. Jones said, "That next to last landing was a peach. If you can just figure out how you did that do 'em all that way." He said I used too much rudder once but I corrected it.

"Always look out for other machines, and stay far away from them, for a side wind might catch you and drift you right into one".

I was very careful near the ground to make shallow banks and long easy turns.

Oct., 15, '17
Nineteenth Flight
Take-off 10:43
Landing 10:52
Time in air: 9 minutes
Wind - W to N.W. strong.
Instructor Mr. Jones
Machine Curtiss JN4D No. 1221

I dislike west winds for we have to take off toward the hangars and make our first turn right over them. We went up to do air work as it was too windy to practice landings. There were seven machines up and Mr. Jones was constantly on the lookout. Once one of the machines started straight for us. Mr. Jones cut off the throttle and dived, coming up on the other side of him. Traffic in the air is rather funny. You pass to the right or dive under or else tip up sideways. Mr. Jones is very careful.

The motor started missing when we had been up about five minutes and I got the highball to go below so I made a landing. They fixed the motor.

Oct. 16, '17
Twentieth Flight
Take-off 8:11
Landing 8:28
Time 17 minutes
Wind - North
Instructor - Mr. Jones
Machine Curtiss JN4D No 1217

We used 17 again and practiced landings. I think I made some really good landings. I had trouble drifting for I did not head directly into the wind. Mr. Jones hardly made a comment. My take-offs were pretty rough and crooked. Near the ground the air was bumpy.

I've decided I might as well have a patron saint. I have a friend whose patron saint is Shelley. Mine shall be Lafayette. I love Lafayette. He represents what I feel we are fighting for. He was the Savior of America. If I can help save France and the same liberty and democracy he loved so, my life shall have been worth while.

Oct,. 16, '17
Twenty first Flight
Takeoff   9:57
Landing 10:12
Time -   15 minutes
Wind   almost East
Instructor Mr. Jones
Machine same no 1217

We did air work which Mr Jones said was, "pre-tty good" meaning rotten. 17 certainly is a dandy little machine. She responds like I imagine the little speed scouts do. I made the landing when we came down. I dove (dived) almost vertically and slowly straightened out. Mr Jones said I made it too fast. We struck the ground about a half a second before I expected and so I wasn't quite straightened out. I also banked several times on my glide. This is the first time I have done this.

I've had 523 minutes in the air. I ought to be soloing soon.

Tuesday morning (Oct. 16, 1917?)
Dear Mother,

I've finished "Mr. Brittling". I was a little disappointed at the ending. Somehow it seemed a weakness in Mr. Brittling to turn to religion in his trouble. Nevertheless I liked the book. Hugh is a rather unusual character. Wells didn't seem to think him an unusual British youth and yet I haven't found his equal in our bunch. Are we less brainy than the British?

Last Saturday I went with Pierce to visit a friend of his in Belleville. They had a little private party for us and gave us a very good time. We got into a discussion of the draft. Grace Baker, one of the girls became very eloquent in her denunciation of the men who had not enlisted before the draft. All us soldiers - all volunteers, strongly supported the draft.

America is certainly coming around. I believe she will justify our highest hopes. Belleville is an intensely German community in ancestry but they are American in sentiment through and through. I love the little town.

Your son,
Edmund

Hqrs Detachment A.S.S.C.
S.C.A.S. Belleville, Ill.
Oct, 16, 1917
Dear Sister,

I never acknowledged receiving that picture. It certainly is good; don't you think so? It is particularly good of Mother. You are the only one of us slighted.

Our work here is not heavy to say the least, so I am making the most of my correspondence. It's lots of fun too. I am writing to about twenty people more or less regularly.

Did you ever notice how differently you write to different people and how you sometimes write the same thing in two different ways to people who may exchange your letters? And then sometimes you experiment with friendship. One of my friends received a rather unusual and shocking letter last week, I fear. I am having all kinds of enjoyment wondering what kind of an answer I'll get, if I get any. And another of my friends is waxing very serious over religion. She fears, perhaps not groundlessly, that I am an atheist. And still another one is giving me her - her life through her letters. Perhaps she doesn't know it. But what she is giving, is making me live much more intensely, much more really, much more ideally. To another friend I write very endearing letters and I get cold, icy, answers. Then once in a while I write to a friend who is sure to read the letter at a periodical gathering where some people I know might be present. Isn't it fun living in the midst of this world and at the same time being very detached eating and sleeping out in a desert.

About forty minutes every day we – I was going to say really live, but I really live all day long now. Anyway during those forty minutes my physical enjoyment is at a maximum for I am flying. O, it is wonderful. Sport? This is the super-sport! My how I love it.

I sent you Mr. Brittling. He has told me his story. It is well worth while. Hear him for yourself. Your brother,
Edmund

[Oct. 18, '17]
Twenty-second Flight
and 23rd and 24th
Two on the 17th and one on the 18th of Oct.
15, 11 and 11 minutes
All one grand disappointment

I got worse and worse – Mr. Jones was pretty much discouraged about me and I was down and out – My faults consisted in landing too fast – leveling off too high – using too much rudder on my turns – not banking enough – not keeping lateral balance when near the ground – not lifting up my tail enough on the take-off.

Something had to be done so I spent every spare minute the next day imagining I was flying and landing and figuring out just how to correct my mistakes. Finally I would sit on my bunk and imagine I had a wheel in my hands and a rudder under my feet. I got so I would unconsciously correct my mistakes.

I was determined that I had got to get that.

Oct-19-'17
Twenty-fifth Flight
Start -   8:15
Landing 8-42
Time   27 minutes
Machine - 17
Instructor - Jones
Wind N.W.

    We did nothing but landings and take-offs. After about three take-offs with Mr. Jones handling the throttle and me handling the wheel - Mr. Jones turned around and said - "Now you take off." So with one hand holding the wheel and the other on the throttle I took off. It was a job handling the wheel with one hand and I was glad when I got the throttle wide open and could use both hands on the wheel. There was a strong wind and it was hard to make my first turn out of the wind. Once Mr. Jones jerked the wheel over. I had not been banking enough - was skidding. Then it was harder to turn back into the wind - you have to nose way down and bank very steeply.

    On my landings Mr. Jones still helps me out - But Pierson said he did that with him until the day he solo-ed. They didn't charge me with my 26th flight so I will skip it and call my 27th my 26th. I did air work and got a "good" out of it. I can see some improvement - (286 minutes to date).

Hqrs Det. A.S.S.C.
S.C.A.S. Belleville, Ill.
Oct., 20, 1917
Dear Sister,
    I didn't realize until after I had sent it how utterly self-centered and, more than that, egotistical, that letter of mine was. I think probably most of it was bunk anyway.
    Mother told me you folks weren't getting any too friendly with Miss Sass. Thos. and I evidently didn't make a very good impression or she wouldn't feel so superior to the sister of two soldiers. How does Miss

Ralston feel about it? Does the male element (Mr. Harnish) vote pro or con? 'Tisn't a case of jealousy, is it?

By the way, I'll bet your woods and hills around Marshall are beautiful now. There is a patch of woods just east of our field here. This morning I was making one loop of my figure 8 over it and as I tipped the machine up very steeply in the "bank" I could look straight out the side of the 'plane directly down into it. It was very pretty - the leaves had taken on all shades of autumn coloring and from about 800 feet we could just get a general view of it. Really it's like getting up close to a mosaic and then getting off a ways. You get the same improvement in effect from the 'plane - more than that - the sun is shining on the tops of the trees and we are looking down at their tops. I think the leaves show their prettiest side to the sun anyway. And at 800 feet we are just near enough to see the individual coloring an just far enough to get it blended in with the whole. You would really be surprised how wonderful it looks.

The other morning when I went out to the field it came over me how modern our surroundings were. Three planes were circling around above us. Seven planes were on the ground either taxiing around or awaiting their pilots. A mechanic "turned over" a propeller on one of them and the sharp staccato exhaust spit out at us. The engines of the three planes in the air gave out a constant high pitched rapid fire. Two motorcycles with side cars shot past us bringing out the major and one of the instructors. The automobile ambulance came to a stop by our side. Two "White" trucks brought barrels of gasoline out to the field. A steam grader was at work on the road. Off toward the horizon a balloon was coming toward the field. (It afterward landed on the field and five men got out of the basket.) the old was giving place to the new.

Selah -
Edmund

Headquarters Detachment,
Signal Corps Aviation School,
Belleville, Ill. Oct., 21, '17

Dear Mother,

I've got into a scrape with myself. I'm going to tell you about it and ask you what I should have done.

It was this way. Two weeks ago Saturday night I went with Mr. Pierce, one of the older men here, to a little party in Belleville. Four of us cadets went. It was to the house of Dr. Huggins, one of the prominent physicians of Belleville. Our host was Elsie Jean Huggins, just a peach of a girl, quite a musician and eighteen years old. Her mother is of German descent. She still has a very slight accent. She is a very intelligent woman. There were several girls there, all musicians and we had singing and dancing etc., just the kind of a time you would enjoy. I had a very pleasant time so when I got home I wrote a little note of appreciation to Miss Huggins. I made it rather general with just a word for the friendship of all the Belleville people for us cadets.

In a few days she sent Mr. Pierce, whom they know better than any of the rest of us, some fudge, and enclosed a box for me. Then I got the note I am enclosing.

Saturday we were dismissed at 1:30 and almost everyone went in to St. Louis. I didn't know where to go so I had about decided to remain here over the weekend, when someone mentioned that David Warfield was playing "The Music Master" in St. Louis, and this was his last night. I decided I wanted to see him so I phoned in to reserve some seats. The ticket office said the house was sold out by the middle of the week. That made me want to go even more, and I wondered if I couldn't take a girl. Well, Elsie Jean is the only girl I know so I decided to take her.

So I got a car and went to St. Louis. It was a wild ride. We had about fourteen fellows piled in the car, a big Chalmers. I sat one one[sic] of the doors and went through the rain and the wind (and it was raining and blowing) at 40 miles an hour almost all the way to St. Louis, about twenty-four miles. I stopped at Belleville and phoned a ticket scalper in St. Louis for tickets. He said nothing doing - he had been trying to get tickets all week. So I went on in to St. Louis and went to the Jefferson Hotel. They told me all their tickets were gone by

Monday. So I went over to the theatre and argued with the box office man. He finally told me he had two seats returned in the last row on the main floor. He had one or two single seats also. I took the seats in the last row.

I took a street car back to Belleville and got there at 5:30. When I called up Dr. Huggins, Elsie Jean was not in, so I tried to see if I could make arrangements with a taxi man to take us to St. Louis and back. He said there wasn't a small car in Belleville for hire and the big ones would charge $20.00 for the round trip. I didn't happen to have that much money on me so I decided we would have to go in on the street car. Then I called up again and Elsie Jean answered the phone. I asked if I could come up and speak to her for just a few minutes. On my way to the house I began to realize that I was trying an impossible thing, a fact I had been fighting all afternoon. Well, I asked her if she would go with me to see David Warfield. She was tickled and she skipped in to tell her mother and coax her to let her go. Her mother didn't know what to do. She had made a rule that her daughter should not go out with any of the soldiers, yet she wanted her to go and she suspected me of being a pretty decent sort of fellow. She wanted me to go to a Belleville show but I told her I wanted David Warfield, not a show and besides I had the tickets. I'm sure when we first mentioned it, her mother knew she wasn't going to let her go. - Then she said she wished she had a ticket. I urged her to come along - said we could probably get single seats in other parts of the house.

In the end she came along with us. We got to the theatre at 8:10. There were about a hundred people standing in line to buy standing room. I didn't know what to do. I was so inherently selfish I never even thought of giving them my seats and getting standing room myself. While I was wondering what to do, Mrs. Huggins had bought a standing room ticket and was assuring us it was all right, she would find a seat that wasn't taken. We went in and found our seats. I was really ashamed of them - way in the last row. I urged Mrs. Huggins to take my seat but she was firm. She said she would find a seat.

At last I sat down and the show started. O, David Warfield is wonderful. I had one of those rare inspirations we get when we get out alone, quietly under bright stars, or when we come on a scene of great beauty in the out-of-doors, or when we see a great painting, or hear a great musician, or talk to a great personality. And I had the added feeling of being with someone I knew, who was enjoying it as much as I was. I think that's the reason a fellow almost always takes a girl to a good performance with him. I hate to enjoy myself all alone. I always enjoy myself just twice as much when I have a girl along who is also enjoying it also.

But all along I had a feeling of discomfort about Mrs. Huggins. I knew I ought to make her take my seat but I hesitated about insisting on it - and then I was selfish too. I wanted to sit next to Elsie Jean. At the end of the second act, I went out and found Mrs. Huggins still standing. She was enjoying it immensely. I talked to her all the intermission. She still insisted on standing so I didn't urge her further and went back to my seat just as the curtain went up. The last act was wonderful. It was very sad yet full of subtle, tear-compelling humor. Elsie Jean was having the time of her life and so was I, you know, you just love David Warfield. You just want to kneel at his feet and worship him. Really, Mother, I think not near all the divine is in heaven. There's a whole lot of tangible divinity in personalities all around us and I want to worship it wherever I see it. David Warfield had more of it than anyone I know.

We got back at one o'clock and I got to the barracks at two. This morning I told Pierce about it. I said, " —I finally got two seats in the last row. --- Elsie Jean couldn't go without a chaperone so we took one with us - her mother ---"

" -- And where did you sit --- "

Now there you are. It still comes back on me. I feel like a Boer. A real gentleman would be horrified at what I did. It isn't what Elsie Jean thinks of me. I don't ever expect to see her again. At most I won't see her two or three times more. Even if she thought I acted perfectly, I know I was an ass.

I did several things like that at Illinois last year - I learned a lot but what to have done in this case I didn't know - or else I didn't think. What's the matter with me anyway.

Two solo fliers came within ten feet of each other in the air yesterday. The major saw it one of them was taken off the flying list *indefinitely.* [underlined twice.] The other was taken off for four days. Are we taken care of here? Are they reckless? More folks are drowned at Lake Geneva in proportion to the number involved than are killed in airplanes in this whole country - don't worry.

Edmund

(Read - Our debt of gratitude to France - Oct - Worlds Work - please.)

Oct, 22 '17
Twenty-Sixth Flight
Start   7:36
Finish   7:52
Time   16 minutes
Machine No 17 - Instructor - Jones.
Wind South

Our first men to solo started to-day. Gothlin, my old roommate at Champaign, and Hole, both of whom started a week after I did - soloed this morning. O, how envious I am. I did landings this morning - made lots of mistakes but corrected everything on my last round. Some of the mistakes were - landing too fast - not leveling-off high enough caused by over-correcting my habit of leveling-off too high, and last and worst - not completely closing the throttle when I shut off to glide.

Oct. 22, '17
Twenty-Seventh Flight
Start   9:15
Finish   9:26
Time   11 minutes
Machine - same, Instructor - same, wind - same

I'm sure Mr. Jones, after seeing Couch turn Gothlin and Hole loose intended turning me loose. He told me, "Now, taxi up to get a good place to take off, take off, make the round, and land. I don't want to touch a thing." But he had to touch something - first in taxiing, then in my take-off, then in my landing. So he didn't turn me loose. I'll be flying under instruction Christmas! but Arnold isn't soloing yet. My jealousy has that consolation.

Oct 23.
Twenty-Eighth Flight
Start   7:29} tailspin   {7:41
Finish   7:32} wreck {8:00
Time total 22 minutes
Machine same - controls are getting awfully stiff on old 17.
Instructor - same
Weather - N.N.W. wind - strong icy turning to snow

We flew while it was snowing. The first round was made by Mr. Jones as usual. During this round one of the solo men got into a tailspin near the ground and piled up - nothing serious but Mr. Jones was scared and wouldn't let me take the machine around. He hung on to the controls on every turn. He got me again for not leveling off soon enough on landings. Snow made us stop so I got only one ride.

[The tailspin wreck was somebody else's flight, but happened while ETA was up with Mr. Jones, so it interrupted his training flight. Looks like they were on the ground from 7:32 to 7:41.]

Oct 24 '17
Twenty-Ninth Flight
Start   7:42
Finish   8:05
Time   23 minutes
Wind   N.W. mild - instructor and machine same
First round by Mr. Jones as usual.

Second round I made a good take-off. My first turn came over the north end of the hangars. I always hate to fly low over the hangars. The air is always bumpy over them and it is kind of skittish business being only 100 feet over some buildings when the air is bumpy. I tried to keep level on my turn and so since I wasn't nosing down, I banked very little and took about 160 acres to make my turn in. my landing was perfect and I think I did it all alone! On one take-off, after I got up about 30 feet, I felt the machine sort of settle - the controls became sluggish. It wasn't like a hole in the air. Mr. Jones immediately took control, nosed way down and brought us around in a steep bank. I asked him, when we had landed, what was the matter. He said the motor was losing speed and I was trying to hold it up in the air instead of nosing down to get up speed. Moral - Never lose flying speed. When you feel her settle, nose down.

That tailspin yesterday kind of took the pep out of the Major. He has issued an order for every man to be taken up and put in a tailspin by the instructor and required to get out of it alone. He says unless the men get into one they'll never learn it. The instructors have started on the solo men. They take them up about 4500 feet, nose up into a stall, twist the wheel to the right (or left) so as to fall into a right sideslip, and push hard on the right rudder. Before you know it, you are spinning around at 250 miles per hour and headed more or less straight at the earth. It is terribly disconcerting but easy to get out of it if you don't loose your head. The student must get the machine out of the mess. He does it by cutting off the motor, putting his controls in neutral, and nosing down. Then the instructor repeats the performance and by this time the student is crazy to do them himself. Then they do two stalls and some side slips. The tailspin is the most dangerous thing in this game because it is so easy to get into and unless a fellow has altitude and a cool head he is liable to spend a session in the hospital. All you have to do to get into one is to use too much rudder on a turn. The instructors do them every day for exhibition. They are as common as looping the loop. We sometimes don't even look up at them. From the ground it doesn't look bad. The machine is nosed down only about 30° of 45° - not near as much as in an ordinary steep glide, the nose of the machine describes a circle about 50 of 75 feet in diameter and the tail follows around pointed almost straight outward. The machine comes down about 200 feet in every turn. It makes about one turn every two seconds. When they come out of it, the machine stops turning and dives 200 or 300 feet and then straightens out. It is so simple - in fact it is just the same to get out of a tailspin as to get out of any unusual position - always - cut the throttle and with controls in neutral, nose down. You'll come out of anything.

Johnson, the fellow who got into it yesterday said he knew perfectly what he was in but he didn't know what to do so he sat tight and prayed to God. And about that time they pulled him out of the splinters. That's what a lot of us do - sit tight and pray to "God". And that's all the good it does us. Last Sunday a crowd came out from Belleville and saw some exhibition flying. There is some rivalry between two instructors here. Mr. Lewis, one of the younger men, - about twenty-two, - won't admit that Mr. Couch, an old exhibition flier is any better flier than he is. To prove it he got up about 4500 feet and did a tail spin in which he turned 26 times. Then he went up again and looped the loop 13 consecutive times like this: He came to within 100 feet of the ground on his last loop.

Mr. Couch answered by tailspinning to within fifty feet of the ground. That was exciting. The ambulance had cranked up and the wrecking crew was started up when he finally straightened out just above the ground. One of them is going to kill himself before they get through. I know I don't want to go up with either when they're in the fighting mood. We all laugh at them.

I am beginning to realize that flying is mostly instinctive anyway. They keep telling me and telling me how to make a landing, and I couldn't even tell when I was diving too steeply or going too fast or too near the ground. It was all Greek to me for a long time - then I began to feel it. I couldn't tell anyone now how to do it. It's a matter of feeling and only when you get the feel can you do it.

Oct 24 - '17
Thirtieth Flight
Start   9:23
Finish   9:40
Time   17 minutes
Wind   N.W.   Instructor Mr. Jones - Machine 17

I taxied out - took off - and did air work. I tried pulling back on my wheel again in making my turns. It certainly gets you around in a hurry, and it's safe - unless you do it too much and so climb on your turns. I am beginning to get a little at ease, especially when I'm up real high. I can sit back and relax and look around a little. I even look around and watch the tail following us. I am mighty careful though, for sometimes when I turn around again I am climbing or nosing down or not level laterally. It's hard for me, now, to fly straight when I am not looking straight ahead. Lieut. Prevost said to me once, "It's foolish to look straight ahead. When you get over to France you will never look ahead. You will fly always looking in back of you". Since my first joy ride my flights have been tense - nervous - I am getting now a little relaxed. I can enjoy it. But I have had 8 hours!

When we were up almost 1500 feet, Mr. Jones gave me the highball to land so I cut her and dove at about 30°. I spiraled around on my dive and made a landing - a good one. All the rest of the stuff I did was poor. We took off again and made a few more landings. Mr. Jones said, "don't stand the machine up on its ear when you're so near the ground. (Referring to my climbing on my first turn) If your motor should happen to stop, you'd be in a h—l of a fix. Use more rudder on those right hand turns and nose down a little".

I was expecting to solo this morning, and was disappointed. I spoke to Mr. Jones, "Mr. Jones, what's the matter with me that I am so slow learning this stuff?"

"You let me worry about that, will you" - rather irritated. "I'll tell you what I wish you would do - get me a list of your time."

So I got him a list - added up all our individual flight records:

Arnold	525 minutes
Allen	475 minutes
Daniels	457 minutes
Lindaur	93 minutes

I was surprised that Arnold has 50 minutes more than I. He isn't soloing yet - But Hole and Gothlin only had 300 minutes when they were turned loose.

Oct. 25 ['17]
Thirty-first Flight
Start   9:04
Finish   9:20
Time   16 minutes
Wind   S.S.W. The machine - 17 - Instructor Mr. J.

This is the first time Mr. Jones has let me take her around the first time. I did it twice and was quite encouraged. I thought sure I was going -

When we came down the second time, he turned around and said, "be careful about that throttle. When you cut it off, cut it clear off. Don't come down with it half open like you did that last time". Then he went on and bawled me out some more and finally ended with, "I'll take it around this next time and show you how to do it right". So he took it up. Then I took it twice more. When I had gotten down and climbed out I stood on one of the wings as usual to listen to my mistakes. "Allen, you're pretty rotten." - and he went on to show me just where I was rotten - and it wasn't one or two places, either. During my wait for my next flight I just moped wondering if I was ever going to get it. I couldn't remember half the stuff he said I did wrong so I didn't try. I just moped.

Oct 25 ['17]
Thirty-second Flight
Start   10:14   (Then - 10:31
Finish   10:30   to
TIME   16 MINUTES   10:37)
WIND   S.W. Instructor J. Machine 17

I took it around the first time and tried so hard to do everything right. Then I took it around again. When we came to a stop, Mr. Jones turned around and said, "Well, Allen, I believe you can do it. What do you think?"

My heart just went mad. I believe he could hear it above the roar of the motor. - "I believe I can, Mr. Jones".

"Now don't say so if you have any doubts about it because I'd rather go around with you than have you fail".

I didn't have any doubts so he climbed out.

O, boy! What sensations I had when I turned on that throttle with no one in the front seat. I began to get lonesome. Funny! Mr. Jones never said a word to me during my 8 hours of flight and yet when he got out I was lonesome. I never before fully realized that I was running the machine. With Mr. Jones in the seat right in front of me I always seemed sort of secondary even when I had the controls. Now there was no primary and I knew I was running the plane. I made a very conservative turn and a 2-point landing.

Mr. Jones came running out and said, "That was good. Do as well again and I'll be satisfied. Go ahead."

My second landing was 3 point. I got my tail down at just the right moment and touched on all three points at once - hardly bounced a bit. Mr. Jones was as pleased as I. He told me, "If you make as good landings as that all the rest of your life you will live a long time. It was very good". I was happy as a ten year old.

Major Rinehart's airedale always chases the planes on their take-offs. He has had several narrow escapes. This morning he jumped right into a propeller turning about 1500 revolutions per minute. It cost $200. - $100 for a new dog and $100 for a new propeller. It cut the dog in two and the propeller in about 100 pieces.

Now I'm on the solo list and fly in the afternoon. It's lots more fun than in the morning. The lieutenant has charge of us. He calls out, "Cleary, take 22 and go up to 2500 ft and do six figure 8's. Allen, take 18 and make four 3-point landings on the corner of that grass plot. We flew till the moon came out and we could hardly see the ground.

It's more wonderful than ever. If we don't get too many delays, I ought to get my Military Aviator and pilots license by the first of December.

Headquarters Detachment, ASSC   Oct. 27, 1917?
Signal Corps Aviation School
Belleville, Ill.

Dear Mother,

I have made my first solo flight. Did you know that Guynemer[6] - the most wonderful aviator who ever lived - was unable to pass any of the tests when he applied for service? He was a consumptive who had but a year or less to live. He finally got into the service through pull or maybe just kicking through red tape. He brought down unofficially about 150 German airplanes. Officially about 75. But the queer thing about him is that he took so long to learn. I feel quite encouraged.

Since I soloed I haven't been up. The weather has been rotten - rainy-snowy-blowy. Once we solo men went out, but they didn't get around to me. It will be a week since I solo-ed in a day or two. I will have forgotten how to fly!

You know, there is one rule in flying which if observed will make a student perfectly safe: "Don't try any stunts under 1000 feet." That is very conservative but I am going to live up to it to the letter. When up about 5000 feet you can do _anything._

This morning after a day of rain and a night of cold, it has cleared up finally and the men under dual instruction are flying. I will fly this afternoon if it doesn't start blowing too hard.

Our bunch from Champaign are the last men to stay here. Two weeks ago about 20 came from a ground school in Georgia and last week 20 more came from U. of California ground school, but they have all been transferred to France or Egypt. The American government has established a flying school right near the pyramids. All that are left here are the original 34 from Boston Tech. and Austin, Texas, and us - 20 - from Champaign. We will probably stay here until December 1st when we will move south - Memphis or New Orleans or San Antonio or Pensacola - probably Memphis.

Isn't it great that our men are in the trenches already. They were just as crazy to get there as we are. And the National Army will be just

---

[6] Guynemer -French aviator killed in 1917. Credited with 54 downed German aircraft — TKO

as crazy to get there next spring if they are not already. Tell me that spirit isn't wonderful! And to think we are relieving the poor poilu[7] who have borne the brunt of it for 3 years. And some of them are 48 years old! At last we are beginning to pay our debt.

Sometimes, Mother, I begin to realize how much harder it is for you to forgive me, than for me to forgive myself. And yet you will do it, won't you?

We raised $15,850. in our squadron of 54 for the 2nd Liberty loan. I took $500.00 worth and we are all doing it out of our pay!

Your son - Edmund

Oct - 31 - 1917

Dear Mother,   It was blowing so hard yesterday we couldn't fly. So I have not been up yet. I will ask Mr. Hill to go up with me before I try it alone. Mr. Hill is the best instructor out here. I may fly this afternoon.

This morning we got up at reveille as usual at 5:30 - had reveille roll call and calisthenics at 5:45 - breakfast at 6:15. I was on police duty this morning, so I spent 2½ hours sweeping, scrubbing, etc. then I took a bath and fooled around with my buzzer the rest of the morning. This afternoon I went to French class at one o'clock -

Wireless at two - and Paperwork at three - after that I have been watching the planes on cross country work. When I am not on police duty I loaf all morning - or read or practice wireless or write letters and in the afternoon I loaf also when we don't have classes. This is an easy life.

One reason I don't learn to fly is that I never have a chance to go up. I guess they think it is better to go slow and let what they give you soak in before they give you any more. In the meantime the Germans are invading Italy and Russia and France is bleeding. England has learned her lesson and has gotten the good out of this war it needed. America remains. There will be a different America after this war, a

---

[7] "poilu" a French infantryman in WW I —TKO

deeper, sincerer, purer, America; an America with higher ideals and nobler purposes; an America in which all the classes will be knit together and working together toward a common end.

Thursday morning - I went to a Halloween dance last night here at the post. I invited Elsie Jean Huggins. I am inclosing her acceptance. I brought her mother along as a chaperone. We cadets had to go to bed at ten o'clock. It seemed funny to say goodbye to the girls at the dance and go home. They stayed and danced with the non-cadets - (enlisted men we call them) and were taken home in government cars. I had a great time. I'm getting to be a better dancer. I only missed one dance all evening - and there were about twenty more fellows than girls.

Three R.M.A.[8] men finished their cross-country flights yesterday. They only have one more test before they finish. Then they will be laid off and we will get a chance to fly every day again. The cross country was a triangle - Avia to Freeburg to Belleville back to Avia - 30 mile. It seemed queer to see them leave an then return in about twenty or twenty five minutes - Some speed -   Edmund

Headquarters Detachment
S.C.A.S. Belleville, Ill., Nov 4 - '17
Dear Mother,

Here it is a week and a half since I solo-ed and I haven't been in a machine since. We are short of machines. Then we have not had many good days - and on the good days they would push the more advanced men. Seven of them now have finished their R.M.A. tests and are regular licensed pilots. Their commissions have been applied for. We will soon be saluting them. The only advantage of this is that now they are through and the rest of us will be given more time. All but ten of our fifty-four have soloed now. I hope we all keep together and are put in the same corps when we get in active service.

Galli Curci in "Lucia de Lammermore" with Campini directing was given in St. Louis last night. I didn't know about it until Saturday and

---

[8] Reserve Military Aviator

found out that a friend had canvassed the city several days before for seats and couldn't get a thing. So I did not take a chance but stayed here this weekend. Yesterday I went hunting in the woods near here. The quartermaster issues shotguns to us. To-day I went on a hike over to Shilo - and on to O'Fallen and Lebanon across country. The woods are beautiful now. To-day was a typical Indian summer day. Most of the folks around here are friendly to us. Everybody is German descent. Some are openly hostile. One farmer shot at a plane flying over his farm. The bullet split the propeller and buried itself in the radiator - stopping the motor. Lewis had to land. We raid their orchards and woods but usually give them a dollar or two for what we eat. We are always joking about how some German is going to poison us all some time. They would get some valuable government property. This place cost almost $4,000,000. and they are planning to train only us 54 fliers this year. $80,000 apiece not counting our pay, Ft. Sheridan, and all we are going to get.

O, Mother, I wish my life was as useful as yours. You are doing something really constructive. You are indispensable in your work.

Do you know, - Miss Gregory said it was unusual when I told her our family was so enjoyable that I just enjoyed being with either my Mother, Sister or Brother, not because they were my relations but just because of their interesting personalities. We are friends, as well as mother and son, aren't we, Mother?

Don't worry about me - I'm going to have a plane of my own when I come home after the war. Then I'll take you up.

I shudder too when I climb up our water tower.

It's entirely different from a plane.

Edmund

P.S. I took out $10000 insurance yesterday. We got over $500,000 among us 54.

Headquarters Detachment A.S.S.C.
Signal Corps Aviation School, Belleville
Nov., 6, 1917
O, Mother,

What do you think has happened? Well, I'll begin with yesterday. You see I hadn't flown for almost two weeks. Arnold, who has charge of us, soloed a week after I did. We went out to the field yesterday and started down the list. They were starting down the solo list giving all the men a chance at a machine without first taking them up with an instructor - all the men had been off flying a week or more. Now there is considerable silent rivalry between Arnold and I. I know he feels it as much as I do. We started flying the same day and under the same instructor. When he (Arnold) came to my name on the list he told the lieutenant - Olds - that I ought to go up with an instructor first. (He went up alone as did almost all the rest - and of course I wanted to.) So I went up with Lieutenant Richmond. I went wild as soon as I turned on the throttle. Somehow it gets in your blood. It's a regular fever - you tingle to your finger-tips and have the most extraordinary feeling of exaltation in the world. It's tremendous! Lieutenant Richmond is so different from Mr. Jones. He handles a machine like some wonderful chauffeur would a car - with two fingers. It's so easy for him. When I got in the machine he turned clear around in his seat and took a good square look at me and said with a smile, "Who are you?" "Allen, sir". He is just a boy - about 21 - The kind of a boy you wouldn't dare be familiar with; the kind you love [inserted above: and worship] the minute you see him. He and Lieut. Olds and Lieut. Harvey have just been assigned to this command. They are all about 21 and wonderful fliers and wonderful men.[9]

I flew entirely different with him than with Mr. Jones. I did everything with smash and energy and initiative. He would point smartly to the left and almost before he got his finger out there I would bank up sharply and turn smartly. I have gotten on to the science of a

---

[9] Robert Olds would become Major General Olds responsible for the first operational unit of the B-17 Flying Fortress. Health issues would limit his participation in WWII. Coincidentally, he was born and would die in the same years as ETA (1896 - 1943).

turn. It came to me all at once. Now when I bank steeply I use lots of rudder and when I bank little I use little rudder with it. It was very hard for me at first to get the right proportion. Well, the Lieut. cut off my motor and I spiraled in and landed. He taxied up and called Lieut. Olds - the field commander. I left immediately but overheard, afterward, Lieut. Olds telling Major Wheeler about me and found out that Lieut. Richmond had told him I had done wonderful work!!!

Then I got in no 22 and soloed. I find it perfectly wonderful soloing. The two weeks lay-off have given me confidence, I believe, and I fly without any effort at all. My landings are rather poor. I was marked 80 - 85 - 85 - 80 - 85 - 85 on the six I made. Lieutenant Olds said they were satisfactory.

Last night the classes were re-divided and this morning I found I was in the second class! A whole lot of men who soloed before I and who have been flying ever since are in the third class. And poor Arnold is in the third class. The new Lieuts. are putting the thing on a basis of merit and pushing the more able men forward. The first class is about ready to do their R.M.A. work.

To-day I went up just once. I made three figure 8's and three landings on which I got 90 - 85 - and 90! The Lieutenant - Richmond - said "very good work". I never felt so confident or secure. I look over the side and enjoy the scenery below. To-day was a perfect day - not a ripple in the air.

About that tailspin: Johnson wasn't reckless. He made a turn and suddenly his tail whipped around and he found himself tail-spinning. He was at his extremity.

The lieutenants are wonderful fliers. Two of them go up side by side in two machines. Then they chase each other all over the field at 100 miles per hour. And they always play safe - no foolhardiness.

I helped assemble a new Curtiss to-day. I superintended the assembly of a wing - the first time I've had a chance to use anything I learned at ground school.

And I'm getting to be an expert radio operator 16 words a minute.

O, but I'm egotistic - your Son - Edmund

Nov. 7 - 17
Dear Mother,

I am enjoying? a rather singular reputation to-night. "Luckiest man alive" - "Daredevil Allen". "The man whom God loves" etc. and I'm paying for it by missing promotion. I'll tell you all about it.

We went to the flying field at 1:30 this afternoon and were assigned to making 8's and coming in on a glide and making an "S" of the glide. Everybody just wiggled a little on the glide instead of banking up and making a real of it. When I got my turn I banked up about 45 degrees and made a pretty good S. I made four just about the same. The last time I came to a stop within ten feet of the mark. The lieutenant told me it was a very good exhibition but I had made several mistakes. Then he said, "Don't be afraid to bank more than you did." I said - "Well, if I go over 45 degrees I have to use my elevator for a rudder and my rudder for an elevator, don't I"?

"O", he said, "That's all right, all you have to do is to pull back on your elevator when you want to turn."

On the next time round everybody did better and really banked up and made a very appreciable S.

The lieut. said, "Allen, here, is the only one who has banked much and he didn't bank enough". He marked me 90 on my stuff, though.

When my turn came I decided I was going to do a 90 degree bank, i.e. the machine instead of this position flies in this position around a turn.

I tried on my first turn and came around very fast and beautifully. So when I came to glide in, I nosed over steeply and banked up to 90 degrees and turned around in a grand S. Then I turned back to straight and banked over to 90 degrees in the opposite direction. Then I found I was very near the ground and was side slipping and couldn't figure out just why I was side slipping. I thought I had got mixed up when I changed my elevator and rudder over to rudder and elevator. All this time I was trying desperately to straighten out and was having an awful time. This was another puzzle. I finally straightened out just above the ground. The next think to do was land. In order to control the machine I had to shove the controls way over (two feet!!) or turn them clear over. It usually required but two inches to entirely change direction. I wasn't a bit worried. I was too busy trying to get my tail down at just the right instant. I did it and made a beautiful landing. Then I gave it the gun and took off again. This time I made another 90 degree bank on my first turn and learned a little about the use of my rudder on it. I glided in about the same as before and as before had a h----- of a time getting straightened out on my landing. Then it all came to me - I had no flying speed. Of course the controls were sluggish. On my landing I suddenly found myself 15 feet above the ground without any speed and my nose pointed upward. Almost without thinking I shoved the controls <u>clear</u> forward with as much force as I could. Then instantly I pulled them clear back to my stomach. And I just <u>touched</u> the ground - beautifully. This was all that saved me and the machine. My third and fourth rounds were practically a repetition of the first and then I forgot how many times I had been around and went around again. Each time I had an awful time landing and I made a beautiful landing. I taxied up to the lieut. He didn't wait for me but came out and before I could get out of the machine said, "Allen, you deserve a h ----of a lecture, I never expected to see you alive again. You came nearer to a smash-up five times than you will ever know. But those were beautiful landings and you deserve a mighty lot of credit for them. You certainly handle the machine wonderfully - only why in God's name do you come out of those turns with your nose up in the air? Those turns were great - only you always lost your flying speed on them. You handle the machine beautifully but for that. <u>I was going to recommend you for RMA right away</u> but now I'll hold you over for couple of days. So I lost my chance for promotion!

    You see, I didn't realize that I was coming out of my turns with my nose high. I was wondering why I lost flying speed. I knew that I had nose over to begin with until I had about 100 miles per hr. and three seconds afterward I had only 50 miles per hour. The answer is that those 90 degree banks absorb speed as it were. So I should have straightened out and nosed over again to get up speed before I banked up in the other direction. What I did was perfectly correct except for about a 10 degree difference in glide. Now I know about it I can do it in perfect safety.

The funny thing about it is that to all but the lieutenant it looked like very very advanced exhibition work.

Everybody in the barracks crowded around tonight like I was some great hero. They told all those who didn't see it about it and it certainly sounds funny. "We all turned white and stopped breathing." "Everybody on the field gathered together in a prayer meeting". "Williams turned over with his face to the ground and said, 'Tell me when he's down' ". "That was the prettiest exhibition I ever saw."

I just talked to one of the RMA men who was standing by the lieut. He says, "You did all right, and the lieutenant knows it. I'll tell you your mistake - flying speed.

I suppose this is very dry to you but it is intensely interesting to me. I am really about two weeks ahead of most of the other men in my class and except for this mistake would be put on R.M.A. tests tomorrow.

I'm sending you my Kodak. We have to send them out of the post. Just turn it to No. 2. I guess No. 1 is spoiled. I have left it turned a little past No. 1.

I'm just wild about flying.

Your son,

Edmund

Nov. 14, 1917

Dear Brother,

I received you and sister's letters and forwarded them to Uncle D.R. with my approval. The only doubt I had was whether sister wanted to go to Rush or not. If she wants to go nothing shall stop her. I can help quite a bit as soon as I get my commission - which will probably be before 1918.

As to the initials, A.S.S.C. stands for Aviation Section Signal Corps - which is what I'm in, and S.C.A.S. stands for Signal Corps Aviation School which is where I am at. It has recently been changed to A.S.S.C.N.A. which means we have been made part of the National Army. Bu we don't care much. It doesn't do any good to care.

Yes, I soloed; and then for two weeks didn't get up. Bad weather and hustle to "give to him that hath" kept me out of the running. The more advanced men were pushed and since we had broken up all the planes, we were in the soup. I was the third man in our bunch of 20 to solo. But of course I had a head start on most of them.

When I got back to flying again I did better than I ever did before. Lieutenant Richmond took me up once and then turned me loose. Then I did some rotten landings and some air work. This started me in the third class. The next morning I found I was advanced to the second class. That day I felt awfully reckless. The lieutenant was cussing about the fellows not making steep enough banks. When I got up I executed some stuff that still keeps the fellows here smiling broadly whenever they pass me. I banked that old boat up vertically on every turn and finally recovered just in time to make a landing. When I got down Lieutenant Richmond bawled me out terribly. Ever since then he has been my best friend. He told the commandant I was the best little flier in the class and I was immediately jumped to 1st class.

Then we started R.M.A.[10] tests. The first consisted of making a 180 degree turn on a glide and landing to a mark. The next was a right and left turn of 90 degrees each to a mark (on a glide). Then we cut the switch at 300 feet altitude and landed to a mark. The next was an altitude test of 1000 feet. This morning we continued with a climb to 1000 feet, killed the motor, and come to rest within 200 feet of a mark. Each of these I had to do twice before I completed them satisfactorily. I was getting discouraged but I had a strong stand-in with Lieutenant Richmond who for some reason or other has taken quite a liking to me, and he would encourage me or pull me out of any trouble with the commandant. I certainly admire Lieut. Richmond. He is a wonderful flier - about 21 year old - young yet dignified - not democratic - rather the justification for aristocracy.

The next test was landing over an assumed obstacle 10 feet high and come to rest within 1500 feet. This is valuable as practice for landing in a field 1500 feet long with a line of trees at one side. Next was the climb out of a 2000 ft. square field to a height of 500 feet. Then came figure 8's around pylons 1600 feet apart.

---

[10] Reserve Military Aviator

We have one more to do yet. We tried it this morning but the clouds were so low we couldn't get up more than 1200 feet without getting all bawled up in them. It is to cut your motor at 1500 to 2000 feet and glide down making a complete right and complete left spiral and coming to rest at a mark (i.e. 720 degree turn). I did it from 1200 feet when I wanted to land after I had been up for a joy ride but it wasn't counted. Then come our cross country flights. One triangular 30 mile flight - Scott field - Freeburg - Belleville and return and one 40 mile to Salem - land and take on gas and return.

The first fellows - eight of them - who finished these tests two weeks ago have left for France (last Monday). So I imagine I'll be here just about three weeks more. I _may_ get my commission before I leave.

After we got thru our tests this morning he sent each of us up for half an hour to do whatever we wanted to. I just had all kinds of fun - stunts and steep banks and glides and stalls. Loops and tailspins are barred or I'm sure I'd have tried them. On the glide in, as I said, I made some pretty right and left spirals. The lieutenant told me they were pretty. Then he sent us up in formation. He told me to take No 1, McAtler, No 2, etc. We had six in the formation like this:

```
1
    2 and 3 were about 75 feet              3           2
    above 1, and 5 and 4 - 75 feet      5                   4
    above 2 and 3.                                              6
```

The lieutenant told me to manage it. We all got lined up on the field in formation. Then I took off. The others followed at 5 second intervals. As soon as I got around once, I throttled down to wait for them. Then we started off at full speed. I tried to keep low so that the higher men would not have to fly in the clouds. I flew with one hand on the wheel and the other on the throttle or out the side signaling to the men who got behind or too far ahead. I very seldom looked ahead. Most of the time I was looking behind watching the others and trying to keep them in form. It was some job. First No. 2 would shoot ahead then No. 3 would get way behind. We were traveling about 100 miles per hour so it only took a fraction of a second for one machine to reach the spot another had left. I throttled down most of the time because the last three men 4, 5, an 6 never caught up to us. I never had such real sport.

You asked once what an aileron was. Remember how we used to hear about the Wright Brother's warping their wings? Well, on these machines they have a little flipper on the back of each wing that is hinged on to the wing, and instead of warping the wings, we move the flippers or ailerons. The control mechanism that works the ailerons is a regular automobile wheel. When you turn it to the right it flops your right ailerons up and your left ones down and so banks you up to the right (lifts the left wing) - like going around a bank on a race track. If you tried to turn with the rudder, without banking up, you would skid just as an auto does on a wet turn.

The yoke on which this wheel is mounted is also pivoted so that you can pull the wheel toward you to point your nose up or push it away from you to nose down.

Today we started instruction in machine guns. This is supplementary to what we got at ground school. We are mounting the guns on planes and are going to practice with them at targets towed by other planes or at captive balloons. We are also equipping a plane with a wireless set. Our new instructor who is just back from Canadian training schools told us of a machine - obsolete he called it - which climbs to 19,000 feet in 15 minutes - forward speed 145 miles per hour. He says German planes now go 175 miles per hour. Some speed!

When I get my commission and get to France I'll get $2800.00 a year.

           Your Brother, Edmund

Nov. 14, 1917
Dear Mother,

 Yesterday I had a very discouraging time. We did R.M.A. tests and I had to do every one of mine twice before I got it satisfactory. The most interesting one was coming down on a glide, just clearing a rope stretched between two poles ten feet high - representing a fringe of trees or a line of telephone poles - and coming to a rest within 1500 feet - supposing the field to be only 1500 feet long. The first time I cleared the poles 75 feet. I wanted to play safe and came to rest at about 1000 feet. The lieutenant wasn't satisfied. He said, "If the field was only 1000 ft. long you would have smashed into the fence at the end of it." My second time was all right. Another test was climbing to 1000 feet - killing the motor and gliding down to a landing within 299 feet of a marker. My first time I shot way over the mark and stopped 1000 feet away. The second time I stopped 50 feet from the old mark.

This morning we tried a new one - climbing to 1500 feet and cutting off and making a complete right and a complete left spiral and landing to a mark. The first man up ran into the low hanging clouds at 1000 feet. He immediately shut off and dove out. So we had to abandon that test till it cleared up.

The next test was climbing out of a 2000 feet square to an altitude of 500 feet. This was easy. Only one man got outside the square and had to try again.

The lieutenant then sent us up for half an hour to "have a good time". I did 90 degree spirals, glides, stalls and everything I could think of. It surely is fun. When I came down I made a pretty double right and left hand spiral and came to rest on a mark.

As soon as the last man was down he told us to go up in formation and told me to lead the formation. Why me - when Hole and I are the only ones of the new men who are in this advanced class and about 15 of the old men are in the lower classes? I guess that spiral I had just made caught his eye. He certainly is good to me now. Every morning when we go out to the field he makes a point of coming over to me especially and saying, "Well how are you this morning, Allen?" I guess he is so glad I got out of that mess I made last week alive that he treats me especially nice - like we imagine we would treat a dead person if he just hadn't died. The lieutenant told us just how to fly. Each man had a number (I was No. 1) and had a corresponding place in the formation. We got out on the field in formation and took off at 5 second intervals. After I had rounded the first turn I throttled down and waited for the rest. It was great sport. I was flying with one hand on the controls and the other out the side signaling to one of the other the men back of me to hurry or slow down o go higher or come lower or move to the right or left.

I very seldom looked ahead. All my attention was fixed on keeping the formation intact. Once in a while I would glance around to see if I was running into a plane but the rest of the time I was looking behind me. The speed of the machines with reference to each other is 200 miles per hour when passing, so it doesn't take long to shift positions. It was some job keeping everyone in place. When I stuck my arm out to the side I could hardly hold it straight, so great was the wind pressure. Tomorrow we may do some more formation flying.

If it clears up tomorrow we will first do our 2000 ft right and left spiral and then our triangular flight and maybe our cross-country. Then I'll be an R.M.A.

Did I tell you - the eight men who finished this test two weeks ago left for France last Monday? So we may leave in a few weeks. Isn't that great! I hadn't hoped to get to France before Christmas.[11]

We have started machine gun practice. They are mounting the guns on our planes and teaching us to fight in the air. This surely is getting interesting. Soon we will have wireless on the plane and machine guns too.

Have you read about the new direction finder that works by wireless? They are planning (actually) to establish a wireless station in the Azores and fly all of our machines across the Atlantic using the "finder" to steer the course straight toward the wireless station. My! I'd like to be in the first bunch to fly across the Atlantic.

I am having a wonderful time these days - hunting - playing football - studying, flying. I am very happy.

Margaret is having a rather unsatisfactory time this year. She should do something else. I quite approve of her going to school again. What does she want? I'll write her tomorrow

                Your son
                      Edmund

---

[11] In the class of 12 cadets preceding ETA's class, which were sent to England to complete training, 7 were killed while still in training. This apparently is why he wrote his mother and requested she give away everything he owned. The average life of a flyer was 40 days.

Nov - 19 - 1917

Dear Mother,

Friday was a poor day for flying. I got up in the afternoon in a Standard and had lots of fun flying it. The Standard is not nearly as alert and easy as a Curtiss. It's lazy and awkward, but as a novelty it is interesting. I quite like them. It was very misty and so - hard to see the horizon - and, you know, we fly by the horizon. I had about an hour flying that day.

Saturday it had cleared up a little so we took our long cross-country flight - 70 miles. We landed at O'Fallon. The flight was very interesting. I will send you the map I made and followed. Isn't it queer how soon an event loses interest - If I had written this letter Saturday evening I would have written pages about that flight. But now it seems so commonplace. I have been over the course about five times to-day and it is a perfect cinch. When I got back to camp I was about 4000 feet up. I decided I would have to do some stunt. It is a shame to waste 4000 feet in a straight spiral. I decided on a stall. I had never done a complete stall before. I opened the auxiliary air valve and shoved the throttle clear forward. Then I nosed the machine up - and up she went - up, up, up. I thought I was going to loop. Soon the motor began laboring so I cut it off and for just a second hung there pointed straight up. Then the machine started to fall, and O, how fast it can fall. It just pulls you right down with it. While it was falling it nosed down and soon I had lots of speed in a straight nose dive. I had dropped about 1000 ft. in my stall. Then I went into some steep spirals - six of them before I was low enough to land. When I had taxied around to the place where we park the machines - Lewis, one of the instructors came over and bawled at me for "landing on top" of him. I had never seen him at all - He was directly under me when I glided in - not ten feet apart! Anyway I had finished my last R.M.A. test.

Saturday afternoon I went to St. Louis and bought my insignia. It cost me $9.81 - I got two embroidered (?) pair of wings with a shield and a gold U.S. This is sewed on the coat over the left pocket. Then my lieutenant bars and signal corps emblem and officers hat cord completed my set. I cannot wear any of this until I get my commission - about the middle of December.

Sunday evening I went to Belleville with Ellis and met a St. Louis girl - Dorothy Winkler who is a peach. I really had great fun. We danced and smoked (I didn't smoke) and talked about how we were going to give "Dot" a ride. I told her I would tie a white handkerchief to my helmet ring and fly low over her house. She was then to run out to a nearby vacant lot where I would land and take her in. I would bring an extra helmet - leather coat - goggles etc. Dot took it half seriously. She is just crazy to fly. This morning I got a Standard about 8 o'clock and flew till twelve with a few stops. I went to Belleville four times but couldn't find a suitable field so didn't land. I took a mechanic with me each time. They are crazy for a ride and now we can take them. I put one of them into a steep spiral that made him wild. He clutched wildly at the cowl, the scardest kid that ever was. The next one was a man of about 25 years. I decided to get right over the Belleville square and make two steep turns in a spiral. On the second turn I didn't use quite enough elevator and started to spin like a top. It didn't take me but an instant to realize I was in a tailspin. We were nosed directly at the ground and traveling about 200 miles per hour all the time spinning. I worked the controls desperately. There was no response. It seemed like some great power were drawing us down to certain death. The centrifugal force produces the most extraordinary sensation. That poor mechanic was terrified. Thank Heaven he didn't turn around and look at my face for he would have died of heart failure - for - I was as terrified as he! I think that fact bothered me as much as the tailspin. I had always thought that when danger faced me I would become calm - but here I was - so scared I couldn't think. I just put my head down inside the cowl and tried to figure out what I ought to do - all the time Belleville was getting nearer at the rate of 150 feet per second - and I wasn't more than 3000 feet above it. I moved the stick to each corner of the cockpit as far as it would go - nothing happened. I felt perfectly helpless. Then I remembered the rudder. It should be straight but it was crosswise - the left hand side way forward. I tried to push the right hand bar forward - It wouldn't budge. I put both feet on it and shoved. Then I remembered that the throttle should be closed. Here my engine was wide open and drawing me down all the faster. But just at this instant the ship straightened out - about 500 feet above the square. As soon as we were leveled out the mechanic turned around in his seat and grinned - a kind of sick looking grin to be sure - but still a grin - Now that's "gutsy" (to use the vulgar parlance of army life) and it's what we admire whenever we see it. I know I wasn't grinning. That grin put me at ease immediately. I throttled down so he could hear me and yelled forward to him, "That was some little tailspin, wasn't it"? He nodded. He doesn't know yet that I didn't purposely get into it and get out just as easily.

When I got back I was greeted with a bit of startling news. Ellis, Alexander, Meister and myself have been picked and recommended and accepted as instructors. We are to stay here at Belleville all winter and teach flying. Well at first, I was terribly disappointed. I had my heart set on France this winter. But then maybe I can help just as much by teaching others how to fly and go to France - and then when I do go I will be a far better flier. And every man in the Squadron envies us - so it isn't so bad. I have often thought of instructing and turned it down because - it's too much like joining the quartermaster department - a noncombatant in war. But, of course, we haven't a word to say. Those are orders. And I guess it won't be too bad.

This afternoon we started in learning to fly from the front seat. Tomorrow I will be given some outside landings in fields all over the country. About the end of the week we will be given classes.

Ellis and I flew together this afternoon. I flew in the front seat part of the time and then we changed. We tried to see which one could give the other the most thrills. Tomorrow I'm going to loop - I [am] going to fly to St. Louis and land on the Country Club grounds. O, it's fun.

Your son
Edmund.

Flying Cadet Squadron,
S.C.A.S. Belleville, Ill
Nov. 26 - 1917
Dear Mother,

We've changed our name to Flying Cadet Squadron. Doesn't it sound better? I am expecting to move this week into a cottage. All the instructors live in cottages - and they don't want us to associate too closely with the men - after we get our com's.

Mother - I realize I should not have described all the horrors of my exciting narrow escapes, to you. None of the other men do. They are very careful what they write to their folks. They never tell anything but what would make the home folks more easy about their welfare. I was making a sort of diary of my letter. I wanted to record as accurately as possible my impressions in my first tailspin.

Since then I asked Lieutenant Olds to show me tailspins. We went up together. He put me into three tailspins and I put myself into one and got out of all of them. That is a very necessary part of our training. Often a pupil will get a machine, in which he is riding with his instructor, into a tailspin. Unless the instructor knows how to get out of it, there is liable to be a disaster. I can now get out of almost any position that the Curtiss can get into. In regard to my inexperience and inability I beg to state that I am considered the best flier in the squadron. That is the reason I was chosen as instructor. To be an instructor in flying - one must know how to fly well - without sideslipping - skidding - stalling or other evidence of poor flying. He must be able to get the machine out of difficult positions. He must be able to analyze poor flying and tell the pupil what was wrong and how he can correct it. I believe that the authorities have seen all these qualities, or promise of developing them, in me. I am a very conservative flier. Especially when I have a pupil am I conservative. For if I did any stunts with a pupil, he would think he could do them alone and he might get into trouble. The young aviator learns most of his flying after he solos.

In regard to danger of being killed (we might as well look the facts in the face) there is some. But not a single flier has been killed yet at this field - and we have done as much flying as at any field in the country. We have had three serious accidents - all solo men who had just had a few hours solo - one man was knocked unconscious - got a slight cut on his leg and got some blood on his face. He is back flying now. The second got his leg broken. He is getting well rapidly and will be flying again soon. The third fell last Saturday and cut his lip! All three machines were absolute wrecks - had to be burned. I have not even broken a landing gear - in fact - I have one of the few perfect records - in that capacity - in the squadron.

Now I think I'm about through talking about myself. Are you satisfied that everything is running nice and smoothly and that I am safer than I ever expected to be at this time - when I went to Ft. Sheridan?

We did have a rather exciting thrill Saturday. Simpson was on his cross-country. One stretch came within three miles of the field and we watched him pass - just a speck in the sky. Suddenly he dropped straight down for 1000 feet. Then he checked his speed for an instant and floated there. Then again he dropped - down down - and disappeared behind trees. The lieutenant and surgeon jumped in no. 1477 and in three minutes they disappeared behind the trees. They returned in half an hour to tell us the tale. They had found - just kindling wood where the machine had dropped - just two struts were unbroken. A big hole in the ground showed how hard the machine had hit. Simpson was in a nearby farmhouse bathing a cut lip. He had tried to fasten his loosened safety belt and had lost control of the machine. The machine nose dived and he was pitched out. He still clung to the wheel. The machine turned over and flew upside down with him hanging to the wheel - for ¾ of a mile. Just before it hit he had drawn himself up into the cockpit. When it struck his face had hit the cushioned cockpit and his lip had been cut.

We flew all day yesterday on account of the quarantine. The first Sunday we have flown! Today it snowed but we flew through it.

            Edmund

Flying Cadet Squadron
S.C.A.S. Belleville, Ill, Nov. 26 - 17
Dear Sister,

To-day it is snowing - it has just turned to rain - and we have discontinued flying for the day. We started out this morning about 6:55 and flew till 11:00 when the snow was flying so thickly that there was danger of collisions in the air because of difficulty in seeing far. I have a class of eight solo men. These fellows have had four to eight hours solo and nave shown enough ability to go up from the third to the second class. I have four machines which I keep in the air all the time. This morning I had them making a 180° turn on a glide and landing on a mark. I would grade them on each glide and landing. As soon as they show marked superiority I shoot them ahead into the first class where they take their R.M.A. tests. It's lots of fun teaching. I get all the men who are not flying around me, and we together judge each glide and turn and landing. I think the men learn as much that way - watching others and seeing what's wrong with them - as they do in actual flying. We call it ground flying. It is simply thinking - thinking about the mistakes you have made - trying to analyze them and figure out ways to correct them the next time you go up. And it's serious business - not only from the point of view of personal safety - (for the better flier you are the longer you will live) but what is considered worse - discharge from the service. Six men have already been discharged simply because they didn't show class in their flying. The men are mighty interested and want to learn all they can. They can't learn much from me for I have had but twenty hours myself, but we together can learn lots just keeping awake and watching and thinking. The men have gotten so they can recognize a poor turn or a landing in bad from as quickly as I can, and they are getting good - all of them - at flying.

O. - I didn't tell you that I'm an instructor, did I? And I'm stationed here at Belleville all winter. It kind of spoiled my plans at first. But maybe I can be of more service teaching a lot of young aviators and sending them over - and when I do get there, it will have to be a good, speedy, Fritz that can get me before I get him. Why, I have visions of being an ace a month after I reach the front. (I suppose you know a man is an ace when he has brought down his fifth Hun 'plane.)

Lieutenant Oldys reminds me lots of Lafayette. He has the same pride - the dignity - yet the humility Lafayette must have had. And he is as great a flier as Lafayette was a general. It seems to me that Lafayette was about 150 years ahead of his contemporaries. We have just come to realize the necessity of fighting for freedom - international freedom - regardless of nation or state. It is that spirit that animates Wilson. More pointedly it is that spirit that animates the fellow who, when he has been so unfortunate as to be turned down by a recruiting officer here, goes to Canada and enlists. It spurred on Lufbery and Rockwell and McConnell and Thaw and Chapman and Prince and Thenault and the rest of the American Lafayette Escadrille[12]. I feel as if I knew those men - (you really should read "Flying for France"). Men like these have seen [obliterated word] France and the French people making a fight for liberty against tremendous odds. They saw the essential rightness of the cause and they didn't forget Lafayette and Rochambeau.

Why is it that our wonderful traditions embodying all that is best, most worthwhile, in this life, are handed down in so few homes. So few children think of DeGrasse as any more than some old Frenchman who got into a little fight near Yorktown; when they should remember his name as that of a man who risked and did more for the United States than any other foreigner. Most of us feel no debt of gratitude to France we would be willing to help pay, if it caused us the slightest discomfort. O, I hope America wakes up before She has lost her opportunity to do this service. And I believe she will. You can see it all around you. Boys and girls are all working for some good cause. All are pulsing with the desire to do good. We are waking up and seeing the part each of us can play in this great act of life. And that is the interesting feature of it. Each one must act a part. (Sister, I believe you have found your part). Whether that part is for good - or for indifference or harm we can choose. But we are responsible for making that part help in the upward tendency of the race. I never before thought that we

---

[12] Lafayette Escadrille - Squadron composed primarily of American flyers

individually had any part in evolution - but we do. There we are different from beasts. We have a deliberate choice. We can play a part that will make for finer, nobler, personality and nationality and humanity; or we can play the part that does not let the life force in us work out that evolution in our lives and in the race. And just now we have come to the place where this life force needs the power to place the interest of the group above the interest of the individual.

But I am getting preachy. I'll have to quit.

We have some thrills here almost every day. Mother will send you the account of the fellow who rode a Curtiss upside down for half a mile or more. Lieutenant Olds was taking up some civilians Saturday. One of them was a 200 lb. man. He flew around the field a couple of times then came in. He had too much altitude so he shut off his motor and just settled for about 300 feet. Then when about 100 feet high he nosed over into a steep dive to get up speed for his landing. When he "went to pull out of it about 10 feet high, he couldn't pull back. The nose dive had lifted the civilian out of the front seat and he had let himself be thrown against the wheel in the front cockpit. Instead of getting back he had just rested there. When the lieutenant tried to pull back he had 200 pounds reposing on the tandem wheel in front. He just braced his feet and pulled with all his might and just cleared the ground. Later the civilian asked, "I suppose you felt a little pressure on the wheel?" in a drawl.   "O - - no." also drawled.

Lieutenant Richmond also had a thrill. He took off and turned up into a vertical bank right off the ground. When he tried to come out of the bank he could't straighten out. His long fur coat had caught in the aileron wires and he couldn't turn the wheel back. He just held it in the bank and went round and round right off the ground until he had gotten up lots of speed - then he banked over almost upside down - jerked his coat out - and straightened up. That was hard work.

We had two machines turn over on Friday. They were landing in a strong wind and just nosed down and the tail came over. No one hurt.

       Your brother,
       Edmund

Flying Cadet Squadron
S.C.A.S. Belleville, Ill.
Dec. 3, 1917
Dear Sister,

I am sending you one of Miss Gregory's recent letters. Please return it to me. I am very glad that she saw you and that I resembled you. It seems to me that I am pretty good at absorbing things usually. Due to your persistent efforts when we were very young - seven and nine - I absorbed a whole lot of knowledge and even before that I absorbed something of your personality. Mother says that I owe my education to you and I think I also owe my character and personality to you. So no wonder I resemble you, for what I have is second-hand - but none the less valuable for the first hand was so priceless.

Dear sister, I've never admitted to you how much I appreciated my sister. We've never been very affectionate and then both of us have had a good deal of reserve and when I was younger I expect I was a little jealous. But I am just thrilled by Miss Gregory's appreciation of you. She is so wonderful - so ideal - herself that her appreciation is a [the word is almost obliterated: criterion?]. It makes me feel very humble that she should say you were very like me.

My class of dual men started last Sunday. They are on landings already. I am hoping to solo them after about six hours. I like teaching very much. My men are up on their toes all the time.

This afternoon I went up for a joy ride. First I climbed about 4000 feet - above the clouds. I flew along about 100 feet above them for a long time. Above was clear blue sky and a bright sun. the top of the clouds was just like the pictures of sunrise on Pikes Peak. Then I shut off and dove down. it seemed queer to dive straight at this solid layer of white. Suddenly mist enveloped me and I could hardly see the tip of my wings. In a minute I came out below and saw the earth. It was all dark and dreary and cold down below - so different from the sunshine and beauty of a few seconds before. I landed and found I was 15 miles south of home - so I took off and flew north till I sighted the hangars. Fifteen miles in ten or eleven minutes! Real sport!

Your brother,
Edmund

[A note to their mother in his sister's handwriting is written at 90° at the top of the letter above his "Dear Sister," left of the return address:

"Please keep this letter very safely, and private. I'm sending it to you because you haven't heard as recently as this - Margaret."]

Flying Cadet Squadron
S.C.A.S. Belleville, Ill.
Dec. 11, 1917
Dear Mother,

My flying days at Belleville are over. The snow and cold stopped flying last Friday at noon and we haven't attempted it since. Yesterday and to-day we have been engaged in taking apart the 'planes for shipment south. We leave for Dixie on the fifteenth. The chances are that the men who have passed their R.M.A.'s will not go away but wait here for their orders for France. This bunch includes all of our old bunch from Champaign so that I will go south with an entirely new bunch of men. However I expect my commission before then and will probably travel with the officers. I am hoping the personnel of the instruction and executive staff will go intact to the new field, for it has been mighty pleasant to be under these men; and I am thinking it will pleasanter to be one of them. We will go to either: -----

Lake Charles - Louisiana
Pensacola   Florida
Miami   Florida
Houston Texas
Waco   Texas
Wichita Falls Texas
Dallas   Texas
San Antonio   Texas or
San Diego   California

I knew you would be curious to know just where - as we are - and so I wanted to be explicit. There are two other possibilities, Memphis Tenn. and Ft. Sill - Okla.

There are aviation fields at all these places.

Houston Tex. Is the most likely place. Even the commandant doesn't know where we are going.

Mother, I wish you would see Wilson as he is - a more far-seeing man even than Roosevelt. What support do you think Roosevelt would have had - had he been president and urged armament and war? It took the United States fully two years to prepare - mentally for war. There are thousands of people who are yet not in sympathy with our war. A year ago the population was all against war - and 2½ years ago we would have had internal strife had a president pushed us into war. You lived among an element peculiarly pro-ally and pro-war. But the masses have been slow to grasp the ideal which was to mean so much sacrifice for so vague and - almost theoretical - an end. It took England five days to declare war. Had she done it sooner public opinion would have been split. It took their people five days - in face of the fact that a definite treaty had been violated - in face of the fact that only the English Channel separated them from the Hun - in face of the fact that the English were better prepared mentally by the constant talk of war with Germany. They had as much provocation for war that first day as we would have if Japan seized a part of Mexico and prepared an invasion. Yet it took them five days - and in that five days almost every faction in the Empire united. While I greatly deplore the fact that our people - all our people - did not see the necessity for entering the war years before we did - nevertheless - had a president pushed us into war with only a part of the nation behind him - we would have had trouble just as sure as this is a nation - of the people, and etcetera -

I fully believe that Woodrow Wilson saw this; and he bent his efforts to bringing public opinion up to his exalted point of view. And I say, "exalted", reverently.

Why is this war horrible, Mother? Were the crusades horrible? What means could be horrible out of which is to come such an end. Out of this war will come a purified, inspired America - an America with higher ideals - nobler aims - greater possibilities. Of course it takes fire to burnout the dross - and the fire hurts - but is it horrible? Is it not merely a necessary purging? And this refinement is not only national. It is personal as well. What other thing (besides a tremendous religious experience - and that is not possible to most of us) could bring out the courage - the unselfishness - the resourcefulness that the war brings out in a man?

McConnell's epitaph is, - "Greater love hath no man than this: That he lay down his life for his friends."

To aspire to be worthy of such an epitaph is my greatest joy.

But why speak of epitaphs when there's arguing to be done? As a matter of fact, I have seen the wonderful improvement in several fellows, already. Lindauer - an old friend of Wallace Pendry, was in Co 1 - 11th P.T.R. when I was in Co 3 - 11th P.T.R. at Ft. Sheridan. I knew him as a most distasteful egoist - a vile snob - a mentally repulsive fellow soldier - unsympathetic, lazy, selfish, uninteresting and disgusting. I went to Champaign with him - in the same car. He affected me exactly the same as spoiled pet - but he was so much improved by Ft. Sheridan that had it not been for his telling me all about how he was going to get a big commission at Sheridan if he had stayed - if he had not told me that - he would have been almost bearable.

Champaign did a good work for him. He flunked one week and so graduated a week after we did - and he got ten demerits during his course. When he arrived in Belleville - two weeks after us, we loaded all the dirty work on him. He was on police duty every day and on kitchen police as often as possible. When he started flying he was going to beat the record - soloing - and make the best flier in the squadron. At the end of a month he was still under dual and so - was discharged. He told me confidentially, "I used to think I was as good a man as anyone in the world - I could do anything anyone else could do. But I see now that I can't and I feel more kindly toward the world". Really by the time he was discharged we had come to like him real well. And we took him to the train with tears in our eyes. And he hasn't even seen a battle yet. I'll wager when he comes back from Europe he'll be a real comrade to any man.

The sox arrived amid great rejoicing - for I had had cold feet for a week. They are wonderfully warm. Wool knit sox are the aviators greatest help for - It does get cold way up high. Thank Eva for me, please. My sweater is great. It's warm without bulk. R.M.A. means Reserve Military Aviator. What is P.S.? - I am well equipped for the cold. Uncle Sam has given me woolen undergarments and leather outer ones.

Edmund

Flying Cadet Squadron

S.C.A.S. Belleville, Ill.

Jan., 1, 1918

Dear Mother,

It seemed so mean of me to allow you to stand there so long on that icy platform Sunday night. I had such a comfortable ride that I thought of that last picture I had of you standing there. The train was but 45 minutes late in St. Louis. We arrived in the Union station at 1:30. I spent the afternoon getting a pair of field boots an overcoat and my suit. St. Louis is quite a satisfactory place to trade. I owe $27.50 on my $27.00 boots and $45 on my $50 overcoat - my bank account is gone and I'm broke. Also my commission isn't here. Only ten cadets are still privates. A bunch of them went to Washington on their vacations and came back with commissions. But my orders are here anyway; they came this afternoon. Six of us go to Lake Charles, Louisiana - tomorrow. We are having a great time speculating on our prospects. Someone recently got a letter from Mr. Lewis - one of our old civilian instructors who was sent there. He says it is the awful-est old hole in existence - rotten field and everything - but we refuse to be pessimistic. We are expecting to be dee lighted. Lake Charles is not far from Houston Tex. We will probably make lots of cross country flights. Possibly I have lost my rating as instructor. I may be put on advanced training at once or may be assigned to executive work among the squadrons. We may be put on duty as officers in charge of flying and do nothing but direct the work. Shields & Anderson, who have already received their commissions and Arnold, who has heard that he was commissioned, and Maupin and Wegener and I, are the six who go in this order. About ten fellows have orders to Ft. Sill - Okla., and ten more for San Antonio and Ft. Worth.

Tonight I am going down to Belleville to say goodbye to my friends. Poor Elsie Jean! She hasn't heard from me for almost a month - I'll have to make up tonight.

Yesterday afternoon at three o'clock I passed the soda fountain in Famous Barr's store and realized I hadn't had a bite to eat that day so I got an egg malted milk - It was scrumptious. Our mess out here has gone to pieces. This morning when we got ready for breakfast at ten o'clock we found some frozen eggs and canned sausages which when fried tasted pretty good.

The new year came in very quietly out here. We went to bed at ten-thirty without raising any rough house.   Next letter will come from Louisiana.

     Edmund -

Flight Instructor Role 1918

Assignment as an instructor to Gerstner Field in Lake Charles, Louisiana, and Love Field, Dallas, followed. This was a traumatic period in his life. Shortly after Eddie moved to Gerstner Field, his original instructor, Mr. Jones, is killed along with his new student cadet. They spend about one month at Gerstner Field going through instructor training and introductory test flying the new Thomas-Morse Scout known as the "Tommy". This is a faster, more demanding aircraft than the Curtiss JN4D "Jenny" and is accepted with enthusiasm by the new instructors. However, they do not get to transition to the Tommy and instead find their Curtiss JN4D's are replaced by the Standard J, perceived to be even less capable than the original Curtiss machines. They then move over to Love Field in Dallas for their main instructor duties. At Love Field, ETA experiences his first wreck which does some facial and dental damage. The damage does not keep him from continuing his role as flight instructor although there is some recovery time after nasal surgery. ETA will spend February thru June 1918 in an instructor role at Love Field in Dallas.

Eddie Allen's introduction to the flight instructor's role was described by Thomas Collison[13]. "Eddie and his pal, Bob Ellis, said goodbye to Scott Field on Christmas Day, 1917, Goodbye to Joe Weiner, later killed in action with the 95th Pursuit Squadron on the Western Front. The two promising young pilots received their commissions as First Lieutenants at Gerstner Field, Lake Charles, Louisiana, there where Major Mitchel, one-time mayor of New York (and after whom Mitchel Field is named), fell from a Thomas-Morse Scout and was killed......Louis Meister, who had won his wings along with Allen and Ellis at Scott Field, joined them at Gerstner.... Meister was killed in 1931 in a Verville Trainer.

Although neither Eddie nor Ellis nor Meister had been schooled in aerobatics, they were, in February, 1918, assigned to Love Field, near Fort Worth, Texas, to teach acrobatics! Eddie's first crash came one day when he was assigned, with Ellis and Meister, to fly to Greenville, Texas, to give flight demonstration before a fairgrounds crowd. It was all in the interest of boosting the sale of Liberty Bonds. Taking the lead, Eddie made a pass over the centerfield of the county fair track. He came in for a landing and promptly washed out the undercarriage. Somebody had forgotten a small item which may be described as a deep ditch across the end of the centerfield! His undercarriage repaired, Eddie did a climbing turn...a chandelle...to get out of the short centerfield. But he chandelled right back to the ground again, landing in some trees! He learned, too late, that his airplane didn't have enough power to perform

---

[13] "This is Eddie Allen" draft biography material by Thomas Collison 1944

such miracles. At that time, it wasn't realized that down-currents are more prevalent over trees than over hot, dry earth. Eddie cut his lip on that one.

One day Eddie and Bob Ellis flew to Benbrook Field near Fort Worth, where the Canadians were in charge, to pay respects to Major Rhinehart, later General Rhinehart. A beautiful white-painted Curtiss Canuck biplane, the Canadian version of the Curtiss JN-4 (Jenny) was staked out on the field. The airplane belonged to Fred Harvey, Jr. Anxious to prove that his Canuck was better than the American Jenny, Harvey invited "Take her up, Eddie." Eddie did.

Years later, Eddie said he was in a festive mood that day; he wanted to show off. It was not easy to put much altitude in the bank quickly. The Canuck did not climb as rapidly as the JN-4. Upside down flying was the latest thing then; Eddie planned to fly across the field upside down, go into a spin, level off close to the ground and land. He would make it short, but spectacularly sweet!

Reaching 1,200 feet, and knowing full well he should go higher, Eddie grew impatient. He turned the machine on its back, discovering too late the Canuck was not built for inverted flying. Oil spewed out from the breather tubes covering his windshield and goggles…but Eddie was in inverted flight, all right, and flying across the field. Then he shoved the plane into a spin. Holding the stick with one hand, he attempted to remove his goggles with the other, but the elastic holding his goggles eluded him. Suddenly he realized he was close to the ground — too close! He had seen men die from trying to recover too close to the ground. That was curtains for certain. And so he deliberately remained in the spin! Young Harvey's gorgeous, white Canuck struck on one wing first, then cartwheeled into a shambles. Eddie found himself on the ground, trying to hold himself up with his hands. An ambulance trundled out across the field after him. The astonished driver of the meat wagon found Eddie laughing. His nose was shoved to one side; he was cut about the face. Hospitalized for two weeks, Eddie was told his flying days were over. Eddie's disfigurement remained with him many years, until he fell into the hands of a skilled surgeon who made a new bridge from one of his ribs."

H.D.C., Gerstner Field, Lake Charles, La.
January, 7, 1918
Dear Sister,

Imagine us sitting around a big stove in the barracks almost freezing - in Louisiana - thirty miles from the gulf. This is where we are and what we are doing. And last night we shivered under five blankets. Still it is fully thirty degrees warmer than Belleville and flying is going on every day. We arrived yesterday.

January 8. This was as far as I got yesterday. Not that I was busy - only lazy. Today - besides sleeping a whole lot. I went out and watched the aere-o-planes. You know them things that go way up in the air? Well I saw a couple of them - yes mam. And they had real men in them. And they made lots of noise when they started up. I really didn't believe there was any such things but now I've seen them, I'm satisfied. This is a two-unit field - about 1500 acres. There are over 300 cadets and probably 2000 enlisted men here. Our mess is some different than Belleville - we have to beat the next fellow to it if we get anything to eat.

Looking back we see that we have left the best field in the whole world. The officers and organizations were just perfect. The mess and quarters were wonderful. Even the climate was favorable. This field will have to go some to beat it! But this is only the seventeenth day of flying on the field. It is less than three weeks old. Things aren't organized yet. I wish they would let us get to work and help organize it.

We have been assigned to the cadet squadron here and have done nothing but loaf since we've been here. I have been expecting assignment to the instructors school every day. They are probably waiting until we get active duty orders from Washington. O, I didn't tell you - I've been commissioned. We received the telegram on the sixth and immediately wired acceptance. The only trouble is this - at this post, the commanding officer takes the Army Regulations literally and won't allow us to consider ourselves as officers until we get ordered to active duty and discharged from the army. We are now awaiting those orders. One of our men, Shields, got his to-day and immediately moved into officers quarters. I have written to the Adjutant General of the Army and the chief signal officer requesting speed.   In the mean time -

        Pvt. - 1st class Edmund T. Allen
                Bids you good night.

    Headquarters Detachment Cadets (H.D.C.)
    Gerstner Field, Lake Charles, Louisiana
    Jan., 7, 1918
    Dear Brother,

I don't need any money. We got paid for December on the second of January - just before we left for here. I have eight dollars left and no debts. As we can't get passes now to go to Lake Charles, I guess it will last until February tenth, next payday.

Gerstner Field is a new place - this is the sixteenth day of flying. On the third day - Mr Jones - my instructor, was killed with his pupil. Next day another man was killed. Yesterday when we arrived, the first thing we saw was two machines sitting on their noses out on the field. There are two fields here in one - 24 hangars - about 300 cadets - and several thousand enlisted men. Most of the cadets are fresh from ground school. They haven't had their first ride yet. I think we were sent down here as instructors. When we arrived we found a telegram from Washington commissioning us. We wired acceptance at once. The catch comes here - The commanding officer at this field considers a man still a private as long as he is still enlisted as such. And as we have not been assigned to active duty we are still reserve lieutenants not on active duty and we live in the cadet barracks and get cadet pay. We are hoping not to have to wait two months more for those active duty orders.

One of the men sent down with us had received his commission up at Belleville. He was wearing his hardware and boots and spurs. When he got down here he had to take it off and put on the white band. It is an awful blow to him. He surely did hate to come down here in the charge of a private.

As nearly as I could estimate, there are 100 planes here. They are all stick control JN4D's! Some new Thomas Morse Scouts have arrived here. They have a wing spread of but 12 feet. They have a rotary engine and go 105 miles per hour. One of the fields here is for advanced flying. Machines are looping and tailspinning all the time. The weather is great for flying.

Things are all bawled up as far as organization is concerned here. I am crazy to get in and organize a little bit but can do nothing as a cadet.

So long –
Edmund

Gerstner Field, Lake Charles, La
January 27, 1918
Dear Mother,
We have no idea of the weather in the north. When it gets down to 25° here we complain terribly forgetting that you are all snowbound up north. Now, it is quite warm, but blowing about 45 miles per hour. All the strong winds are south winds and as our quarters are on the north side of the field we get the benefit of all the dust that has been scraped and plowed and rolled up on to the field. In coming back from a cross country yesterday, we could spot this field from 30 miles away by the dust. There is no dust anywhere else but here. You see, in smoothing off the field all the grass was plowed under and the ground loosened and pulverized. The results of a rain last but a few hours.

I am very glad that Thomas is happy and contented. Do you suppose he will see much of Dwight in camp?

The work here has started in earnest. Officers roll-call is at 5:50 – breakfast immediately – then flying till noon and from one till dark. After dinner we have lectures. Until further notice flying has been stopped on Sunday. We are all delighted to have the day off.

Our course is almost over in the instructors school. We graduate on the 29th and start teaching. Yesterday we had some fun. It was an exercise in forced landings. Mr. Vernon the head instructor would take-off and go away. In two minutes when he was just a speck on the horizon the first pupil would take-off and follow. We all followed at two minute intervals. Mr. Vernon would land in some farmers back yard and taxi around so the next man could not see in which direction he had landed, or where the good ground was.

When we had all landed he would talk to us, pointing out any mistakes we had made. Then he would take off again. We landed in sixteen fields all around within a radius of ten miles. Once I landed across wind and almost turned over.

Tomorrow we fly the Thomas scout. We have had several lessons in starting the motor on the ground. The motor is more powerful than the Curtiss and the 'plane is but half as large as the Curtiss 'plane. These scouts have made 150 miles per hour with the wind. We are expecting lots of fun.

Edmund.

Gerstner Field, Lake Charles, La.
January 29, 1918
Dear Sister,

I am sending you some of Thomas' letters. He is evidently in Ft. McPherson and not Ft. Oglethorpe - evidently enjoying it also.

The candy came from Mrs. Ralston. It was really wonderful candy. I had to fight to keep enough for myself to really get some satisfaction. Then to-day I got another box from Marshall. Did you send that? What in the world do you call it? Sea foam walnut divinity fudge? M-------m!

Rain stopped flying at noon to-day and it has been raining "off and on" ever since. Our front yard is a lake a hundred feet across and a foot deep. Water is the main feature of the landscape - I should say waterscape. This morning I flew the Thomas-Morse scout. Did I tell you about it? The engine is a 100 horsepower rotary engine - i.e., the cylinders revolve around the pistons and crankshaft. It is very hard to control. You sit in the cockpit and turn on the gas. Three men hold each wing and a seventh man cranks the propeller. The engine immediately races up to 1200 revolutions per minute. The men on the wings until you give a signal, - then they let loose and the little scout darts forward picking up speed very fast. At around 100 feet forward it has gained 75 to 80 miles per hour and it leaves the ground. Then you

Thomas-Morse S-4 Scout "Tommy"

can pull it back and it will go up at an angle of 45° if you want. In fact, unless you hold the stick forward with considerable effort it will go straight up. Of course it couldn't go straight up very far and a stall would result. Then you push the nose down to horizontal and you pick up speed very rapidly till you go about 105 miles per hour. If you want to go faster turn and go with the wind. Then you'll have 105 + speed of wind. If that isn't fast enough nose down and gain about 50 miles per hour on a dive. With a good strong wind you could go 200 miles an hour. There's lots of things to watch. The main one is - other machines. It is necessary to keep weaving from side to side so as to always know what is ahead of you. Then there is your gasoline pressure which must be kept between 4 and 5 pounds all the time - and your tachometer which should register just 1200 revolutions per minute - and the air speed indicator - compass - gasoline and oil gauge - and oil pressure. When doing stunts and controlling the ailerons, rudder, and elevators one is quite busy - yet you have time to enjoy the rush of air past your goggles and cheeks - the tremendous speed at which you are traveling and above all the sense of power which is making you pass the Curtiss planes as if they were anchored. The scout is so little it all seems a part of you. When they equip them with machine guns and wireless apparatus we will be busy.

Edmund

Love Field, Dallas, Texas
Feb 12, 1918

Dear Mother,

I have been so tired every evening when I come in the quarters that I just pile into bed without even the ceremony of making the bed. Every minute I sleep is so much toward catching up and feeling rested again.

You never saw a bunch of fellows so glad to get away from Lake Charles as our bunch was. We "fought" for the first auto and got in to town as fast as possible. I came right through to Dallas stopping but five minutes in Houston - to change cars. Our train arrived in Dallas at seven A.M. and after piling out of the sleeper I got a good breakfast and hurried out to Love Field - only six miles north of the center of town. Dallas is a wonderful town - rather wild but enjoyable ne'er-the-less.

The jitney drivers only soak you two bits to come out to the field. This makes it handy for the officers can get into town every night.

This is the first time I have reported to a new field as an officer. A motorcycle sidecar met me at the gate and took me to headquarters. While emoting to the Adjutant-Major Styles, I heard a gruff voice behind me bawling out the cadets. Then all of a sudden "Why - hello - Allen". I turned around and there was Major Wheeler our old commanding officer at Scott Field. He is an officer in charge of flying here. Of course, it put me in solid with adjutant right away to have Major Wheeler greet me like that. I stepped up and put out my hand with "Hello Major Wheeler". We had a good chat about old times at Belleville. Then he told me I was to be put on advanced flying here. Ellis and Meister, old Belleville boys, arrived the week before and we three are assistant officers in charge of flying. We are in charge of all advance flying.

I got to work next day and mapped out a course of instruction including stunts - forced landings, battle formations, and night flying. I got Ellis' and Meister's approval and took it to Major Butts. He said "Go ahead. You have a free hand. Do whatever you think best.

We will turn over about twenty cadets a week to you for advanced instruction. When they finish it to your satisfaction they will be commissioned".

We started out hunting ships and lined up eight pretty good ones. Our first class of twenty reported and started their stunts today. All the officers on the post, Major Butts included, now want to take our course. Ellis, Meister and I work together well. Today I spent six hours in the air - most of it at 6000 feet.   - Edmund

Love Field, Dallas, Texas
Feb 16, 1918

Dear Mother.
I did receive the $95. and thank you very much for it. I did mention it in the first letter I wrote after I received the money. Since then events have happened so fast that I can hardly catch up, and so I failed to speak of it in later letters. The money was so entirely unexpected and I had so reconciled myself to do without my overcoat and bedding that I really thought at first I didn't need it at all. But I bought an overcoat and am now getting my bedding. It would have been throwing money away to have bought blankets in Lake Charles and mail orders are always unsatisfactory. Army blankets are at best cold propositions and now they charge $10. for a hard, smooth piece of cloth that wouldn't keep anyone warm. (I'm afraid I must have exaggerated my need in my letter.)

Here at Love Field they issue to officers - blankets on memorandum receipt. I have four government blankets and two of my own. I bought a very good - cheap - mattress for $2.00. But it is warm - with my six blankets.

I am very glad that you are so comfortable now. It certainly is rotten to have an uncomfortable bed and a cold room. I'm sorry I worried you so about my comfort. I'm afraid I could never be so concerned about anyone else's comfort as to lie awake working I guess I am so selfish myself that I can not realized anyone else's unselfishness and altruism. Come to think of it - my letters to you never do mention anyone else but Lieutenant E.T. Allen, do they?

A letter from Thomas came yesterday. Also one from Margaret - in fact eight letters - forwarded from Lake Charles - came in one mail. Thomas's letters are always interesting. He seems to get so much out of life - Every event has a new interest. He always has something interesting to say and he says it so nicely - so effectively that it gets across - always. I am going to try to copy his style.

I am sending father's watch to you. The crystal does not fit. The last time a new one was put in the wrong size was used.

My affairs are as follows - I have $100. in the Champaign Ill. 1st National Bank. My bank book is there at the bank

I have $3.40 in Illinois Bank of Champaign

I have 10 $50. Liberty loan bonds - on which I have paid $200. $50. a month goes from my pay for them. They will be forwarded automatically to you next September - when $500. has been paid. The government has a accident statement for each flier. They will telegraph all concerned in case of an accident.

I have $1000. insurance in Travelers. You have the policy. They have been notified to mail all information to me at 315 Dayton St., Hamilton in care of you.

I have $2000. insurance in N.Y. Life Company. I have not yet located the policies (2 of them). They have been notified to address at Hamilton.

I have $10,000. insurance with the government. The policies should be in your hands by this time. They are usually sent home by this time. The premium is automatically taken care of on my pay voucher.

My pay now is $166.67 per month (plus 25% for flying - if Pershing does not succeed in having that revoked). Actually my pay now is less than that as a cadet. I never realized this until I found a bill for $50. for January mess and officer clothes bills and excessive laundry bills and Saturday and Sunday hotel bills etc. The higher in rank an army man goes the more it costs to live. If I ever get to be a captain I expect it will be the same thing. I'm glad I have to save $50. a month. I'll continue that at least. I am making a regular budget only from month to month, but it includes every cent. Shall I send it to you?

An announcement of Dorothy's wedding on Dec 7, 1917, came today. I didn't know she was married. Did you know about it all the time?

I heard from Mrs Price in Champaign. Mr. Price is off diet - eating anything he wants. He is gaining weight - has a good job in the library - name on library stationary and all. She says "Drysdale is Sport Editor of the Illini. His head is so big it's about to bust - the poor idjiot"

A letter from Owen tells of his trial at San Antonio. He is having as much trouble learning to fly as I did. I have to laugh now at my diary. I took it all so seriously. Some fellows just picked it up without any trouble. But I am convinced that if I had not gotten down to business I would not have learned in the required nine hours.

Some of the poor birds we are trying to instruct in trick flying have had 15 hours solo and have never done a vertical bank. We were doing them after 45 minutes solo.

Things are being rushed to the limit here. Washington is insistent that everything be speeded up. We are working twice as hard as we did at Belleville. Flying from 6:00 A.M. till 6:30 P.M. No stop for noon.

        Edmund T. Allen

Love Field, Dallas Texas
Feb,. 19, 1918.
Dear Mother,

I have just returned from a movie at the Y.M.C.A. here on the post. I imagine there were two or three hundred men there. It was very interesting. The fellows did not know we came in and we sat at the rear so they were free to make remarks. Some of the wittiest remarks - comments on the pictures - were made I have ever heard. The funniest ones were those applying military and aero terms to conditions depicted.

A new order from Washington takes away all the Curtiss planes from this field. We swore we would not stunt anything but Curtiss planes and we fixed up for ourselves six of the finest planes on the post. We had new wires and new parts and new motors. They would climb to 5000 feet in 25 minutes. And then they come out of spins so easily and handle so readily in unusual positions that we like them very much. To-morrow they are all to be taken down and shipped to San Diego. I don't know what we will do. Ellis and Meister have refused to stunt the Standard plane and I rather dislike doing it although I have not refused. The Standard is a big, unwieldy, ugly, blankety - plane that everybody hates. It has an unreliable four cylinder motor in it that goes on the bum about every fifteen hours of service. They say it takes two mechanics to keep each motor running. And they are going to make this a "Standard" school. Nothing but Standards and Dayton-Wrights here. The Dayton Wright is exactly the same as the Standard except that the parts are not interchangeable. But I suppose in a few days we will be stunting Standards as if nothing had happened.

I suppose you read about the rotten accident here the other day. I saw the whole thing and it was rather horrible. Too bad - if he had only been put in a tailspin before - and knew how to get out of one it wouldn't have happened. I regard that part of our work as the most valuable. It teaches the men how to get out of tailspins - quickly and easily. Only they should have this work sooner for it is in the first ten hours of solo that most of the men accidentally get into tailspins. A tailspin is a queer thing. Any unusual position always ends in a tailspin and by applying the controls one would, on first thought, apply to get out - one only holds in it the tighter. Do you remember my wails of woe when I was still under dual work that I did not pick it up faster? - that I did not fly by instinct but had to figure things out? Well - that first tailspin I got into would almost certainly have ended in disaster if I had been a flier by instinct. The instinctive thing to do in a spin is to try to pull up the nose and this is what almost every fatality is caused by. The sensation of the spin adds to the danger by paralyzing the thoughts. In my spin after I had finally overcome my paralyzing fear - my thoughts got to work and I figured out what was wrong. Instead of pulling up I must push down farther and straighten out the rudder. Suddenly I stopped spinning - fortunately still above ground. I now realize that that was the closest call I ever had - most fatalities occur in the first spin. Of course now a spin is as much a part of the day as mess - more - for we spin ten times a day and have mess but three.

A class in French has been started by a woman in Dallas who has hired a native French speaker and is giving lessons twice a week to officers of this Post gratis. I think I shall start parley vooing next week.

We are having a lazy time here. To-day there was no flying on account of the cold and the high wind. Yesterday afternoon we flew steadily from one o'clock until six-thirty. I soloed three men on stunts. One of them did a loop - rather difficult - with this type of machine. One of my men, Stevenson, was climbing up to 5000 feet - We got to 4000 feet in thirty minutes. I made the signal to turn around and he thought he would show me that he could make a vertical bank, so he cocked the machine up to vertical and - - - - hung there. In about a second we were side-slipping like thunder. I sat tight wondering if he was going to pull out of it. He finally did. I thought this was as good a time as any to teach him vertical banks so I shut off the motor and hollered some instructions to him. He then proceeded to exhibit his technique. In about a minute I looked at the altimeter and read 2000 feet. We had come down 2000 feet in a maneuver in which you are supposed not to loose any altitude! And it would take 15 precious minutes to climb back up. I madly grabbed the controls. He signaled for me to take them back and then proceeded to hang his head over the side of the fuselage and shed his lunch on the villagers below. The vertical banks had got his goat. He will be all right tomorrow.

I do not need any more wool knit socks. - emphasis on the underlined words. Those I have are very fine but I have plenty of them. I like your plan of taking work at Oxford. and the course is great.

               Edmund

Standard J

Love Field, Dallas Texas
Feb. 24, 1918
Dear Mother,
How are you standing the warm wave? Or aren't you having a warm wave? Last Tuesday it was so cold we called off flying, and Sunday the temperature was reported at 93°. It is very warm today in spite of the strong wind. This afternoon the wind is blowing up all the loose earth and we are having a regular sand storm. Sometimes we cannot see a hundred yards ahead. Flying has been called off.

I had my first wreck the other day. My motor died when I was about 100 feet from a line of trees and just a little above them. I knew I couldn't glide over them and I was too close to the ground to turn so I dove and tried to zoom over them but I hadn't enough speed and so I settled right into them.

I crashed through and hit the ground on the other side - wiping off the landing gear, smashing a wing and twisting the fuselage. Neither myself or my student were even shaken up. I requisitioned a near-by automobile to take me to the field four miles away and reported. We got the machine in that evening.

Next day one of the lieutenants - not knowing who I was, told a bunch of us that the commanding officer had said, "motor froze up - hell! It was nothing but ___ ____ boneheadedness". I had reported that the motor had frozen up. (i.e. got hot and stuck). I immediately went to the repair shop and got the men to working on the motor - taking it apart so I could find out what was the matter. I found one piston and cylinder all scarred - the aluminum melted and running into the grooves. I showed it to Major Butts and Major Wheeler and all the engineering officers and I have it saved in a locker to beard the lion in his den with when he returns. I may be court-martialed for disrespect to a superior officer but I surely will clear myself. I'll try to be as respectful as I can. He thinks that I was trying to make a landing and misjudged my distance above the trees and so hit them. I laugh! Me - (excuse me) who has taken bricks off of chimneys in Belleville!

Sorry I haven't anything interesting to tell about this time. Did I tell you that they are shipping all the Curtiss 'planes away? We are all seriously considering absolutely refusing to stunt the Standards. I don't know whether that would be considered "cowardice in the face of the enemy" or not. We have entirely caught up with the list of students awaiting advanced instruction and are now taking men temporarily out of the cross-country classes. They would rather do our work than any other part of flying.   Edmund

Love Field, Dallas Tex.
Mar - 8 - 1918
Dear Mother,

When I opened your last letter and pulled out the clipping, I really cussed for I thought it was one of the kind all the fellows here are getting from mothers and sweethearts about "Aviator at Ft. Worth killed", and then when I found out what it was I was doubly delighted. Thank you for sending it. It was mighty interesting. I like to get clippings like that, especially if it contains word of dear Peg o my heart

[14] I am getting to like medium-sized towns. One doesn't get nearly so lonely as in a big city where everybody is a stranger. Dallas is just such a medium-sized de grandeur moyenne burg.

Last night we entertained Major Butts, our officer in charge of Flying, and his wife and her sister. We got a box at the Majestic and saw "Turn to the Right". The play is certainly worth while; and with the cast we had last night the show was wonderful. We all enjoyed it immensely. We got the big Cadillac "8" to call for us and take us down and bring us back. Major Butts is a West Point man - a very strict disciplinarian. All the cadets are scared to death of him. But he enjoys a good time and he likes us. Last Sunday the Cadillac called for us and we went out to dinner - his invitation. We ate at one of the finest homes in Dallas. Before the meal everyone got out in the dining room and had an appetizer - Sazurac cocktail. Then afterwards a Cardac New Orleans drink. I felt rather uncomfortable at first but no one paid any attention and so I got by without any scene. When we were all seated in the living room talking, I seemed entirely inarticulate for some reason. I couldn't talk at all. Surely I never used to be shy or timid, did I, and the Army should take that out of one. All I do is listen and grin. Then the Butts went calling and took us along. At every place they would serve cocktail of some kind and the ladies all took some too. Some more uncomfortableness! Ah well - just so I don't get the holier-than-thou attitude.

All our Curtiss planes have gone and we have started work with the Standards. We three, Ellis, Meister and I have written letters to the Commanding Officer stating the unsuitability of Standards for advanced flying. We recommended a Curtiss JN4"H" which is especially built for our work. We are not expecting anything to be done.

Did I tell you about our lecture? We got Major Butts to get all the cadets in the school building one night; and then we lectured to them

---

[14] "Peg 'o' my Heart" was a popular song. He was probably referring to his cousin Margaret, whom they called Peg. She must have been the subject of some article in her hometown, Hamilton, Ohio, paper. if it was an article about his sister Margaret, also called Peg sometimes, the hometown would have been Marshall, Ill. I wish the clipping were still there! — MRB

on "Safety in Flight". The major was pleased. I got a job next day of giving three lectures a day on "Air Studies". I got away with it all right and gave them an examination after fifteen lectures. I now have 242 exam papers to correct - (or throw away). It's lots of fun lecturing to cadets. I'm not at all shy there. I wonder why?

Edmund

PS - I suppose you mean by "that splendid flier at Ft. Worth" - Vernon Castle - Well, he could fly, but not splendidly - that was mostly press agent work. - - - the government isn't reckless. They simply had to increase their instruction staff from 60 to 500 and do it quickly. We teach a man to fly here in two weeks.

The following are the notes for a talk by ETA, most likely in March of 1918 given to aviation cadets. This may be the lecture on "Air Studies" referred to in the previous letter. It certainly is an excellent primmer on aviation and "aircraft" in WWI. The notes are in pencil on 18 "note card" sheets.

To most civilians the word "aircraft" conveys a very inadequate and muddled idea. Aircraft naturally falls into two classes. Those which are lighter than air and rise of their own buoyancy such as balloons, and those which are heavier than air and rise by being propelled at an angle to the air as when a boy throws a tin disk boomerang.

There are two main types of balloons. The old captive balloon which was such a novelty twenty years ago is, strange to say, still an important feature in modern warfare. It has only changed its shape. However, long before the anchored balloon came, we had free balloons which float before the wind and are operated by throwing out ballast or releasing gas so as to lower or raise the balloon into the air current which is going in the right direction. These free balloons have been of little value in war and have been used mainly for sport, meteorological research, and training balloon observers.

Early in the war the ordinary anchored balloons such as those which have been on exhibition at our state fairs were recognized to have several distinct disadvantages, the most prominent of which was its uneven swaying. This motion would make even old sea captains seasick - a seasick observer would be of little use to an artillery unit whose fire he was directing. The first development came from the Germans who made the balloons in a sausage shape. The British added two lobes at one end which resembled elephant ears and the French put on a rudder resembling the trunk of an elephant. This is where the name "elephant balloons" originated. We have recently greatly improved on this design and produced at Akron O. a kite balloon which is very stable and has practically eliminated motion. These kite balloons have been of inestimable value in the war. They are anchored about five miles behind the lines often to a motor truck sometimes to a stationary windlass. The observer is in constant telephone communication with the battery below. On clear days he can see clearly from twenty to thirty miles behind the enemy lines. So important is this observation that the enemy will often send over a squadron of airplanes to risk their lives and planes in order to shoot down that kite balloon. And it is at great risk that a pilot attacked a balloon. The development of defensive measures has outstripped that of offensive tactics, and so certain pilots are specially trained for anti balloon work. Such a specialist was Lieut. Luke of the American Air Service who has shot down some 21 enemy balloons.

There is yet another class of aircraft in the lighter-than-air classification. This includes those machines which are lifted by gas and propelled by engines; the blimp and the Zeppelin. The dictionary name for blimp is dirigible balloon but it is not known otherwise than plain blimp in the air service. These craft are mainly used by the Navy for spotting submarines. They have a speed of about 60 mph and can spot submarines submerged 100 feet deep. They carry depth bombs with which to attack the under water craft; but usually they work in conjunction with destroyers with which they are in communication by wireless. The Blimp sees the sub and tells the destroyer where it is.

Zeppelins are used mainly for long distance bombing. They were of great value at first but defensive measures have been so perfected that they are now of little use. We all recall the last Zeppelin raid of England when five of the 7 Zeppelins were brought down before they could get back.[15] When we consider that these craft cost Germany about $750,000 each we see why she did not try it again. Zeppelin type dirigibles will undoubtedly be used to great advantage in the future for passenger and mail service. [The Zeppelin is a very large sausage shaped gas bag (500-300 feet long) enclosed in a metal framework. Suspended from the framework are the compartments which hold the bombs, the pilots and the engines. Zeppelins are usually "pushed", the propellers being in the rear].

We now come to a consideration of heavier-than-air machines. It has been stated that the first flying machine was the boomerang with which the Australians became so adept. And Australia has claimed the honor of originating aviation. But no one disputes that Wilbur Wright was the first man to rise from the ground in a self propelled airplane. Airplanes as used for war purposes are classed as Scouts, two-seater fighters or reconnaissance machines and bombers.

Scouts are offensive machines. They are built to hunt out the enemy and fight him. Scouts carry but one man. They are usually very small and fast and maneuverable. Their two machine guns fire straight ahead and are so synchronized with the propeller that they miss the blades as they revolve 1500 times per minute in front of the muzzles. Everyone has seen two sparrows fighting a crow. They fly around him as he is flying at top speed. While one of them is working the crow's head, the other will fly straight at his tail and peck him. These sparrows are very much like two scouts attacking a heavy bomber, or a two seater. If those pecks were streams of lead which would penetrate to the heart of the enemy and the crow had some lead in reserve for defense, the analogy would be complete. The two best known scouts are the Spad and Nieuport, others are the Camel and the S.E.5, and the Sopwith.

---

[15] The Zeppelin blitz of WWI was not militarily effective but took more than 2 years for England to develop a level of defense. London, Liverpool, and surrounding towns were attacked in 1914-1918 resulting in more than 500 civilian deaths. More than 30 Zeppelins were shot down or lost in accidents.

Two seater fighters were built to obtain a greater range of fire and keep as much maneuverability as possible. The pilot still has his two guns and in addition has an extra gunner behind who can shoot in any direction. These machines are used mainly for reconnoitering behind the enemy lines. The Bristol fighter which was such a failure when built in this country, is the most successful 2 seater in France. The De Haviland which we built at Dayton and Detroit is also a 2 seater fighter.

Bombing airplanes are usually very large. The Handley-Page and the Caproni are the best known types. They are from eighty to one hundred thirty feet from tip to tip and carry about a ton of bombs each. Bombers are usually slow and un-maneuverable. They are strictly defensive machines, avoiding fights wherever possible. But they are well armed. There are three gunners to fire at attackers. Usually also they are accompanied by Scouts as an additional protection. These bombing planes sometimes go as far as 600 miles behind the lines, but most of their trips are 100 to 200 miles.

Airplanes are also classed aeronautically as monoplanes, biplanes and triplanes. Monoplanes have been used with some success from time to time, but on account of the difficulty in making them strong enough they have not been very successful.

Biplanes which have the two planes trussed together with struts and wires are the most easily and strongly built airplanes. Practically all the types in use, scouts, 2 seaters and bombers are biplanes.

Triplanes are not used very much. The Kirkham Triplane which was reported to have a speed of 168 miles per hour is the only successful triplane we have developed.

Love Field, Dallas Tex.
Mar 14, 1918
Dear Mother,
Thomas wrote me that you are planning to go down to see him soon. Do go; it will be well worth it. And as it is probable that he will leave in a couple of weeks you could go right away. And then you could go on to Florida. Florida must be beautiful now - orange blossoms and

magnolias! I'm sure you will enjoy it. And you need it, Mother. You need travel and interest. It is more important than we realize. If you confine yourself to your work there, your viewpoint narrows, doesn't it? I find most people think that way. You can be of more service if you get away - like Jane Addams does. One of the things that, more than anything else, helped reconcile me to the farm life was a speech by a successful Tennessee farmer who said that the farmer who did not manage his farm so that he could leave at any time was a poor manager. Couldn't that apply to other professions?

We are doing so little here that there is nothing to write about. The weather is rotten for flying. Either the wind is blowing 45 miles per hour or the dust is so thick that we cannot see the ground from 1000 feet or it is misty or cloudy or rainy or muddy. The result is that we have put in very little work. I was up day before yesterday practicing the "wing over wing" and Immelmann turn, two very simple but useful maneuvers. They are quite hard to do on the Standard plane, the only one we have here.

Last night I went to a dance at the country club. It was given by the Dallas hesitation club for officers of Love Field and Camp Dick. Most of us went alone and met unescorted young ladies - - - I enjoyed myself very much.

At the dance I met three old Scott Field pals, Pickerell, Williams and Clark. They had just been sent down to Camp Dick from Fort Sill where they finished the course in Reconnaissance and Artillery observation. They told me that all the fellows who had been sent across from Scott Field went down on the Tuscania.[16] I guess most of them were saved. That must have been some experience. In that bunch were Cameron, McDavid, Bird, Branshaw, Potter, Beauton and Ferris. I haven't found anyone else in that whole Scott Field crowd who was as lucky as Meister, Ellis and I. And of the three I am luckiest for I finished after they did and now have the same position they have.

---

[16] [The SS Tuscania luxury liner transporting 2397 American soldiers was torpedoed February 5, 1918, by the German submarine UB-77. The liner went down with the loss of 210 lives.] — TKO

We three now have one hangar to ourselves. We have nine planes of our own with the privilege of taking any plane on the field or that is sent in. we absolutely have the cream of the work and privileges. During this rotten weather I am feeling ashamed that we cannot do any work for noblesse oblige, you know.

This morning we swiped some rifles and shotguns and went out on the range. It was great fun but I surely need practice.

Edmund

Love Field, Dallas Tex.
April 7 - 1918
Dear Sister,

Miss Gregory has asked me to send you her address so that you may write to her if I get sent to France or have a serious accident or forget her address or something. It is Allene Gregory - 1008 Lincoln Ave., Urbana, Ill. You won't forget, will you?   I am going to ask another favor, Margaret. Will you promise to turn to me if you need anything for yourself? I want you to know that you can rely on me for love and understanding under any circumstances, and I'll do my best to be worthy of your reliance.

I wish you could visit me here in Dallas. This is such a wonderful town. I am sure you would enjoy it. And when the weather is good we do lots of interesting flying. One day last week we had 94 machines in commission, and most of them were in the air all day. We have soloed every man in camp and most of them have had a good deal of solo work. Almost a hundred have graduated and been commissioned. They are all crazy about stunt flying, and regard it as the most important work they get. They like the Immelmann turn best of all. It is a very spectacular stunt consisting of rolling over laterally onto your back and then coming out as from a loop.

Another stunt I am just perfecting is that of making a loop and when upside down on top of the loop straightening out and flying upside down for a second - then turning over laterally and flying straight. Now a combination of the two would seem in order and so on Friday I made an Immelmann as far as (b) and then instead of coming out as in (c) I completed the roll and came out straight going

in the same direction as (a). Another stunt we are working on is a series of backward and forward stalls like this.
If we can do this on the front they never will be able to hit us. They'll be laughing so hard they can't shoot straight.

How is school work? Are the children waking up to the bigness of the time and work?

Edmund

April 16, 1918

Dear Sister,

I am sending Tom's letter to you. Your letter came on the same day that this came. Did he send you Florence's picture? She is very pretty - don't you think?

I got a card from Dwight from Camp Merritt, N.J. I expect he will be going across soon.

I just heard from Mary Jane. Did you know about her accident? That must have been awful. I hadn't heard about it at all. She wrote a wonderful letter and called me her "big brother". I was always more or less embarrassed when out with her last year and I think that led to considerable misunderstanding. She is a very wonderful girl but I hardly know her at all. It seems queer that she endured me last year. I was so evidently a lame-brain.

I met Owen Price over at Camp Dick the other day. We flew over Dallas with Liberty bond literature and landed in Camp Dick, which is a concentration camp for cadets and flying officers. He had just arrived from San Antonio and he had no idea I was at Love Field. He just came over to see who was in the machine and - we met. He has been out to Love Field and flown the Standards. San Antonio evidently gives a poor course for he has had almost 100 hours of flying and no individual instruction since his dual work. We are making that a big point in our course. I had a man today who hardly knew how to fly. How he ever got to advanced work I don't know. I had to practically teach him how to fly. He was quite sick when we came down. First he got dizzy and then nauseated.

We have some doctors here who are interested in carrying on experiments to overcome hypersensitiveness to dizziness and nausea. They are doing it by the constant use of the "dizzy" chair. A man who gets sick easily is revolved in various positions - daily - until he is immune. One of our graduates has evolved a scheme of giving certain exercises in calisthenics which tend to accustom the cadet to unusual positions. Forward, backward, and sideward somersaults, rolls, spinning, etc. are used.

I don't understand your idea of cowardice, Sister. You are very courageous to consent to put off your marriage. Your "yellow streak" is true blue. It is Miss Gregory who has made me tolerant - if I am tolerant. She is the most wonderful person I ever knew. She is so very tolerant and friendly with the world. She is like a wise doctor who does not condemn a man because he is sick but seeks without prejudicing him to find a remedy for his ills. It is so easy to prejudice people by condemning them, and that does more harm than good. Jesus brought that out so clearly in his treatment of the sinful woman that I cannot understand how the orthodox churches seem to miss the point so entirely. I am almost intolerant of orthodoxy. But its tendency is upward anyway.

Good bye, Sister,
Edmund

May - 7 - 1918

Dear Mother,

I am so full of uncertainty and disappointment that I must unload to someone. Are you willing to be the goat? Do you remember I told you in my last letter that Meister and Ellis had applied for assignment to Boston to take the course in Aeronautical engineering at Mass. Institute of Technology? Well, Meister has received his orders to go and Ellis' order will probably come through this week. I also applied in spite of my lack of qualifications. About twenty graduate mechanical engineers were turned down so I didn't take a chance on turning in my application with the rest but went in to see the Commanding Officer personally. I told him I was lacking in educational qualifications but I could study and I was very much interested in that kind of work. He said,

"Do you know what they want to make out of these men? -- They want assistant engineering officers to test out ships. -- Jobs like Lt. Showalter had here. You have a good chance by sticking where you are to be Officer in Charge of Training soon. With your flying experience you could be of much more use as Assistant Officer in Charge of Flying or Flying Officer than studying Calculus or having charge of hangars. But if you want I will recommend you for this course."

Well that put a new light on the subject and I decided to take his advice. But before leaving I asked him if he could suggest something for me to do to improve myself - to make me of more use than I am. I told him I thought I was standing still. He told me, after a minute, that if I would spend an hour a day reading and digesting all the War Department Stencils on Aviation - that I could soon be the best informed officer in the Signal Corps.

But Meister and Ellis are going and that breaks up the trio. That's the worst part of this Army life. Just as soon as you get to really like a fellow you are busted apart and never see each other again.

That puts me in as Officer in Charge of Acrobatics. I am surely going to make my influence felt if I have to rearrange the flying schedule. There are lots of things that could well be rearranged. We are going to have an auxiliary flying field for stunts and I will have charge of this field. Just now we are doing double work. We take up every cadet as soon as he solos and show him how to get out of spins and sideslips. He then goes back to solo work for fifty hours and just before he graduates he comes again to the stunt class and gets his advanced flying.

Friday I received a wire from one of the Greenville girls announcing that the Red Cross squad would be in Dallas to visit us Saturday and Sunday. They came out to the field Saturday noon - that afternoon we went down to Camp Dick and gave an exhibition of trick flying over the Fair grounds. Other machines from Fort Worth were there but the crowds were unanimous in deciding that we "put it all over" them. It was the first time I had done any stunts since my work at Greenville. I made quite a hit with the populous when I climbed out of my ship and grinned. They all knew about my "wreck" and my missing tooth. There were about ten thousand people in the grandstand and in front of the track to witness our performance. We were complimented by Major Rinehart and Captain Harvey and the famous French Ace Lemaitre who came from Fort Worth.

Saturday evening we went downtown and took the girls to a vaudeville. The girls are Lela, Zula, Inez and Nina Cantrell - daughters of the Greenville physician and a Miss Harwell, a Miss Rab, a Miss Harrell. They are good sports - out for a good time - in love with all soldiers - but absolutely above board all the way through. After the show we went out to a country home and had a wonderful time generally. Pat Maloney brought us back in his big car so we could get some sleep before morning. Sunday the girls came out to the field and ate at the Officers Mess with us. We went riding all afternoon and I got Owen in the party. Altogether I think we gave them as good a time as we had at Greenville.

We are going to have a dance here at the post next Saturday evening. It is to be a very formal occasion. We have a 40 piece military band engaged for the music. The girls are to come from the Southern Club - the best girls in Dallas. I am inviting Mary Hobson, a very interesting Southern girl. [We will probably come out in her Packard.]

Owen and I have been together quite a bit lately. He comes out to Love Field about once a week and I often see him down town at night. I guess he is enjoying himself. Camp Dick is not a very enjoyable place but he gets away almost every night.

I am making the N.Y. Life Insurance Co a cheque for my premium. My Travelers premium must be almost due. I hope it does not go too long again like it did last year. I am quite sure it is $16.95. I am enclosing a cheque for the amount. If you have not already made any arrangements about it I wish you would send them the cheque. I am afraid it is overdue. I will soon mail you a cheque for the $95 you loaned me in Lake Charles. I am catching up now - I bought $500 worth of 3rd Liberty Loan Bonds. They will be paid for Jan 31st, 1919.

Margaret Byard's picture has the place of honor on my chiffonier. I was struck speechless [and writeless] when I received it. My darling Peggy is wonderful!

I have given instructions that my 3rd Liberty bonds be sent to Sister when they are paid for. I want her to have this $500. My first $500 is for you.

Did I tell you my J.M.A.[17] request was disapproved. That means no promotion for some time. Do you want me to return Sister's letters to you? I have saved them all - Edmund

I am sending a few pictures of the Greenville wreck. Please return them - I am making an album.

---

[17] [J.M.A. - Junior Military Aviator - a rank preliminary to full Military Aviator] — TKO

[Margaret Byard (Peggy) is his first cousin on his mother's side; Owen Price his first cousin on his father's side. His mother, Abby Dyer Allen's, sister was Margaret Dyer Byard of Hamilton, Ohio. His father's sister was Louise Allen Price. "Uncle D.R." Byard and "Uncle Enoch" Price were both much appreciated by their wives' nephews and niece after their father, Edmund Turney Allen, Sr., M.D., died. Both offered homes and loans. Loans gratefully accepted were dutifully repaid.] —MRB

[Dwight Moore was Mother's fiancé. On the very day they went to get their marriage license, heeding her mother's advice, they decided to wait until he returned from war – just in case he didn't return. He did, but she was too straight-laced to accept that he had been what all – or most – soldiers had been at the end of the fighting in France – accepting of many forms of gratitude/hospitality of the French.

Owen Price was their first cousin. Owen's mother, Louise Allen Price, was "like a second mother" to my mother, she said. Mother (Margaret Allen) was living with that family in Morgan Park, IL (now part of Chicago) when she and her cousin Allen Price graduated from Morgan Park HS in 1915.] —MRB

May 28 - 1918

Dear Mother,

I am almost entirely caught up on my letter answering. Only twelve remain to be written - You and Sister first - then an old pal Alexander - Mrs Ralston - a Dallas girl - a former pupil - Mr. Bryant - Dwight - Margaret Byard - Miss Gregory and Della Frazer. Maybe I can finish tonight - maybe not. I quite enjoy writing letters now. usually they are terribly boresome and colorless but sometimes they are enjoyed. Miss Gregory says I have ability in writing but then I have always taken great pains with my letters to her. Mary Jane knows I have no ability for we simply "kid" each other.

My request for leave has been forwarded to the Southern Department and my Medical Certificate has followed it. It will be some time before it returns. I would like to go up to Champaign and maybe to Chicago - but I can do that later. After June 25 my railroad fare while on leave of absence will be one cent a mile.

A nose examination this morning showed one side to be open and the other side open up almost to the top - where the septum is bent over and almost closes the passage.

Is Thomas in France? I have not heard for almost a month. Owen left about three weeks ago for France. He will probably get his advanced training over there.

This post is improving wonderfully. We have a C.O. who is all on his toes to get new ideas for improving efficiency.

I may see you in a week from now.

     Sincerely,
      Edmund

    [LOVE FIELD
    DALLAS, TEXAS]
    May 28 - 1918

Dear Sister,

  Your letter was very interesting although I'm sure all your compliments are misplaced. What can I need of the money. The government needs it and then you will need it. It's logic - nothing else. I'm glad you enjoyed the candy. If it will make you fat I'll send a box every week. Is that a bargain?

  Falls are a matter of gravity not of amount of flying. We fall every day. One day I fell 8500 feet; but as I happened to be up about 9000 feet to begin with I stopped in plenty of time and not too suddenly. It's not falling that hurts a man or a plane. It is the sudden stop that does the damage. Now it happened that on one of my falls recently I had about 100 feet less altitude than my fall required so I stopped suddenly just 100 feet short of my objective. Another time I failed to notice that I was falling almost twice as fast as ordinarily and so I reached the stopping place sooner than I expected and naturally stopped quite quickly. Now my 'plane stopped before my nose, in fact, my nose kept on going for several inches after it reached the stopping place resulting in a serious dent in the cowl of my ship and incidentally a broken nose - But enough of this.

  I am coming home by way of Marshall next week, I think. I am getting a month's leave of absence (not furlough) and I'll stop off and visit you a day or two.

Tried to get the folks here to let me fly home but as I have orders from the surgeon to suspend flying for a time, I was unsuccessful.

Mary Jane is abused, poor dear. But I've written and promised to be good in the future.

See you soon --------- Edmund

[No place; text implies Love Field]

Dear Mother,

I am back home and mighty glad to be here. Now what would you think if I should walk in and say Hello to you some day soon? The folks here are very insistent that I take a rest and so I applied for a leave of absence and it was promptly granted - for ten days. The surgeon however claims that ten days is not enough and so I am waiting until a request for thirty days is sent to the Department Headquarters and returned - a matter of probably a week. Dr. Riley examined my nose and found that the - (what do you call the partition between the nostrils) was bent and would probably have to be removed as it obstructed the air passages. I have a scar across the outside of the nose where the doctor sewed it up. It looks rather ugly but it may disappear. A cut over my eye and one at the corner of the eye completes the disfiguration. I did not want to walk in on you and scare you., so be prepared. If I wait a week or so the scars may be gone and you would not be able to tell I had an accident.

Everything is changed here. I am quite reluctant to leave. This is to discontinue as a primary training school, as soon as we finish the cadets we have now under instruction, we will instruct only officers who have completed their primary training. I should not be surprised if Love Field were made the best, most advanced training school in the country. We are all enthusiastic, of course.

If I come home I will stop at Marshall and see Sister for a few days. I wonder where I can stay in Hamilton. It is a shame to impose on Aunty again.

Goodbye Mother, I'll wire when I leave -

Edmund

[Only date found on telegram is "3RD" after "DALLAS TEXAS."
Context of Love Field sets year 1918. Reference of "JUNE NINTH" sets month.]

LOVEFIELD   DALLAS TEXAS 3RD
MARGAIRE MALLEN
*   MARSHALL ILLS*
MY LOVE [transmitter error? Perhaps "MOVE"] *WAS DELAYED I STOP IN ST LOUIS FOR NASAL OPERATION/ARRIVE IN MARSHALL ABOUT JUNE NINTH I WANT TO RUN UP TO CHAMPAIGN FOR A DAY ON THE TWELTH THEN BACK AND ON TO HAMILTON BEST WISHES TO THE RALSTONS AND MARY JANE SORRY I CANT BE THERE FRIDAY*
*   EDMUND T ALLEN*

Excerpt from "Boys' Life "-March 1931   Ten years later, ETA recalls his "crack-up in an article for the Boy Scout magazine.

"Before I left for England, after spinning into a crack-up on Benbrook Field[18] in the early summer of 1918, I was given further physical examinations to see how much my injuries had affected me.

Among these examinations was what we called a "rebreather" test designed to show how well your brain would work at different altitudes. While you did three different things — adjusted the brightness of a light with one hand, regulated the pointer on a dial

---

[18] Airfield near Fort Worth established in 1917 and decommissioned in 1919

with the other, and worked a foot pedal to obtain a particular volume of sound — the amount of oxygen that you had to breathe was gradually diminished. Under those circumstances, as in flying through high altitudes, there comes a time when, without your being aware that any change is going on — into complete unconsciousness.....The verdict that was given after the crash was reversed by these new examinations. The opinion now was that I could fly, under certain limitations. My heart had developed a bad skip, and should not be put to any undue strain. The rebreathing test indicate that I should not fly above 8,000 feet.

That bothered me a good deal. What flying could I do if I had to keep under 8,000 feet? But it was at least better than being ordered never to fly again.

As soon as I could, after I got to England, I borrowed a pursuit ship for a solo flight. It took me a little while to get used to the plane; everything seemed at first a little strange, after not having flown for so many weeks. Then, as the feel of the air became familiar again, I put the ship into a real climb. I went on up until the altimeter showed 18,000 feet. That was higher than I'd ever been before. I felt no bad effects. That flight eased my mind a good deal." — ETA

Instructors Survey England / McCook Field   1918-1919

In the summer of 1918 ETA is detached with a group of ten instructor pilots to England to study British flight instruction techniques as developed by Henry Tizard and Frederick Lindemann. These men taught the technical disciplines and scientific approach to flight testing that ETA would champion throughout his career and could well have shaped ETA's approach to flying at NACA and his independent flight test activity.

American participation in the ground war is at a peak; the battles at Chateau Thierry and Belleau Wood in June and July, and the Meuse Argonne in September through November hit with American casualties in major numbers for the first time. The final 100 day offensive resulted in the largest loss of American lives of any single battle in history. More than 25,000 lives and 67,000 casualties over all. Although the first American participation in victories in the war had enormous cost, the battles of the summer of 1918 were climactic and proved to be the turning point in WW I. The final battles at Meuse Argonne convinced the German command they could not succeed.

Eddie travels to Farnborough and Martlesham Heath in mid-August, weeks before Meuse Argonne. They are in England during the final 100 day traumatic ground battle of WW I. They depart England on October 22 and on November 18, 1918, the war ends.

Returning to the U.S. he is sent to McCook Field at Dayton, Ohio, which has become the Army's aviation engineering and test center. ETA was assigned to a special testing squadron where he spends the rest of his Army career. Major Rudolph "Shorty" Schroeder is Eddie's boss while he is still in the service. "Shorty" will become well known for his high altitude flying and work with turbo-supercharger development. "Shorty" is actually 6 foot 4 inches compared to ETA at 5 foot 8 inches. It is interesting that one of ETA's letters describes a high altitude contest as part of an army aviation event which Eddie celebrates for his record-breaking achievement. Later letters will note that "Shorty" would be ETA's only real competition for the NACA position should he elect to apply for it. As Eddie says "then we would have some tryout". Eddie and Shorty maintain communications in later years.

Most of his time is spent at Langley Field, Virginia. He is decommissioned as a lieutenant in August 1919 and immediately signs on with the NACA as a test pilot sponsored by his mentor Edward P. Warner, the chief physicist at NACA.
Note that in this period ETA's letters address two separate "wrecked plane" incidents although the letters may initially lead one to believe they are related. Obviously, air transport still depends upon a pilots skill to make "forced landings". As ETA said

"two open fields are bigger than one" and so he chose both and went through the fence with his single engine DH-4 in route to Washington D.C. The "wreck" in the Philadelphia area apparently involved a twin engine de Havilland DH-4B.

[Two postcards with postmarks almost obliterated. Context suggests July 29 & 30.]
   [First one, postmarked New York N.Y.]

Dear Mother,

Spent the day with Sister on the Hudson. Had wonderful time at Poughkeepsie and Vassar. I went up on the Hudson to see her and then we both took a trip thru the Palisades returning by train at midnight. Stayed at hotel - saw sister in morning in her uniform - then returned to N.Y. on the NY central down the Hudson. Sister is well and happy and taking a wonderful course. -----Edmund

EDMUND AND MARGARET 1918

[This was during the time that Margaret Allen went to Vassar, where she took a course in nursing for college graduates. She wanted to do her part for the war effort, as her two brothers were doing in the military. The program ended with the Armistice in November and she went back to teaching, in St. Paul, but she used her nurse's training when a mother, and we, her children, benefitted in numerous ways. I still remember her teaching me how to make square corners with bedding. Also little songs for sick children she learned in nursing school and sang to me while a thermometer was in my mouth.] — MRB

[Second, postmarked Grand Central (Station)]

Dear Mother,

Another day - They surely pass fast in New York - Back from Poughkeepsie I met Bob - saw my baggage on board - then went out to Mineola where I met several friends. That evening I went out with two girls whom Bob knew. We saw "Tiger Rose" one of the best shows I've seen - then went to a cabaret, got in at 3 a.m. up at 7:30 and at work again - Don't worry!

Edmund

[across top:]

I had Bob send you a box of my belongings. They went Express Collect.

[...and it's off to England...]

[no letterhead, address or date, though he's in Patterson, NJ, probably June or July, 1918]

Be sure to put 1st Lieut. Edmund T. Allen A.S.SIG.R.C. and you may address me after a few days -- A.E.F. via N.Y. City.

Dear Mother,

We have now come under the military censorship and are forbidden to write almost everything including port of embarkation, time of departure, description of boat, trip, scenery and people aboard.

Of course, I am not aboard yet. I am staying at Ellis' home in Patterson N.J. It is a two hours ride on the subway and electric car and ferry to Hoboken where we have to report every day.

I have been very fortunate in the matter of equipment as we have to have a lot of stuff. I went up to the treasure and trinket club and saw Mrs. Bartlett who gave me an order on one of the largest stores in New York for a $35 trench coat and a $30 bedding roll and mattress and cot. All of these things are very good.

My orders to leave Washington came on Monday afternoon and I left for New York Tuesday morning. Monday night Floss and I went for a long car ride after a good dinner at the Washington Hotel. Tuesday I went up to see Bob at his office and he told me that an express package had come for me at the house. I had just an hour to make the train but I went out to the house, said good bye to Ruth and got my raincoat. Then I came down and saw Floss who presented me with a very wonderful writing case. I made the train with still a minute to spare. This was the first real fast train I was ever on. We went 65 miles per hour in places and picked up water on the fly out of the trough in the center of the tracks. It was a dirty trip. My ears and eyes were full of cinders when I got to Philadelphia. I had lunch there and then got a train to New York.

The Penn station in New York makes the Northwestern station in Chicago look like a small town depot. It is certainly a wonder. But New York itself is simple. A blind man could find his way around down town. Famous Fifth Avenue is just "State St. Chicago" and Broadway is Madison St. The blocks are all only half a block long or less, so in reality it isn't very far from 23rd St. to 198th St. The tubes are a real improvement over anything else. Riverside drive and the Hudson are great. Several large battleships and some French Cruisers were anchored out in the river. They lend a real warlike atmosphere to the city.

I will write you every day from now on and so when you miss a daily letter you will then wait about two weeks and get the card announcing my arrival abroad. From then on you should get letters quite regularly except for a wait between boats.

Mr. Ellis is a blacksmith - a very fine and educated man, rather rough in some ways. He is English, his grandfather having been horseshoer to the king and queen. Mrs. Ellis is a kindly old soul who worries to death about her three sons in the service. Bob is the oldest son and he is idolized in the home. He treats me just like a younger brother and I rather resent it. His brother Wilford came home to see him today and Bob went to his room and went to sleep. Everyone waits on him and follows his orders willingly. I hate to go around New York with Bob because he goes around forgetting I am there and expecting me to follow him. Yesterday he was feeling a little bum and I had lots of pep so I took the initiative and walked about two steps ahead of him all the way over to Hoboken and back.

Has Peggy returned to Hamilton yet? You may have time to get one letter to me at

"U.S. Military P.O. Office of the Mail Censor Hoboken N. J." If it arrives after I leave it will be forwarded.

[No signature. The next letter was written in England Sept 20.]

[Martlesham Heath, England]
Sept., 20, 1918

Dear Mother,
I left New York some seven weeks ago and I haven't heard from you yet. Now I can imagine what Thomas went through when he left the States with two loved ones from whom he was to be broken off for so

long. I hardly believe, however, that he had to wait so long, for his mail would go thru but one post office besides his own mail headquarters in France. Mine on the other hand, if addressed simply to A.E.F. goes first to England, (probably,) then to this same mail headquarters in France, then, if they know where I am, back to London. In London it goes first to general headquarters then to aviation headquarters - (BS #3., S.O.S., A.E.F. London), then to the American Officers Inn, then to Martlesham Heath. Besides the letters which were sent to Love Field and forwarded from there to Washington, then to Hoboken, and probably came across on the same boat I did, besides these, I have had but one letter, and that one was addressed to Hoboken on Aug 3.

We did not expect to be here at Martlesham so long; but there are difficulties we did not anticipate, in the way of our getting the information we want. So we shall be here three or four days longer. At present we expect to be back to New York before November 1st. And unless I am much mistaken we will be in Hamilton soon after we return. That will delight me very very much. I hardly believed that two months after leaving New York I should ache to see it again.

Our station is in quite a picturesque location. From the air one can see the coast. There are forests about quite thickly. The heath is treeless but is hardly as good a flying field as Love Field. Our quarters are not exceptionally comfortable either altho we have an old rug, a chest of drawers, a wash stand, on which is a china wash bowl and a pitcher, which our busman fills with warm water twice a day. A little stove in one corner adds to the scenic effect. It certainly doesn't add to the warmth - we cannot get any coal until winter. I have not been warm since I arrived in England; since, in fact, I left New York harbor. Three days ago when for about an hour in the afternoon the sun struggled out and the thermometer managed to get up to 80°, these Eskimos would greet you with, "It's beastly hot today, isn't it." It rains, rains, rains, or just clouds up and blows and blows.

Completing the furniture of our room are two very uncomfortable cots and our two trunks which we have put one on the other and used as a table for writing letters.

There are about fifty officers here. Several of them are Canadians, several South Africans, a few Australians and one a Scotchman. The rest are dyed-in-the-wool Britishers and pretty nice chaps at that. Most all of them are home on sort of leave from the front. Two Canadian boys I have been with a good deal have spent six months in France and each one has shot down five Huns. They are only twenty years old now.

I come into closest contact with all the officers at mess. In the morning after breakfast some of them go into the ante-room where there is a fire and a lot of comfortable chairs. There they read the morning papers and pick out on the big map just where the latest thrusts have been made. Most of them are pretty familiar with the roads, ridges, rivers, etc. along between Dunkirk and St. Quentin, having been there themselves, and so the war news means much to them. After dinner or luncheon they smoke and read until 1:30. At tea everyone gets together for an hour and after dinner they read or play cards or listen to the gramophone and drink whiskey and sodas until bed time. It is an interesting crowd to be with -

Edmund

Excerpt from "Boys' Life - March 1931   Describes ETA experiences in England in 1918.

"Most of the English flyers that I came in contact with were men who had been on the western front for a year or more. They regarded a chance to work in England for six weeks or so, instructing students or testing planes as a "cushy" job, given to them to help rebuild shattered nerves.....

One night at one of the big English training camps where I was sent to observe testing methods, I got into a poker game with an officer who wanted to have a bonus of one pound, nearly five dollars, paid to anyone who held a royal flush. That seemed pretty high, but we all agreed to it. In the course of the evening this chap himself held a royal flush, and we all paid up.

About that time there was a report that a German seaplane had been sighted twenty miles off the coast. A lot of planes went up to ward off an attack, but they didn't find anything.

The next day I was talking with a couple of other officers in the mess hall, just after lunch, when there was an explosion that nearly knocked us down. It broke every window in the place. We thought of course it was a German air raid. But it wasn't. It was a big English bomber that had crashed on a test flight, while carrying three six hundred pound bombs. Bombs, gasoline tanks, and all had exploded when it struck, hardly a quarter mile from the field. The pilot was this chap who had held the royal flush the night before. I picked up a piece of his knee, quite near the door of the mess hall.

That incident occurred only a few weeks before the Armistice was signed in November.

The war came to an end before I got a chance to try out my flying knowledge at the front." __ ETA

Excerpt from "Boys' Life" - March 1931   Describes ETA experiences prior to being discharged from the Army in 1919.

"Nineteen eighteen, the year in which the war ended was also the year of the greatest epidemic in modern history - influenza.

When I got back to America I was ordered to Wright Field, at Dayton, Ohio. There, with the sudden change back from war to peace, everything was disorganized. But before I had much chance to think about it, I came down with the flu, and the mere matter of living or not living was the only thing that seemed very important.

I was in the hospital about a month. In the bed next to me was a mechanic, who had been caught by a propeller he was cranking. The blades had thrown him around in a complete circle, breaking almost everything about him in the process. Both his arms were broken in two places; the bones in both hands were broken. Although he wasn't as near death as I was with the flu, he suffered a lot more.

Just before Christmas I was discharged from the hospital and began flying again as a test pilot, trying out new airplanes. But there weren't many new planes to try out. The war department was canceling orders for experimental planes as fast as it could.

The very thing that would have meant a lot to me during the war came to me now, when it was too late. I was put in charge of a squadron. There were about 150 men in it, but their flying enthusiasm had mostly died when the war ended. With no chance of going to France and getting killed as a reward, they were pretty fretful about doing Squads Right or having to salute any expurgated officer.

I applied for my discharge, but it didn't come through until the end of July, six months after I had been transferred from Wright Field to Langley Field on the Virginia coast. Out of the army at last in the middle of 1919, I stayed right on at Langley Field for a couple of months flying as test pilot for the National Advisory Committee on Aeronautics. " — ETA

Langley Field, Hampton, Va

March 23 - 1919

Mother dear,

I have your letter and realize that I have not written for a little over a week. My last letter was written from New York, wasn't it? I told you of my pleasant trip and my visit in Boston. That visit was very enjoyable and I do not regret a bit that it rather depleted my balance this month. After returning to New York I was very anxious to find out what to do with my wrecked 'plane. I could not return and leave it there and I could not ship it away without authority. Washington did not answer my wire and in the meantime my leave was expiring. In the afternoon I sent a rush telegram to Langley requesting extension of leave until I could dispose of my wreck. I received the extension that night and so instead of returning to Langley, I went to Washington to find out why they had not answered my request. I found Kenneth Leggett working in the training department. (He was one of the ten who went to England in July.) He told me instructions had come thru for me to ship the plane to Mineola, New York, for repairs. That meant for me to go back to Philadelphia and supervise the shipment. Someone had told me that the motor messenger corps provided cars to take officers around the city on official business, so I called them up and ordered a car to take me out to Bustleton Aerodrome, seventeen miles out of Philadelphia. As I was leaving the booth a canteen worker asked me if I could use a ticket to the opera that night. Could I use it?! Later I found that the last seats had been sold two weeks before for this night.

The boulevards of Philadelphia are beautiful. We went twelve miles out on a boulevard three hundred feet wide and the other five miles on the Lincoln Highway. I called up Mineola and asked them to send a wrecking crew down in a couple of trucks to take my machine up there. It is about 125 miles by road. Yes. They would send them tomorrow morning early. I went back well satisfied and heard Caruso in La Boheme.

That night I stayed at the Navy Club, a beautifully appointed club maintained for the convenience of officers in Philadelphia, and I stayed there all the rest of that week, for when I went out to the aerodrome the next afternoon, Mineola trucks had failed to arrive; and when I called them up again I found they had not started and would not start without express orders from Washington. When I told them I had orders from Washington they said they would call Washington and find out. That's the last I heard of them. Next day I called up the motor transport corps in Philadelphia and asked them if they would haul my airplane to Mineola. Some Captain said, "sure" but when he talked to the Colonel about it next day he said I had better come around and talk to the Colonel myself. This I did and was told they had no authority to go outside the Philadelphia district. He would wire Washington and ask for authority. While waiting for authority from Washington I was having a good time at the Club. Social affairs, shows, dancing, billiards, libraries etc. were all at our disposal. By Friday I was thoroughly disgusted with the delay so I asked for a truck to help me load the airplane on a car and ship it to Mineola. The Motor Transport Corps loaned me a truck and the Inland Warehouse loaned me seven men and by seven o'clock Friday night we had the DH4 up on top of a high side gondola freight car all ready to let it down. Then to add to the bad luck one of our planks broke and the bus fell over on its side and then upside down outside the car. We could do nothing more that night but next morning I got a wrecking crew from the railroad company to come up and lift the plane into the car with their steam wrecking crane. I had to see the superintendent of the road to get the crane and then sign a contract paying them $12.00 per hour for it. But at last the thing was done. The Bill of Lading was sent on to Mineola and my ticket was bought back to Langley. That afternoon, some Naval Ensign asked me if I would take his girl's sister to the dance. We went together and enjoyed ourselves very much. After the dance and dinner we went to see "Friendly Enemies" a play depicting the change of heart of a German-American during the war.

Monday morning I went to see how my job was getting on during my absence and I was surprised to find all the wing sections in place and the two side fuselages up too. I was very hopeful of a speedy finish, but this week's work has spoiled it all. It seems as if there had not been two days work done on the machine in the whole week. Fifteen men ought to be able to turn out a lot of work in a week. This week they loafed on the job. I am too easy on men. I make a poor boss, lacking the power to drive and get work out of them. I have placed a sergeant in charge of them and leave the details to him. But I don't check him up enough. The men all like me - but that doesn't get the work done.

On Saturday we got another of the motors installed. I am hoping to get the remaining one in on Tuesday. Then by Saturday I will know how much longer this job will last. By three weeks from today she ought to be in the air.

The Handley-Page that flew to Ellington came thru here today. They were carrying eleven men with their baggage, besides about a thousand pounds of spare parts and tools. It doesn't look very big beside the ship I am building.

I have finished Mankind in the Making by Wells. The book expresses the ideal toward which we are tending - of perfect births - perfect development mental and physical and better finer lives for men and women. And Wells is so concrete that we could really see the ideal.

The book you sent is here and I have read it. It represents a particular type of thinking, doesn't it? I never answered that letter of yours of about a week ago. It isn't easy to write about such things. I don't suppose it was easy for you to write me. I am very glad I have freed myself from the bonds of fear in thinking these things out for myself. I have made a mistake in talking so freely about my ideas, and yet it is only by discussing these things that we can get new points of view and additional food for thought. No, I do not think it would be all right for me to indulge in sexual indulgences. But not because I want to be an innocent child. I cannot believe that moderate sexual indulgence impairs a mans efficiency or manhood. In fact Paul believed it unwise to waste an undue amount of nervous energy in fighting down below a natural minimum our natural desires. Yet there is something else, a subtle something that only he "whose seed abideth in him" can attain to. It is this that I want - a wisdom, an insight, an understanding, a spiritual life that perhaps has little to do with orthodox religion. It is this that will make of me, if I can attain to it, a source of good such as it is my ideal to be. And an innocent child cannot be a great source of help.

    Perhaps I have made a mess of this explanation. I have never tried to explain it before - not to anyone.
    Edmund

    Langley Field, Hampton, Va.
    May 4 - 1919
    Dear Mother,
    Yes I expect to be here a month longer - probably more than that. This promises to be a permanent job. General Menober, the Director of Air Service has taken a lot of interest in the work and will probably push it. That means that I can remain here as long as I like - until September even. I don't know what I will do about the best man business. Guess I'll have to get a leave of absence. Last week I started up to the Conference in Washington with General Menober but I never got there. About half way the motor stopped dead and I started looking for some soft place to hit. As there was no field big enough to get into with a DH4, I made use of the old axiom "Two fields are bigger than one," and I picked out two fields and used both of them. I didn't know how much damage the barb wire fence in between them would do, but I did not have much time to waste so I took a chance and my gallant DH4 galloped right thru the fence, knocking down the posts and breaking the wire. On investigation we found that the cause of the

trouble was a stripped magneto gear. It took us three days to fix it and by that time the conference was all over so I came back to Langley. The little town of Rainswood near which we made our unwilling descent contains, as far as we could ascertain, two families. The Jones family runs the store, post office, church, primaries, Liberty Loans and educational institution. The other family just lives and takes up room. The main amusements are horseshoes and dominoes. Naturally an airplane was far more important than the other duties of life, and as a result during the three days, folks just camped out by the 'plane and examined it from top to bottom, never failing to put their feet thru the wings at every opportunity. Mr. Jones is a very strict orthodox Baptist. He told me when I offered him a cigarette, "Young man, I have lived fifty one years and have never smoked, chewed, drunk, gambled or danced or sworn in that time." A very remarkable record! Mrs. Jones' people were all Methodists. When she went into the Baptist church they disinherited her. And we had an idea religious prejudice was dead in the country.

Edmund

Langley Field May 13, 1919
Dear Mother,
Our aerial circus was a sad failure. It rained from the time the crowd started to come until they left. We had to omit part of the program on account of the low clouds and the rain. The people were very much interested in the various exhibits. I wish you could have been here. You would have seen and had explained to you all the instruments used by pilots and gunners, The compasses, indicators, gauges, meters; the guns and gun mounts; the cameras and photographic equipment; the various types of engines and airplanes in use from the little Thomas-Morse Scout to the Multiplane. Lieutenant George our ace who is now stationed on the field got a British S.E.5 scout from Washington and did some very thrilling stunts.

One of the stunts on our program was the destruction of a German battery. This battery of six guns was stationed out on the field. At a certain time they started to fire on Langley Field. This was represented by smoke puffs which were set off at intervals of a few seconds out at the battery. About this time an American Battleplane of the type used by our pilots overseas took off and bombed the battery. Two bombs were dropped and the resulting explosions left two large craters where the battery had been. About this time the enemy sent up their little scout, represented by Lieutenant George in the S.E.5. He engaged the American De Havilland and after a combat of some five minutes he succeeded in shooting away the controls of the larger machine. The De Havilland lurched heavily to one side and sideslipped down behind some trees. It was great fun. I was flying the De H. After slipping down behind the trees I flew low skirting the field until I could come in and land.

We also had a sausage or captive balloon. Lieut. George attacked the balloon and the two observers jumped out in parachutes. It was quite realistic.

We sold about $14,000 worth of bonds - rather less than we had hoped to do. Still this country is not rich. The farms are very poor and most industries are very small.

On account of the fact that we disappointed so many persons on Saturday, it was decided to repeat the circus on Monday. The crowd Monday was almost as large as that on Saturday and the circus was a great success. Many civilians got rides by buying $1000 worth bonds. The circus is to be repeated again on Decoration Day. We are trying to interest people in the Air Service to such an extent that they will force Congress to make us big and give us lot's of money.

I was going over my courses at Columbia for next fall with one of the men here who has completed four of his six years there. I take English (Introduction to English Literature), French (beginning), Philosophy and Logic (a year course), History (Survey of modern European History), math (trigonometry, algebra and analytical geometry) and physical education.

This comprises a pretty stiff course. There are 35 hours of class work a week which require 24 hours of preparation. This makes 59 hours or ten hours a day including Saturday. But I know I could handle it. If I worked as hard as I did at ground school I could graduate with honors. The point was brought up as to whether or not I could take the same subjects at Illinois for the first three years and get my A.B. there. Then I could get my Engineering at Columbia. Expenses at Columbia will be almost twice those at Illinois. If I get a scholarship at Columbia I could make it for $600. a year. That means that I could go to Illinois for three years and then go two years to Columbia and come within a year of finishing. Summer work might give me enough to finish the last year.

    I am writing to Columbia for more information.

    Your letter came today. I am glad of all the information about Hamilton. Have written to Mr. See and Uncle D.R. Have been reading Lord Chesterfield letters - Very interesting.

                                          Edmund

Langley Field
June 16, 1919

Dear Mother:
    I have not had such a busy week since the armistice was signed. It was mighty enjoyable to have every minute full. Everyone on the post was busy last week getting ready for the Field Day Saturday. In addition we had a dance Friday night and on Wednesday the adjutant informed me that I would run it. I have never done such running around looking for an orchestra at the last minute, getting the hall decorated, the floor fixed and proper cooling arrangements. The punch itself was a big item but the ice cream and cake too, with plates and napkins added to the rush. Then there was publicity among both army and navy officers - getting girls for fellows who didn't know any - persuading everyone to come - arranging for transportation (motor cars) and reception - collecting the money and seeing that everyone met every one else. I was a busy man , too busy to think about the big event of Saturday.

At noon Saturday every plane on the field went up and put on a welcome "jazz" stunting party. I had a De Havilland with a Liberty motor. Seeing a little French scout I dived on it and in a minute he took up the challenge and we fought for a half hour. I was tired out throwing my big ship around but he could still maneuver all around me. I got my observer a dozen good shots but if he had had ammunition I doubt if the fight would have been very unequal.

As soon as we got down I took up a T.M. Scout and attacked a sausage balloon and a dirigible, forcing the observers to jump with their parachutes. We scouts then stunted. I had the satisfaction of being told by the navy pilots that I had done a "roll" better in my T.H. scout than their pilot had in his French "Hauriot", a much better machine.

The big event of the day was an altitude trial in which we tried to beat the Langley Field record of 23,000 feet. I could not find a passenger for some time as I wanted a pilot who was very light. Finally Lt. Yaeger consented to go. We had a time getting our oxygen apparatus to work and the masks on properly, but finally everything was ready and we took off. I climbed to 10,000 feet with the motor throttled and then opened it up wide. The time to 10,000 feet was 11 minutes. It took 20 minutes to climb to the next 5000 feet and I didn't get to 20,000 feet for 45 minutes. It was slow work. At that altitude the motor was not giving much power and the plane struggled to get higher. I started to take oxygen at 15,000 feet and began to notice the need of it at about 18,000. At 21,000 I was yawning a great deal. My altimeter stopped about 21,000 feet and I continued climbing, depending on the altigraph to register the height. At the end of an hour I was about ready to quit. My breath was coming in little gasps and my eyes hurt. It was getting mighty cold, too. The water steamed out of the radiator with the engine wide open and congealed of the cap. I was hoping the gas would run out so I could come down with a clear conscience believing that I had reached the ceiling of the ship. An hour and a quarter - and I was ready to quit. I couldn't understand why the engine continued to run when the men said they had only put an hours fuel in it. Minute by minute I stuck it out. The big plane was hanging in the thin air, unable to climb with the reduced engine power. Finally at the end of an hour and a half I decided I couldn't stand it any longer. Nervousness, more than anything else bothered me. I began to see strange things floating around me. Strange tales we have heard about the ether fish that inhabit the upper air. I began to believe them. My head was about ready to pop. Then I thought about Lt. James' record of 23,000 feet. I knew I hadn't climbed much since the altimeter stopped but I thought I must be pretty close to 23,000 feet. Another minute and I would come down. No. I'll stick it out for one more. I was almost ready to say enough when the motor started popping and with an extra little gasp of relief I nosed her down and started a "dead Leaf". It was five minutes later that the folks on the ground were able to distinguish the Silver Ghost "falling". In ten more minutes I was on the ground. Everyone crowded around asking how high I had been. All I could say was that at 21,000 feet the meter had stuck. We pulled out the altigraph and opened it up. The ink had stopped flowing at 23,000 feet. No! on a little way it had started again and made a little mark at 25,000 feet. It was worth it - a Langley Field record - a De Havilland record - a Liberty motor record - an American two seater record.

         Edmund

Langley Field, Va
July 12, 1919

Dear Mother,
 I am writing with pencil because it will be faster and I am rushed for time. Returning from Washington I find my table stacked with unanswered letters - twenty three to be exact. After disposing of two business letters that I had to go thru immediately, I am settling down to the real business and yours come first. I have been slow about writing to you lately. It has been for one reason that I've been sort of torn up lately mentally. But I hope to start a regular schedule soon.
 My trip to Washington was fine. Espy and I arrived after a long slow journey and got a room at the "Y". Mrs Espy came up on the boat and

stayed at her cousin's house. For three days Murry Espy[19] and I hustled around - going back and forth way out to Chevy Chase to the Bureau of Standards for the Wind Tunnel tests and Walter Reed Hospital where Mrs. Espy went for an operation - and to the Air Service Building for instructions. Finally Espy went back to Langley - but not until he had taken me to the Kappa Sigma House in Washington for dinner. I met all the fellows, who are going to Washington University. They are the

---

[19] [Murry and Ruby Espy remained long time friends with Eddie. Even after his death my mother maintained "Christmas Card" communications with the Espys. Ruby died in 1958 and Murry passed away in 1978] — Turney Oswald

most agreeable bunch I have ever met, a finer lot of fellow than those at the fraternities at Illinois.

On Monday I received your telegram and decided to go up and see Tom. I wired him that evening that I would be there Wednesday morning. He answered Tuesday and I got the midnight train out of Washington. (Just as I was leaving Wilson arrived in Washington. Crowds jammed the entrance to the White House - Pennsylvania Avenue and the Union Station. There were 5000 people at the Union Station <u>after midnight</u> to see him get in!). Tom had left the Pennsylvania Hotel and so I went out to Camp Dix which I thought was a little way out of New York. I got there at 10 A.M. and spent two hours in a vain endeavor to locate someone in authority who had the faintest idea where Lt. T.D. Allen M.D. could be located. While eating in the casual officers mess, Tom came up behind and blindfolded me. He was already discharged and ready to go. So he decided to go back with me to Washington. He has probably told you how I bribed him with the promise of an airplane ride to delay his getting to Hamilton until I could get a chance to talk to him.

What worried me was the proposition about which I wrote you from Washington. I was to invest $500 in the company to show my interest, and was to get $5200 for the next year for flying and running a field for the company besides getting my share in the profits of the company. The proposition looked awfully good. I thought it was worth it to delay my school a year for a sure $5200 and possible profits. I wanted to talk to Thomas about it and get his point of view. He gave me this idea - that of the two possible paths - going to school or making money I want to <u>go to school.</u> That was the most important thing to do. I might regret turning down this proposition but I will surely regret delaying school. And so I am going to school - and to Illinois

Tom stayed with me until I could get away from Washington and then he came back in the JN4 with me to Langley. We made the trip in a little over an hour and a half which is good time in a Curtiss. We ate our supper on the way and then wrote a letter. Tom seemed to enjoy the trip very much. On the way down when we arrived over the field we did a few Immelmann Turns and a spin. This was enjoyed by all. Tom stayed all night and hustled away first thing this morning.

I just received a wire from Hamilton — not all of Hamilton — saying, don;t keep Thomas longer than necessary - needed here. Well he's gone — no amount of keeping would keep him.

If I can rake together about $4000 I am going to invest in a Curtis and take it to Champaign. On Saturdays I can make enough money to keep me in school and can pay back the $4000 in about two years. Do you know of a capitalist who wants to risk that much in a wild cat adventure for philanthropic purposes?

Edmund

Home for the wedding!

Eddie Allen During NACA Test Flight

NACA / University of Illinois 1919 -1920

Eddie joins the fledgling NACA at Langley Field in Hampton, Virginia, for one month summer employment prior to his re-enrollment at the University of Illinois in the fall of 1919. He is discharged as a Lieutenant in the Army the same week he is hired by the NACA and thus becomes the first civilian NACA pilot. The NACA has two Curtiss JN-4H "Jennies". The bureaucratic maneuvering to get Eddie onto the NACA payroll for less than one months work as a test pilot is almost comical. Even then they manage to lose his 2 week paycheck of $125. ETA will return to fly for NACA the next summer.

"The first flight research project NACA embarked upon involved flying the Jennies to determine how their actual flight behavior differed from that predicted by wind-tunnel tests of models. A secondary goal was determining the control forces required to keep the plane in longitudinal trim. The trials were, to say the least, interesting. Edward Warner and his assistants discovered that, although nominally alike, the two Jennies differed greatly from each other and from published official drawings! One airplane had a badly warped propeller; the wings lacked uniform camber and failed to follow reference drawings……

Having established the baseline characteristics of their test aircraft, Eddie Allen and F.T. Norton embarked on a study of the structural loads experienced by a Jenny during maneuvering flight. This study, undertaken at the request of the Navy, involved taking readings on an automatically recording accelerometer during takeoff, landing, loops, abrupt control pulses (such as suddenly pulling the stick back, then immediately releasing it), rolls, spirals, and spins. At times Allen would deliberately pound the airplane on the ground during landing; during one landing, the faithful JN-4H groaned through a 5.25 g impact. Abrupt 3 g spins and 4 g rolls were commonplace."[20]

Eddie returns to the University of Illinois only now his major is changed from agronomy to engineering. He also renews his relationship with "Miss Gregory". His second year at U of I is apparently challenging, both scholastically and socially judging by his letters.

---

[20] Richard P. Halion "Test Pilots" Smithsonian Institution Press 1981, 1988

At some time during his second year he meets Richard Lewis Aldrich[21], a fellow undergraduate who would also become a lifelong friend. "Dick" was also associated with Carl Rahn and would attend the free spirited events in Baraboo, Wisconsin, which Eddie describes in letters to his sister Margaret.

During the spring he contacts Edward P. Warner, the Chief Physicist at NACA, who had become his mentor and key proponent for employment at NACA. Warner, who subsequently became a professor of aeronautical engineering at MIT, again pursues Eddie for test flying at NACA for the full summer (1920) and then convinces him to enroll at MIT in the fall. Allen contributes the flying work that produce two NACA Technical Reports which he co-authors with F.H. Norton - TR No. 99 "Accelerations in Flight" (1920) and TR No. 112 "Control in Circling Flight (1921). Eddie will continue his relationship with NACA during 1921 and 1922.

Eddie and Allene Gregory are married on August 7 of 1920 and move into an apartment in Boston. The marriage incites controversy amongst the family which Eddie strives to ameliorate but without success. Eddie is 24 and Allene is 32. The marriage eventually dissolves in 1923 although the divorce does not take place for several years.

Excerpt from "Boys' Life" - March 1931

"Although I had been flying for nearly two years, part of the time as stunt instructor and as instructor of instructors for the army, that test flying for the NACA showed me for the first time how little I knew about really accurate flying. When you want to observe your plane or your instruments in a ten-degree bank, you have to make a ten-degree bank, not an eight-degree or one of twenty degrees or fifteen or eleven. A right-angle turn has to be a right-angle turn, not a little more than a right angle or a little less. I had to learn flying all over again.

---

[21] Richard Lewis Aldrich would become an expert on Oriental, as well as pre-Columbian and Hispanic American cultures and would teach many years at the University of Miami (1947-1962). He achieved his Bachelors degree from the University of Illinois, followed by studies at Harvard (1921-1923), a Masters at the University of Arizona,(1936), and a PhD from Michigan (1942). In his papers on file at the University of Miami, Aldrich says that his close friend Edmund Allen assisted him financially and enabled him to complete his many years of study. Dr. Aldrich died in 1976.

This test flying was also my first contact with the real science of aeronautics. It made me want to become an airplane designer. It made me want to know more about aerodynamics. It made me realize the importance of college training, if I expected ever to become more than just another pilot - as one might become a bus driver or possibly a railroad engineer. When college opened in the fall of 1919, I went back to the University of Illinois, shifting from Agriculture to Engineering. The next year I transferred to Massachusetts Institute of Technology at Boston, where a course in aeronautical engineering had already been established. I was there for two years. During the summer I went back to my work as a test pilot for the NACA." — ETA

Langley Field
Sept. 3, 1919

Dear Mother,

Your letter arrived yesterday and I have put off answering it until late to-night. Why not put the $500. into the Building Association? Isn't it as safe as you would like? The only reason against putting it into bonds is that Thomas might need to borrow it. Still I will probably be in position to loan him money if I get vocational training. Even if I do not get government help I will not have to touch my $2000. for over a year. I have enough outside of that to carry me through the first year - unless the cost of living is as high at Champaign as elsewhere. And next summer I should be able to lay by about $200.

If you could get together $20,000. by the time you wish to give up work that will bring you in an income. If you ever do want to do that, and get 4% on it, that would mean $800. a year. You see I am counting on the Florida property either some day actually paying dividends or being sold for a reasonable figure. $800. would not leave you so cramped if you wanted to spend an economical winter in

Florida or visit Tom and I once in a while or have Sister come to see you. Of course if the Citra land gets flooded or dies of ennui and the Hamilton Bank fails, Tom and I can still help you out to about that extent after we get a start. - if you give up your work. But I know you would rather not have it so.

    I was very much interested in the clipping about the tennis tournament. I was surprised that Holstein got third place. I didn't think he was much better than I when I played him, and I am certainly very poor here. I cannot beat any of the six men ahead of me in the rating. I play about every night. We play doubles often with Lt. Finley, Capt. Scott and Major Lackland.

    Your timely warning about falling in love was needed. I have been thinking about getting married and married life so much lately that I have almost got into the mental state of preparation for it. It is perfectly unreasonable because it is out of the question. I feel that school should be so much the business of the next years that my mental attitude should be one of preparation for a work. I want to feel that I am preparing myself to do a real work in the world. It sounds like the old return to idealism of 20. But it is bigger than that ever was. There is something in me that is much bigger than I am, something which has an effect on the people around me. I can see it working out. I am giving a strength, an idealism, a truth to one or two people with whom I come in contact. It is my one idea to develop this power until perhaps in years to come I may energize a people to freedom - a newer freedom - and faith - also a new faith.

    I do not quite understand your point in saying you think I should not let anyone influence me "in planning my life". Did you have reference to the vocational officer and my contemplated course in Plant Production Engineering? I am not at all sure that is what I want eventually but I am safe for three years in taking up my AB work at Illinois. Did you think aeronautical engineering the more indicated field for me? You know Mother it may be the humanities after all. Do you remember me telling you where my interest lay back in '17. But it will be with direction now. If it is Economics - history or politics it will be from an industrial point of view. My work I believe will be with the worker. I have enough of the aristocrat in me to keep me from fanaticism. But I sincerely believe that my sympathies are all with the working class.

No I am doing nothing for money, though the present position I have with the Nat. Adv. Committee is the biggest temptation I have had yet. It is not the money (though the $3600. is no small amount): it is the wonderful experimental work, the opportunity to do something big in aeronautics. We are finding out things about stability of airplanes that were not known before. We are planning tests that will tell us how to build machines that will be as safe as airplanes ever can be. The possibilities of the committee in development to the size of the British Committee will make me the biggest reputation in the country as an experimental pilot. I know I can keep this job against any and all competitors unless perhaps Major Schroeder decided to ask for the job. Then we would have <u>some</u> try out!

I guess I told you about Mr. Hermans offer. He wants me to go to Daytona with him. He is to furnish the boats and give me 50% of the receipts, which would amount to $200. a day in season and about $30. a day out of season. But I wouldn't leave this job for that. The only thing I would leave it for is school.

I have just finished "The Call of the Carpenter" and I will warn you about it. If you are afraid of social democracy on the one hand, or Jesusism as <u>contrasted</u> to Paulinism - on the other hand, do not read the book.

$100. is a tight squeeze for a civilian outfit.

My first suit is costing me $50. I want to buy another suit probably after I get to Champaign - an overcoat, hat, shirts, collars, ties, socks, shoes, gloves, underwear, and a typewriter, a leather bag, a dictionary and a ticket to Champaign, Ill. via Hamilton about Sept. 19th. Registration days at U of Ill are the 22nd and 23rd. If I leave here (Friday morning or) Thursday night I arrive in Hamilton Friday night. Then I have Saturday and Sunday with you. Leaving Sunday night I get to Champaign Monday in time to register before all the classes are full.

Back to Expenses

If I leave out the leather bag the ticket should be about $30.00

Dictionary	2.00		
Typewriter	50.	In bank	627
underwear	5.	Pay Due	214
gloves	3		___
shoes	14.		844
socks	3.	Deduct	373
ties	5.		___
collars	6.		471.
shirts	20.		
hat	5.		
overcoat	50.		
2 suits	100.		
Board to the 23rd	60.		
other expenses	20		

	373.		

Will that carry me a year with expenses the way they are?

Langley Field, Va.
Sept. 4, 1919

Dear Tom,

Mother has sent me one of Flos' letters and I gather from that that you have an apartment in the same building with Bob and Ruth. Another morsel of news was to the effect that you are taking care of Dr. Wilder's office and practice in his absence and married life seems to be all the anticipation called for? I want very much to see you. Champaign is not far off and perhaps I can come up for Thanksgiving dinner.

I was discharged on the 23rd and began work on the same day for the National Advisory Committee for Aeronautics with the result that

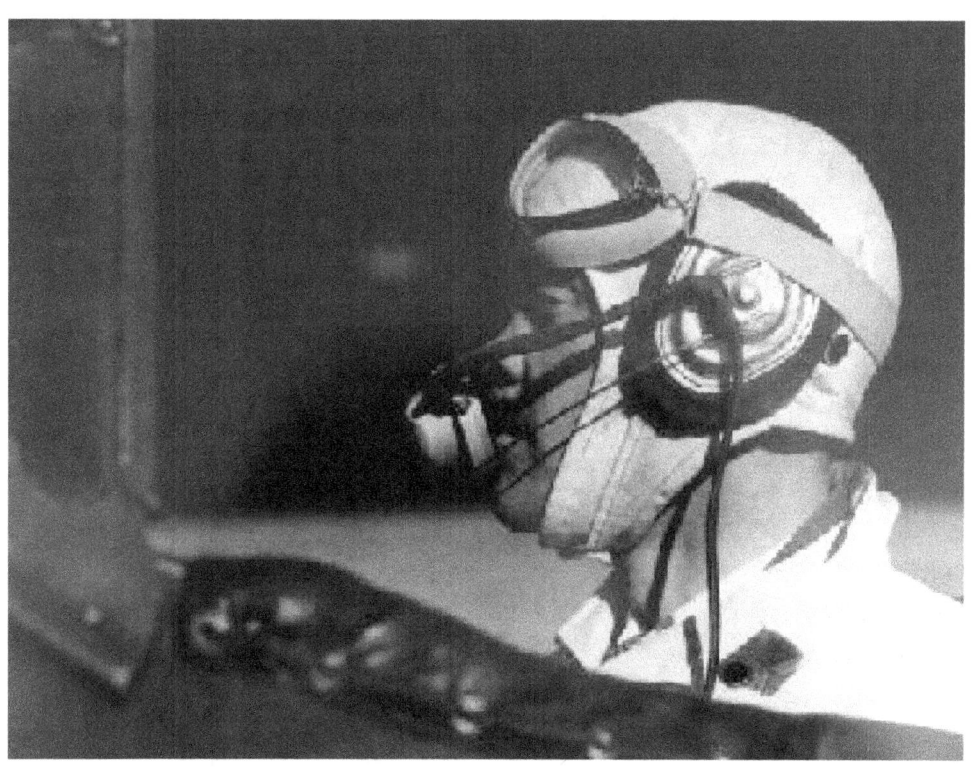

Eddie Allen prior to nasal reconstruction

the 23rd was a lucky day - my wages being $16.71 for that day's work. I like the new job about 3 or 4 hundred percent better than the army. I fly about an hour a day or less and am learning very very much about flying as well as about experimental aeronautics. It wouldn't take much to persuade me to stick to the new job. They offered me $3600 to stay and it sounds about twice as good as the $5000 of the New York concern. The National Advisory Committee promises to be the big thing in Experimental Engineering for Aeronautics. I am rather pleased that I have an option on this job when Mr. Warner has 14 applications including one from Meister and from six other test pilots from McCook Field.

The 18th is my last day at Langley. I arrive in Hamilton on Friday night and leave Sunday night for Champaign. Monday is registration day. I think I shall buy most of my civilian outfit in Cincinnati. A list of my needs adds up to $293.00 including two $50 suits and an overcoat and a $50 typewriter. I wish I could do my buying in

Chicago. I would rather buy shirts and shoes and a hat and clothes there than anywhere else. Are prices very high there now?

I surely enjoyed hearing even indirectly from Flos. Please remember me very warmly to her and remember that I always wish you the finest happiness on earth.

Edmund

718 Prairie Ave
Champaign, Ill
Oct -12-1919

Dear Mother,

This is a beautiful, chilly, sunny Sunday, the first of its kind so far. It cleared up last night just before the moon came up. Allene and I were out on the river in a canoe in the afternoon. The weather had even been rainy for a long time. Just as we finished supper, it cleared off. We have been having lots of good times, she and I. We've planned to take a couple of all day canoe trips on rivers around here. Yesterday she told me a good deal about her father, and at my request, let me read a few chapters of the book she is writing. All the time I kept thinking how very much like my grandfather Dyer her father was. I began to see where certain qualities in her came from - mainly a capacity for a more intense kind of friendship than I ever thought possible. And then I saw in a new way how much of what is me came directly from my parents and grandparents. I am not far enough away from you yet to appreciate in an objective way your gifts to me thru heredity. But I want especially to know more about your father and your mother. Grandfather's personality has always interested me very much - his religious and ethical standards, his ideals of character and business, his intellectual interests and his home life. I believe it is from him that those qualities in myself which I admire, came. A better knowledge of him may help me to understand myself. It is this heritage - this background from which I came that I wish to strengthen for my children.

School work is getting on quite well. Economics is the most interesting subject this year. Drafting is getting endurable - shop is mediocre. The trouble with shop is that I am not at it long enough to get started. The pattern making should be quite interesting but just as soon as you get the red tape unwound, it's time to quit. In math - geometry is very easy. Algebra is difficult because I have forgotten all my elementary algebra. I really don't know how to add or divide algebraic quantities. Mr. Price is helping me out with that. He is working in the library now and Mrs. Price is teaching school in Champaign.

Last Sunday I packed up and went to Chicago. I got off at 115th St and went over on a car to Morgan Park. Aunt Louise and Hugh were in the kitchen. Hugh is very large - almost as large as his father and he has acquired a lot of skill in getting around in his wheel chair. After looking over the clothes up in the attic I decided there was nothing there but an old blue serge suit of Tom's. Aunt Louise had followed my advice and given away everything of mine except a pair of "pumps".

I got a suitcase full of old stuff - one old shirt, a few collars and wash ties and this suit, and took it down town. Tom was just finishing up the business of Saturday at the office. In the afternoon we went out to Rush and held clinic. We arrived in Evanston and found Florence in the kitchen with dinner almost ready. Tom is certainly not going to starve if the meals I had were representative. Florence is some cooker! Tom and I argued about strikes and Socialism all day Sunday very much to Florence's amusement. Saturday night we went down town in Evanston and got a suit and overcoat all for $65.00. Sunday we went to Morgan Park for Sunday dinner. Uncle and Jessica were there, of course. Uncle seems to me a much more lovable person than ever before. Jessica is a lovely little body with, it seems to me, wonderful possibilities for development. Owen is in New York. He is expected home next week to say good by just before he sails for the Balkans with the Standard Oil people. Allen is teaching in Evansville and Lilly is here in Champaign. I haven't found out where yet but I am expecting to find her this afternoon. The Prices enjoy Florence very keenly.

I have been talking to Dr. Rahn about my course and crystalizing my ideas of a career. At present I seem to want a pretty thorough technical training in machine shop practice and machine design with a minor in economics and sociology. I want to get to work in the industrial world as an efficiency man in production work with a view to the production of labor saving machinery. Later I expect to get into the labor side of the problem with the point of view of improving production and conditions of labor.

Edmund

Oct. 26 - 1919   *718 South Prairie*
Champaign

Dear Mother,

   I have been wondering this week how you thought about my last letter. Perhaps it may have struck you as being a little out of key with the main purpose of my college work. I want to tell you the difference between college this year and three years ago. It has changed. I am no more interested in the elaborate Pep-meetings - to stir up Illinois enthusiasm. The football games do not interest me. I find interest in other directions. The typical college life doesn't appeal any more. I believe I told you that this year seems more like a post-graduate year than anything else. Economics occupies a large part of my time and thoughts - especially as it deals with labor problems and industry. I find an absorbing interest in the papers - in events about labor troubles and international muddles. From the American Society of Mechanical Engineers I am getting the Engineers' viewpoint, which has it all over that of the farmer, the politician, the metaphysician or the teacher as a point of attack on the problems of life and of society. This engineering point of view is absolutely honest. It doesn't try to fool anybody or to fool itself. It calmly and dispassionately looks over the material with which it has to work and the work to be done, sets to work and does the job and then asks, - What next. With this in view my shop and drafting and mathematics are very interesting and valuable.

Outside of studies, themselves, I am learning a whole lot about the art of friendship, about the adjustments that are necessary between individuals and between individuals and society. I am finding a little of the general friendliness with life which is possible and concretely works out when one looks into the eyes of his fellow men. This is not Campus bunk. I appreciate the difference between real life as one finds it in London and in the Army and in an ordinary family relationship. And I know that there are possibilities of human relationships I little dreamed of before - that life can be beautiful and creative and expansive as well as sordid and commonplace and ingrowing.

At present I am out of sympathy with the aggressiveness of which we spoke. It seems to me that the ideal aggressiveness would express itself in a very quiet, deep strength and a very perfect tolerance together with a calm assurance that what one was doing was absolutely right and would ultimately be done against all opposition.

It seems to me you are right when you say Jesus should not be blamed for the faults of the Church. When I first broke with the Church and threw over the whole teaching I had been given, I was so much absorbed over insisting to myself that Jesus was not a divinity sent from heaven to save sinners that I lost sight of the fact that whatever his parentage or whatever he became, he did promulgate a teaching greater than that of any other teacher, that his personality was one which I must worship whether I believed it divine or human, and that in him are embodied all the ideals toward which I strive. Please return Bouck White's book when you have finished it. There is also another book of Miss Gregory's "The Madras House" which I left at Aunt Margaret's last year. Do you know where it is?

I have never heard from Margaret in answer to my birthday letter. I think I shall write again and ask her to return the letter of Miss Gregory which I sent her.

It must be lonely for you without the Byards? Do you miss them very much? Mrs. See is such a good friend to have. I am very glad that Uncle D R is improving and becoming interested in activities at Evansville.

Tom will be very much disappointed that you cannot visit him.

Ruby is working hard every day and is very tired at night. I am afraid that when we move she is going to have too much to do. Murry has obtained a loan from the University to stave off a couple of creditors - his vocational check has not come as yet - neither has mine.

Edmund

*718 So Prairie
Champaign Ill
Nov - 18 - 1919
Dear Mother,
Happy Birthday first thing -*

Next about Thanksgiving - Tom wrote me that you were a little blue about not coming up to Chicago this time. Since sister cannot come - I think I shall not go up because we have classes on Friday and Saturday as usual and I am not in a position to cut. So we will make it a reunion at Christmas or New Years.

It must be awfully lonely for you in Hamilton now, Mother. I am sorry that Uncle had to leave but I guess it is for better all around. Do you find Mrs. See a companion? Of course you have all your old friends but Uncle and Aunt Margaret (Byard) seemed to fill a place no one else did. Their home was always so tolerant and homelike.

If you haven't yet sent my sheepskin coat please wrap up my army boots with them [sic] and send them along. I could use them on the hikes that I take almost every Saturday. The woods and country here are beautiful during this Indian Summer. The leaves have turned color slowly, some of them hanging on to their green tenaciously while others have had to let go at various stages so that every hue from bright green to dull brown may be seen in a veritable mosaic of color. But most beautiful of all are the trees, with part of their leaves gone and just the suggestion of their intricacy of branch work against these Indian Summer skies.

Mother these are the days the old air fever grips me again, and I dream at night of flying over a moonlit sea of clouds and wake in the morning soaring toward the sunrise with the pure icy air sending the blood racing thru my veins - not minding frozen cheeks if could I but taste the clouds and feel the sunlight - a sunlight never dreamed of on earth - permeates every atom of my body and soul. It's in the blood, Mother. I believe I could never lose it if I tried.

[Written sideways in the margin: "This was his strong feeling for flying. He felt at home in the clouds – chained to the earth. M.A.R."]

Tomorrow night we are going to organize the silver button men of the University into a club to arrange for having entertainments. May Rolfe, who is the chairman of the Campus Service Committee, and who is mothering the infant organization, has asked me to be toastmaster. We are to have Lorado Taft the greatest American sculptor speak to us - also Dean Kinley and possibly the president of the University. Other talks and musical numbers will be given by students and local celebrities. The first two meetings, tomorrow and two weeks from then, are to be suppers with the talks afterwards. After that, it will be up to the men to decide what they want. Evidently we can have whatever or whoever we want.   Edmund

[No place or date given -
hard to pick the right time -
U of I academic subjects - Probably 1919-20

Dear Mother
I don't know when I have been doing so many things at the same time. I don't get much opportunity to sleep at all. And it seems that everything is going well. I had half A and half B last quarter in my school work, A in English and B in some little courses in Engineering Contracts and in Psychology. O, yes, I had an A in engineering English also.

And with it all I am happier than I have been in many years. Life has brought me much in understanding, in a perspective of human living, in man's place in the order of nature, and in the intricacies of man's own nature. I feel that I understand myself as very very few men

I have known ever get to know themselves. Life opens up with such an understanding.

I have many intimate friends whose friendship is a great source of strength.

It seems as I look at life from a certain detachment even though so busily engaged in it, that my fellow men are as a rule blundering, as far as a personal resolution of life's mysteries and difficulties are concerned. Even the most intellectual (perhaps these in particular) are bereft of a conception or philosophy or even an insight as to what it is all about, they cling to this or that creed or popular dogma, knowing full well that the position is untenable if examined with truth as the only aim. They are full of conflicts, - desires and impulses pulling them this way and that as the wind blows. And when they die we say some pious words. and more are born, and so it goes on in an endless stream.

How few there are among the millions who ever achieve any sort of balance inwardly. Jesus, Socrates, Gautama Buddha, Plato, Aristotle, Marcus Aurelius - and many of these we crucify or attempt to do away with - as if humanity were jealous of those who discover any "why" of life!

I am having lots of enjoyment out of my psychology classes and am planning to use the material in my thesis - an engineering-psychology thesis.

Sorry I couldn't get down to Florida but I'll make it some day. Wait till I get my degree!!!

               Edmund

Strange letter
Edmund Allen to his mother Nov 30, 1919

(This letter needs context. This was during the time when the "Suffragettes" were active. The movement had been slowed by WW I, but by 1919, had fired up again. Feelings ran high on both sides. Obviously, Edmund had been hearing all the arguments about the inferiority of women – this despite the fact that his mother was an M.D. His letter of 11/30/1919 was written less than a year before enough states had ratified the 19th amendment (Tennessee, August 17) to make the required majority of three-fourths of the states.

Here's a little from Wikipedia: The campaign for women's suffrage began in earnest in the decades before the Civil War. During the 1820s and 30s, most states had extended the franchise to all white men, regardless of how much money or property they had. At the same time, all sorts of reform groups were proliferating across the United States–temperance clubs, religious movements and moral-reform societies, anti-slavery organizations–and in many of these, women played a prominent role. Meanwhile, many American women were beginning to chafe against what historians have called the "Cult of True Womanhood": that is, the idea that the only "true" woman was a pious, submissive wife and mother concerned exclusively with home and family. Put together, all of these contributed to a new way of thinking about what it meant to be a woman and a citizen in the United States.

In 1848, a group of abolitionist activists–mostly women, but some men–gathered in Seneca Falls, New York to discuss the problem of women's rights. (They were invited there by the reformers Elizabeth Cady Stanton and Lucretia Mott.) Most of the delegates agreed: American women were autonomous individuals who deserved their own political identities. "We hold these truths to be self-evident," proclaimed the Declaration of Sentiments that the delegates produced, "that all men and women are created equal, that they are endowed by their creator with certain inalienable rights, that among these are life, liberty, and the pursuit of happiness." What this meant, among other things, was that they believed women should have the right to vote.

Suffrage leader Lucy Burns (1879-1966) was imprisoned at the Occoquan Workhouse in Virginia, probably in November 1917, after she and others were arrested for picketing the White House in support of a federal amendment granting women the right to vote.

On Election Day in 1920, millions of American women exercised their right to vote for the first time. It took activists and reformers nearly 100 years to win that right, and the campaign was not easy: Disagreements over strategy threatened to cripple the movement more than once. But on August 26, 1920, the 19th Amendment to the Constitution was finally ratified, enfranchising all American women and declaring for the first time that they, like men, deserve all the rights and responsibilities of citizenship.

… and there's a lot more…) — MRB

718 So. Prairie, Champaign, Ill.
Nov. 30 - 1919
Dear Mother,

I was very glad to get your letter with Sister's enclosed last Monday. Sister is having a lot of development this year, isn't she? I am very glad that she has this opportunity. This friend of hers sounds interesting. How much, I wonder, do you agree with her. I felt that I wanted Sister to know that some of her ideas were not generally accepted in the family. I have been doing some thinking about relationships between men and women, myself, and my instincts have led me to a different point of view. Sister wonders why it takes her a year to get friends. It always seems to me that those that are worth while are worth a year of getting and those I get in less time are often not worth while. I have always liked Stevenson's "We are little islands shouting to each other across the sea of misunderstanding," and while we must shout awfully loud to those whom we feel are worth while we must expect that it will take a long time to get across because those seas are mighty wide and deep and tricky.

Her talk about religious discussions revealed a definite change in point of view in the last few years. And Mother, isn't it a more beautiful thing that she has worked out? She says, "He is the good in everything - in me, in you, in nature; that which awakens our best, and makes us love." That is evidently something that she has not gotten from anyone but has found in her own heart. Of course experience may lead her to find something different. But even if that is not a possibility, I think she has found something that is more beautiful and true than any doctrine or creed. Of course her friend would have to laugh at her argument for bodily care and development. Her only other alternative would be to agree and she evidently doesn't want to do that.

It is a wonderful thing for Sister to go to these operas. She has needed that so much. You need it too, Mother. I believe they fill a very important place in the lives of people of her and your type. If you can see the beauty of a personality that will actually go hungry to buy a ticket to presentations of art, it is a great achievement. We are too apt to laugh at it as impractical. You are going to get the benefit of some of these things if I have to go hungry, Mother. We'll see.

But the thing I really wanted to talk about was her discussion of a woman's relationship with her fiancé. That affair of her friend's was awfully distasteful. Any love that I would have for a woman would die if I realized that she was attempting to give me my ideals. Believe me, I should never let my partner do more than half.   Mother, you know that a man does not look to a woman for his ideals. In the first place I think that men's ideals are higher than women's. They are different. That is certainly not an indication that the women's ideals are right and men's are wrong, or that women's are higher than men's. They are higher in the big broad issues of life and also in the little points of honor of every day living. A gang of boys in a city often has a finer system of honor than women ever work out in their lives.

I am afraid that I am ranting - and that spoils the strength of anything I might say. But I do feel strongly that the ideal of the average man is higher and finer than that of the average woman. The woman may more nearly live up to hers because her ideals are those of her world, while the whole of our rotten economic and religious structure tends to make the fight harder for the man. Another tendency that way is in the handicap that our crooked moral ideals have made out of a man's sex life. The woman's sex handicap is physiological. She is entitled to and usually gets a large measure of the consideration that is due on this account. A man's is psycho-physical and the woman makes it all the harder for him by misunderstanding and misinterpreting. [Tom will probably rave at this.] This is abstract and may go wide of the mark in that account. I am going to take a chance on the understanding in our family and make it concrete. During the five days in which the periodic return of the time of subnormal physical activity recurs in a woman, a man's attitude toward her is usually one of consideration and forbearance. If he thinks the thing out he sees that she is fulfilling a very important function from the point of view of the race. If he doesn't think he feels the same thing. At any rate he does not take the attitude of cynicism toward her physical inferiority - much less does he condemn her for weakness or wrong physical living that might have caused the weakness - altho he might justly do that in some cases where the function had been aggravated by high heels, tight skirts, etc. - . During the corresponding period in a man which comes as a mental and nervous tension - a feeling of physical superiority - the whole thing very definitely connected up with sex and bringing with it always a lot of sexual imagery - this last definitely aggravated by our Methodist Church ideal of sex repression, a policy that is bound to spread it out under the surface where it is ready to boil over at these periods, - during this corresponding period what is the woman's attitude toward the man? She is immediately on her guard against his aggressiveness. She watches that he doesn't get "fresh". She interprets his whole attitude as moral degeneration, as evidence of mental rottenness. She is intolerant of his actions and condemns unmercifully his failure to live up to her ideals. She doesn't consider that he is fulfilling just as important a function in the life of the race as she does. She thinks of sex as realized in his life as an evil (the Church is to blame for that) but in her life it is all right. She thinks that his problem is no harder than hers and is much the same! And she doesn't know that he needs to be taught - and has no one to teach him how to sublimate his impulses so that this may become a source of power for him and for the race.

No wonder we have so little understanding between men and women. The sex problem is a prolific source of misunderstanding.

The woman's ideal for the man cannot be nearly as fine as his own ideal for himself. What woman ever saw in vision a man like Chaucer's knight.

 A knight there was, and that a worthy man
That from the tyme that he first began
To riden out, he loved chivalry.
Trouthe and honour, fredom and courtiesye,
Full worthy was he in his lordes werre.
And thereto had he riden (no man ferre)
As well in Christendom as in hethenesse,
And ever honoured for his worthynes.

. . . .

And though that he was worthy he was wyse,
And of his port as meke as is a mayde
In al his lyf, unto no maner wight
He was a verray parfit gentil knight.

This is not every man's ideal but it is not far from that in most gentlemen. You will not easily find out what a man's ideal really is. He is much more reticent about saying it than a woman would be. Would you think that a man's ideal of a woman more fine than her ideal for herself? It works both ways. Sister is forgetting that biologically the woman is the conservative element of the race and man is the leader; she is trying to reverse the process. The ideals that are finest, it seems to me, must be worked out for both by both together. Men and women tend too much to grow apart as sexes. Women tend to become too specialized for sex and men too aggressive. The woman's job would become then not the promulgation of ideals as a moral leader in the family but the finding out what the man's real ideal for himself was - what kind of a man he really wanted to be and helping him to attain that. And in conjunction with that, interesting him in her ideal for herself so that he could help her work out her fullest development.

And as for the children, they and their development belong half and half to each. A man will resent his wife doing all the guarding of their ideals just as Dr. Monihan must resent his wife's attitude toward the daughters economic welfare.

Edmund

[before June 20, 1920 to sister]

Dear Sister,

You have certainly been neglected in the matter of correspondence from your brothers this year. I am very glad of a little interim in which to renew acquaintance. My last examination is over and I am on my way to Langley Field. I hope the apartment will be rented this summer so as to help out on the rent. So far we have had no success but the summer school crowd has not yet arrived.

Your plans for the summer sound very good. I think Mother is hoping you will come down to Chicago in August and that you and she can go down to Champaign for the rest of the summer and have your visit there.

I was awfully sorry to hear that we would not have you with us next year. But I suppose the prospect of $1800 looks better than anything in the world now when you have been in debt for so long. That certainly makes a difference if you have some money in the bank. That is probably the reason $100 a month was such a temptation at Wheaton, when $3000 didn't disturb my decision last fall when I had $2000 in the bank.

Mother was the most disappointed of all. She was mainly concerned, besides of course not having you with us, with your going on with preparation for bigger things than you can get just now. She thinks that once you get into university teaching you would find a bigger field for the ability we are all just beginning to realize you have. I believe you would like to teach university folks better even than your high school people.

Tom sympathizes with your not wanting to come down such a long way in salary. He said he was almost to the point where he could make up the difference to you. He would like to do it, but I expect that as long as he is second to a chief in the office he will not get much more money than enough to get by in Evanston with a youngster or two.]

I confess to an entirely different reason for having wanted you in Urbana next year. But first I must tell you about my visit with Helen Walcott. Mother saw her last week some time and told me where I might find her and when. And so I went to the Baptist Miss. Training School on Monday afternoon. We had quite a long conversation about you. She spoke about your teaching "genius" which was recognized in Denison. She thought you would find in University ultimately the life and the work you really want and can be most efficient in. She is enjoying teaching J.S. graduates very much.   [Helen Walcott was a beloved Denison College friend of Margaret Allen.]

She told me about her teaching and the problems she was meeting; race problems, civic and religious ones. And then we talked about the G.O.P. convention and international relations and ultimate world problems of religion and race and culture. She is a most interesting person and so wonderfully well informed and openminded.

I think one of the greatest difficulties in the relationship that Mother and I have been building up during the past months has been a tendency for each of us to look for the source of thoughts and opinions of the other. I have tended too much to regard Mother's ideas - even the most liberal ones as being stereotyped by the old school of thought and religion. I have felt that Mother was asking herself often, "Now, where did he get that idea? I wonder if Miss Gregory believes that, or That must be an idea of Dr. Rahn's." She has been suspicious of Dr. Rahn from the first and has believed his influence a bad thing for her son. But then it has not all been like that and Mother has done an awful lot toward overcoming that tendency and has seen my point of view often when I haven't seen hers.

Mother and Miss Gregory have been very good friends and very good for each other. And this, Sister, is the reason why I wanted you here so much. I wanted you to know Allene Gregory. Ever since 1917 I have wanted you to know her and as these last few years have gone on and I have seen more and more deeply the character of the woman this wish to have you know each other has grown. As I see it, you each might learn by just knowing the other certain things that would enhance the two most wonderful women in my world. One of the things I love most about Allene Gregory is her wonderful capacity for friendship. I have learned that it is possible to come as close to another human being as one is to oneself. I have really learned what love is. And now Allene and I have decided that if we continue to feel about each other as we do now, that when the time comes we will go thru the ceremony which takes care of the legal end of our spiritual union. Does this spoil it for you - to say it this way? You see, there is nothing to announce. Allene refuses to be engaged and I have never yet had the opportunity to propose an engagement. It all seems to have grown toward this, slowly developing.

But I do want you to know her, Sister. I think something wonderful might grow from such a friendship.

      Edmund.

This letter and the information is strictly confidential. Allene and I feel that this is a matter which concerns no one but our two selves but we wanted to tell Mother and now you -

Edmund.

[The letter reflects a very progressive view of the role of "marriage" and implies "we" are sub-renting our apartment. This may be the source of the family controversy as this was certainly not a Baptist family's view on marriage in 1920} — TKO

Langley Field, Va.
June - 20 - 1920
Dear Mother,

Well, the baby is nearly a month old. I'll bet he looks a lot different than he did when I was there. Be sure to let me know of any developments like changes in features or a sudden bulging out of the forehead or the appearance of a mustache.

Are you still undecided about next year? I have not heard anything from Urbana so I expect the apartment has not yet been rented. I wrote to Sister, telling her just why each of us were anxious for her to come to Urbana next year. When she goes down with you this summer perhaps she will change her mind.

I expect you to be very much interested in my work here - even the technical dry stuff. Please don't mind it. The new wind tunnel is not quite ready for testing because of some difficulty with the apparatus for generating current for the big motor driving the fan. I hope to get some work in the tunnel as soon as it is ready. There are about ten physicists here now. I think I told you that Mr. Warner was going to Europe this summer. He has left here and gone to Boston for a week or so. Mr. Norton is in charge of the Laboratory. He and I have a suite in the new officers quarters. Do you remember the very large new building - half built near the entrance to the field? It is completed now and the officers have moved over into it. Mr. Warner and Mr. Norton had one suite and I took Mr. Warner's place.

I have been doing a lot of work with an instrument called the accelerometer. It measures the load or strain put on a 'plane in various conditions such as a porpoise landing or a pancake or a series of stunts. We have some very interesting results. The instrument is photographic recording a record on a revolving film. In one result we found some information never obtained before which is very important. The manometer is rigged up on the other Curtiss plane we have here and measures pressure distribution along the tail surfaces. This is being done to aid designers of new types of 'planes.

There is a lot of interesting work lined up for the whole summer ahead, and Mr. Warner wants me to go to Tech. with him next winter.

     Goodbye, Mother. Say greetings to my nephew -
     Edmund.

Edward P. Warner (1893-1958)

(Chief Physicist of NACA will become a professor at MIT, Editor of Aviation Magazine, Assistant Secretary of Navy for Aviation. Warner will be a strong supporter of E. T. Allen throughout his life. He hired Allen into NACA, encouraged him to enroll at MIT, supported his glider activity in Europe, and helped publish his many articles and papers in Aviation Magazine. Dr. Warner will state following ETA's death that Eddie Allen made more contribution to aviation flight testing than any man other than Orville Wright. —-TKO

[sometime before August 7, 1920]

Dear Tom,

Have just been down to Hamilton with Allene. We were awfully disappointed that we did not see you there. Allene had so counted on it. She is not at all well now nervously, and was really not up to the trip. At Hamilton she decided and we all agreed it was best that she return at once to New York city to her sister while I stayed on here and finished up my business in Champaign. We will sure see you in Boston this year some time.

I am coming up to Chicago about Sunday or Monday and will see you then for a good visit. I am taking care of some things of Allene's here in Champaign. We certainly enjoyed Florence and the baby and Allene especially liked Mrs. See. Mother looks very well I think.

Your brother
Edmund.

Please forward any letters arriving for us to Allene Gregory care Martha Washington Hotel New York City until I arrive in Chi.

[Edmund and Allene Gregory were married on August 7, 1920. There are many missing letters between November, 1919 and this one. Only one, from June 20, 1920, is in this collection. Presumably, his letters were shared around among relatives and never returned to Abby for her safe-keeping. My recollection of my mother's (MAR's) comments of many years ago is that the family was quite dubious about the marriage, and, again according to Dr. Smith, "This union did not last; it apparently started coming apart in 1921. There was apparently a reconciliation of sorts in 1922; but by 1925 it was finished." – MISCELLANEOUS LETTERS by Edmund T. Allen (1896 – 1943) 1914 – 1929 – Richard K. Smith] —MRB

Langley Field, Va.
August 27 - 1920
Dear Mother,

I did not get this letter written before the 27th. I have written to Lida Pendry, to Aunt Louise, Aunt Margaret, Aunt Mable, Dr. Rahn, Mrs. Storke, Mrs. See besides a daily letter to Allene. I received a mighty fine note from Tom in answer to my letter. And now everything is happy. Allene wrote to Tom and Florence this week also. Allene is still in Boston awaiting her furniture. We have a beautiful little apartment, i.e., it will be beautiful when we get it fixed up - near Tech. Allene is enjoying fixing things in it immensely.

I expect that today you are having your hands full with all the things to be done in Urbana. I received your letter with the check from Miss Barber. I think that I have written about the disposal of everything in Urbana. Of course, you know that any arrangement you make will be perfectly satisfactory with me. Whatever you wrote that you have done is all right and I will fall in with it.

I am having a perfectly beautiful time loving Allene long distance. I am afraid we will not be able to see each other until school starts. But next fall will certainly be wonderful.

In regard to my work here, it is going slowly. There are many instruments that require infinite adjusting and re-building. I'm afraid we will not get much farther this summer than this stability and control work. I am having all kinds of offers lately. A new company is just being formed to build a sport plane. They want to give me one if I will fly for them next summer. They also are planning to build an entry for next years Gordon Bennett which they want me to fly. But we shall see; we shall see. I am going to be an aeronautical engineer - not a chauffeur.

      Love -
       Edmund.

September 1st 1920
Dear Mother,

I was very glad to get your letter of the 27th and read of your decision to get a home. It seems to me that that is a very wise thing to do. You have often pointed out the mistake of grandmother in giving up the big Hamilton home and the wisdom of Mrs. See keeping her home. Of course we shall always love to have you come to us and live with us as long as you like - but I think you would feel happier if you felt you did not have to live with us. In another part of your letter you say "It is up to me to carry out mine [ideals], to live my own life". I think you are very right. It is the privilege of every person to have an individuality of his own and not live thru other people. Mrs. Gregory tried to live through Allene's life when she had nothing in common with her and had broken every tie except the blood tie. Allene and she could have gotten on beautifully and friendly if both had lived their lives independently.

We shall love to come to your home and live with you temporarily. I think it would be loads of fun. And yet in visiting you we leave you free of us in every way except love. And this will give love an opportunity to grow rapidly. I think I can help you to almost any extent financially. The prospects are the government will pay my expenses and I have some money in the bank. I like the idea of your getting near a University town.

How are your teeth? Was the extraction painful? I have written to Aunt Margaret some time ago. Yes we will want some army blankets. I think Allene and I can get to Chicago about the 19th or 20th of September. Will Thomas and Florence be there then? I wish Sister could know Allene. But then she will next year. And Allene will be well next year. She is really in rather poor health now. She arrives in Hampton tomorrow from Boston. We live at the field for two weeks and then start for Chi via N.Y. if those plans suit everyone at Chicago. We can change them any time. I need four or five days in Champaign to settle up business -   Edmund

Langley Field, Va., Sept 6 - 1920
Dear Mother,
This is the last day of our honeymoon. We were married on Aug 7th and now it seems as if we were just at the beginning of a very wonderful friendship. We always look forward to the time when we will be "old lovers". But we don't have to look so far for a beautiful vision. As soon as we get back to Boston we are going to build together a home just as we want it. We are building our own place where we can have the friends with whom we can start the kind of relationships we want. There is nothing in the way; nothing we would have different. If we cannot make a success of our loves under these conditions we wont be very good artists. The "success" we are after is inward success - the only

kind really worth while in the long run. How little it does matter after all, Mother, what important position I work to in my profession, how honored and respected I am in the community, how many friends and few enemies I have, how much property I own and operate wisely and justly. After all, there are many such - many such who after fifty years find they haven't made much progress in the inward life. And so if as life draws to it's close Allene and I are in poverty, obscurity - or if I am an outcast or in prison; if we have hardly a friend on earth - and we know that we have walked on the way we have started to walk, that we have succeeded with the inward success which means everything to us now, - if this is our condition at the end of fifty years, we will be content. This sounds like the idealism of 20, doesn't it. But I happen to be about as old as most men who are starting their profession and to have had a good many more years of knocking around among men. In addition I already have one profession in which I have reached "success" and am in preparation for another which promises just as much and as prompt success as my experimental piloting.

But leaving such mundane considerations as "success" aside - we are not looking forward for our happiness. We have it now. I think we often get in the habit of just looking around the corner for happiness and forgetting that of the here and now. Allene is here. She arrived on Thursday from Boston and is staying with me here at the field. We are having a perfectly beautiful time - with everyone being very good to us and an attractive place to stay. We are not going to be separated again for a long time either. Our plans are to leave here on the fifteenth for Cincinnati via Washington - we will hit Hamilton about the seventeenth and stay for a few days for visits - Then about the 20th, I want to be in Champaign - school starts there about the 21st. Allene will be in Chicago. I think we will go to Chicago first and then I will come down to Champaign for about a week. We can have a good visit with Tom and Florence and leave Chicago about the 26th for New York. We have some of Allene's friends in Washington and New York to visit. Allene's sister from California will be in New York the last week of September. Tech opens the 4th of October and we want to get in our home about the 29th. This gives us a lot of visiting and traveling. I think we will be ready to settle down then.

I am glad to know about Dr. Wilder's overcoat. I had always thought it would be impossible to make anything that would look well out of an army overcoat. The other things - blankets and pictures and cot I can make a box for - and send them express. Where did you leave the things that I want to take with me?

Wherever they are I can get them on my way thru from Hamilton to Chicago.

I am very glad you are getting this opportunity to spend the winter in the South. DeLand is such a delightful place. I am sure you will enjoy getting back there. I am also glad you are getting the work that will give you interest and an opportunity to work out your ideals. This responsibility will be much less trying than the work with the little girls of Hamilton, and it will give you time to yourself - to think and grow and have many interests in your life -

Allene and I send love to our mother.   Edmund

[Knowing as we do that his beautiful vision of his marriage was to last only about a year before "beginning to come apart in 1921," according to Dr. Smith's research, we are sad that the vision must fade.

We also know from another source* that this move his mother made to DeLand, which Edmund remembers from his childhood there, resulted in her position of Dean of Women at Stetson University in DeLand.

*My Mother's Book: After Father's death, she insisted we all three finish college. She was "House Mother and Physician" to the Junior Boys at "Morgan Park Military Academy" for 2 years. But her greatest pride was the work she did with and for "her girls" at the "Hamilton Training School." These girls had been through the courts for various reasons, from truancy to very distressing home conditions. As "Sup't. and Physician" for 6 - 8 years there she changed many lives. She was so proud of them. And we were proud of her! ... I forgot to record Mother's position as House Mother for the Academy Girls, then Dean of Women at Stetson University in 1924. Here she got her A.B. degree. – Margaret Allen Reynolds

And she already had not one, but two M.D.'s, one in Homeopathy and one in Obstetrics. In those days, instead of just becoming certified as a specialist in a field of Medicine, an M.D. took the required course work and received another M.D. This Abby had done in Omaha, where her husband was on the faculty of the Medical School. Dianne has her several diplomas.] – Margaret Reynolds Broussard

MIT Course Work / Glider Competitions in Europe (1920-1922)

Eddie returns to full summer work at NACA Langley Field following completion of his second year at the University of Illinois. He will co-author two NACA Technical Reports with F.S. Norton as a result of the summer work (1920). Edward Warner, leaving NACA to become an assistant professor at MIT, convinces Eddie to enroll at MIT in the fall of 1920 and pursue the study of aeronautical engineering. MIT credits Eddie's two years of course work at the University of Illinois as meeting the first year MIT requirements and leaves him with a full 3 years left for him to complete his degree requirements. He will partially complete 2 of those 3 years of academics.

He is an active student and is elected as the President of the Aeronautical Engineering Society at MIT and he continues a close relationship with now Professor Edward P. Warner. During the 1921-22 academic year at MIT, Warner promotes participation of a MIT team to compete in the European gliding events during the upcoming summer. Eddie enthusiastically joins in the building of a pair of gliders and leads a group of students in preparing for a competition in France in the summer of 1922. One of the glider team members is Otto Koppen, who will become a professor emeritus at MIT and will teach his entire career at the institute. The glider effort will be partially sponsored by MIT alumni, however, the team must provide most of their own expenses. After arrival in Europe, the effort will be additionally sponsored by Chicago newspaperman Floyd Gibbons.
The MIT group of four students take two completely different gliders to Europe. At the French competition one of the gliders is damaged and the second cannot be erected in time to continue in the meet. At the time of the damage, the MIT team is leading the competition. Eddie's three compatriots must return to MIT to start the fall quarter. Eddie decides to continue alone and take the remaining glider to the German competition at Wasser-Kuppe in the Rhone Valley. While flying his glider he is injured in a crash and his competitors assist him and take him to the hospital. Eddie does not realize it but many of his competitors are from the 1922 class of Hanover University, all of whom he will later call "brilliant" as they became the leading aeronautical engineers in Germany. Eddie's association with some of those students will be maintained in his later travels to Germany. Even with his injuries, he manages to travel to the gliding competition at Lewes, England. He returns to Germany where he is stricken with appendicitis and requires further hospitalization.

His wife Allene travels to Germany to assist him early in 1923 and they return together to the U.S. in March of 1923. Eddie does not return to MIT to continue academically. He does present the results of the glider experience to the MIT Aeronautical Engineering Society. His marriage to Allene essentially ends at this time.

Allene and Peggy send you their love.   881 Mass Ave Cambridge
Nov - 11 - 1920

Dearest Mother,

I have let a long time go by before writing. Your last two letters are before me. Mother, dear, you speak of your life as if it had been a failure! You say, "Perhaps I have learned more from my mistakes than I could have learned from successes. But your life has been a success. You have achieved in a truly great measure what you started out to achieve. You have met tremendous obstacles with very real handicaps and you have not turned aside but have gone straight through. Money was never a goal. You did not start with that in view - and so please get this clearly - that your financial condition can not influence the proposition of your success - whatever that condition is.

If motherhood was part of your ideal, in that part, I should say you are a remarkable success - with one son who is already a leader in his profession, a good man who will always be admired and loved in his community; a daughter whose ideal has had to keep jumping to keep ahead of achievement, a teacher who has a vision in her heart of service that when it permeates through the lives of mankind will mean the coming of the kingdom of heaven in all our hearts; and another son who is having some measure of success in working toward the ideal in his heart.

Think, Mother, what you started out to achieve and see that in other parts of that dream as well as this part success is yours.

You are loved by many many people. By many whom you hardly suspect of it, Carl Rahn and Lorinda Perry for instance. Sorry you have had so much work with sickness. Can't you demand better food for the girls.[22] Demanding does an awful lot sometimes when we least think it will.   E.

---

[22] "her girls" at the Hamilton Training School

881 Mass. Ave Cambridge Mass.
Nov 21 - 1920

Dearest Mother,

I have not heard from you since my last letter, but I think I owe you two anyway because I received two from you before I wrote last time. And then I want to write to you.

I wish you could visit us, Mother. We have the pleasantest apartment you could imagine. Something like this.

The furniture is all in and the pictures hung. We are going to get the lighting fixed soon, and then we will have most of the settling done. I built the book cases in Allene's room in my spare time and am now at work fixing up my room.

I wonder if I gave the impression in my last letter that I thought your work, accomplished with a high degree of success, was done. I know that you know it is not, but I want you to know that I also see before you as Dr. Rahn put it "twenty years of service". He once told me "Your mother has had the vision in her heart put to a terrible test. The constant hammering of the social order has almost succeeded at times in its effort to obliterate it." A world in which the love of Jesus Christ is extremely nominal and almost unknown in reality has sapped at that dream for years but has been unable to drain it out. You have still the same character, the same beauty of personality of the girl who went to South Carolina to teach the darkies. I love that girl. Everyone does. Allene says, "Everyone that comes in contact with Mother Allen loves her.

        Good night, dear.
                Edmund.

881 Massachusetts Avenue
Dec 8 - 1920
Dearest Mother,

I am having a very busy time preparing for examinations. This is the last week of the first term, and a large amount of material is to be reviewed. I am surprised at the enormous amount of work we are expected to digest in ten weeks. In Physics we touch on a great subject like Inductance for two days and are expected to be able to solve almost any intricate electrical problem containing inductance formulae. In mechanism we spend a week on gears and we must know how to design, construct and operate gears in any possible combination. At the U of Ill we would spend a whole term on the subject, or what is more likely we would not be expected to know much about it.

Allene and I are having a gorgeous time in Boston. This is the great city of opportunity along literary and artistic lives. On Saturday we are going to see Shaw's Fanny's First Play given by the Henry Jewett Players. I have seen them before in Lady Frederick.

Allene is taking some work at the Curry School of Expression and is enjoying it immensely. She is getting the artistic interpretation of Literature also some fancy dancing. She is very anxious to begin teaching but realizes that what she is learning is going to make her a much more wonderful teacher than she ever was before. Next year she will go back to it with a new zest. She is really very very different person from the one who was in Hamilton. The finding of a work and interests which lead her mind outward instead of inward has proved a wonderful remedy. And at the same time there is the stability in an inward sense that comes with the finding of one's self, a stability that I have found strangely lacking in organized religious expression. I wish you could know her now.

Dear Mother, I have had an overwhelming sense of the utter fruitlessness of most lives when looked at from the point of view of eternity - or even from the point of view of the aviator who at 10,000 feet looks down at the swarm of people each one "caught" by a myriad of little things - each one so hopelessly unsure of life's end. How sordid then seems the ideal of "each for himself" - of "America first". There is after all but one character of true excellence in life and that is the one which bears the sanction of eternity. How beautiful then are those lives which are consciously aiming at such a character. It is for this reason that I am loving you.

Edmund

Sunday Dec 11 - 1920
Dear Mother:

I have had a card and two letters from you since I wrote last but I imagine you have had two letters from me so with this I am not so much behind. Our letters cross so much that we can't hope to wait for an answer before writing and get a letter written every week.

I was very much pleased with your card - especially the part about it taking your breath away. I read it to Allene just for deviltry and was quite amused when she didn't understand, -especially where you said your "son has a most remarkable wife". A and I are having lots of fun laughing at each other and with each other. Between us we have quite a sense of humor.

I am awfully glad you are having a good time with the young people and with your old friends.

Peggy has been here twice. - Once when she went to a dance with a Tech man and again when we all four went. I am getting more and more in love with Peggy. She is very lovely to look upon and just as lovely to pal with. We have tried to have her here oftener but she is very busy and can spare us a weekend but seldom. Allene is very much devoted to her also and hangs on the receiver for hours trying to get Pomeroy Hall where Peggy lives. She gave Peggy a key to the apartment so she could get in if she ever came around when we were out. Last night we had tickets to the Henry Jewett production of Fanny's First Play by Geo. Bernard Shaw but at the last minute heard that Peggy couldn't come so we went alone. That play certainly does get across. Shaw knows his middle class to a Tee. Did you read it when you were reading some of Shaw?

O yes we did get out to Aunt Mabel's - but I have had so much to say, I simply did not get that said. It was not an eventful visit. Allene enjoyed Aunt Mabel very much and strange to say, she enjoyed Uncle George too.

At Tech I am a full fledged Sophomore taking some Freshman work as well - Double (i.e., Freshman and Sophomore) Physics. Next term I am expecting to take some more double physics and double math (integral) and differential calculus. I am enjoying Calculus immensely. Is Professor Smith still at Stetson? Thank you for sending the clippings. I knew that Schroeder was out of the army. He invented an airplane instrument that he is selling for automobiles.

I've not had a fur collar put on the army overcoat. It looks well dyed and I think it will do this year. Next year if fur is lower I will get a fur collar. So far I have not worn anything but B.V.D.'s for underwear. If I can get away with it I will wear them all winter. They are very comfortable. Allene has had her coat remodeled a little. It cost $175 seven years ago and a man here wanted $155 to remodel it. She compromised by having it shortened and re trimmed.

I'm so glad you are having some pretty clothes. I've often wished I might give you a million dollars just to see how you would spend it. - If we were unhampered by poverty we would act differently, wouldn't we. And yet St. Francis' way was better. He was nearer to finding the center of life. Those who are unhampered, who have a good middle class income are most often the ones who are "threading life's tangled maze without life's clue".

I don't know if you can read this at all. - Sorry I haven't read Wells Soul of a Bishop. Wells has fallen short of greatness I think. He has seen too far to be excused for blindness and yet he did not see far enough. He lacked courage at the most fundamental point!!

           Edmund

881 Mass. Ave., Cambridge, Mass.
[Possibly Friday, Dec. 23, 1920]
Dear Mother,

I have delayed this letter very long - I'm sorry. This has been a very busy time for me. Our term ended last week and I finished my examinations somewhat exhausted. It seems inconceivable to me that we have covered so much work in ten short weeks. But there are eight more terms of this kind ahead. I really wouldn't care if there were eighteen. This is life now, full, joyous, wonderful. I do not look forward to living after I finish; I am living right now. And all life will be a learning process.

Sunday Allene and I took another hike out to Norumbega Park. I am enclosing some pictures we took. The one of us both is the best one of me and an awful one of Allene. She has so much sweater around her neck she looks choked. On Monday I built another bookcase we needed and some shelves and a music stand or rather a telephone stand. We are going to take some pictures of the apartment and will probably send them in my next letter. I got Allene a small cast of the Winged Victory. There is something about the tremendous freedom and control in the strong lines of that body that means much to us both. I do want women to be free - much more than I want them to be beautiful, - or loving, - or feminine. I want them to be free of convention, free of what people will think, free of things - comforts, free of the hampering effects of their own physiological make-up and most of all free of love. I want men and women both to be free of the internal physiological processes that make them slaves of their own bodies. And with freedom, control. The joyous conscious control of the acrobat - that wonderful control of every muscle and nerve always energizes me and makes me thrill with admiration. In the air the thing that makes me love flying so is the marvelous control of a delicate machine which the pilot has. And here within our own selves we have the greatest of all machines, the most delicate of all mechanisms, and most of us never even realize the glorious opportunity of becoming master of the machinery and not being run by the machinery.

We are going to take some flash lights of all our rooms. Monday night we went to see the Servant in the House, a play of C. R. Kennedy in which Walter Hampden took the part of Manson. I have never seen such a powerful play. The theater was packed and the audience would not go away until, after a dozen curtains, Walter Hampden came out and spoke. That play surely did get me thoroughly.

Tuesday night we had Murry and Ruby Espy to dinner. Murry is in Harvard this year.

Wednesday night we had Mr. Warner or rather Professor Warner to dinner. We enjoyed him very much. He and Allene made a very real contact in a field in which I thought Mr. Warner would be quite ignorant - Literature and Political Economy. Last night we went over to Professor Mac Innes house to dinner and we all went to Hamlet.

Walter Hampden is considered the greatest living Hamlet. Allene and I read the play twice in preparation for seeing it. Queer - I was disappointed. I enjoyed the Servant in the House much more. Hampden made Hamlet too much bowed down with grief. I wish I might have seen Southern before he retired.

Did I tell you we saw Shaw's play "Fanny's First Play"? Tonight we are going to a reception at Professor Norton's house. (Mr. Norton's father.)

We are not going to Chicago. It is too big a trip and too much of an expense for so short a time.

We wish you could be here with us this Christmas to take a hike with us.   Love from us both.   Edmund

881 Mass Ave Cambridge Mass
Dec. 27, 1920

Dear Mother,

Your gift and letter and card all came to-day. Thank you very much from both Allene and myself. We are particularly delighted with the olive wood fork and spoon; it is almost the only thing we need in the way of utensils. But what are the wooden slab and the wooden brush for? We wondered if it was a bread board or cake board or a crumb tray or a salad dish. We are very keen to use it if we know how.

I don't know just how to address this except just to Citra. I wish I were with you there. I should love to see the grove and the Spanish moss and the hammock. I hope you find things booming in Citra with the prospect of a big income.

Allene and I went out to Blue Hill yesterday with Duncan and Gene MacEnnes. We climbed to the top of Blue Hill which is the highest hill in this part of the state. From there we could see for miles and miles. The blue haze kept us from seeing the ocean which is visible easily on a clear day. We walked fifteen miles and played and danced and rolled down hills.

To-day it snowed - almost the first snow we have had.

Allene and I have been reading a very interesting research into the source of the legends of the Holy Graal. It is called "the Hidden Church of the Holy Graal" the argument being that there is a hidden meaning in the mystic elements of these legends. We have also been reading Hawthorne's Tanglewood Tales, and liking them more than ever.

Lots of love to you and to Miss Valentine who first taught me to love Tennyson.

Edmund

881 Mass Ave. Cambridge, Mass

Dec 31 - 1920

Dear Mother,

I am enclosing the pictures of our apartment rooms. I think the pictures are very good; don't you? The building as seen from the outside is rather pretentious in size but not more than all the buildings near here. Massachusetts Avenue is the street you see in the picture. It runs directly from M.I.T. to Harvard and for that matter clear through Boston and out the other way for thirty miles. There is excellent street car service. The subway runs right under Mass. Avenue with a station three blocks from our place. Fortunately we do not have stairs to climb. There is an automatic elevator which usually works. And since our apartment is in the back of the building we get no street noises at all.

In No 2 you see our reception hall or "music room". There is a frieze of musicians around the wall above the doors on the side you do not see. You enter right by the music stand at Allene's right. This picture was taken from the dining room. You can see our dining room table turned up to form a settle at the right.

No. 3 is the view you get as you enter Allene's room. Her desk is closed on the left with the little Joan cast on top of it and behind it a bookcase that goes to the ceiling and has a curtain in front because most of the books are paper backs and papers that are unsightly. The grate under the mantle is a gas grate but it keeps us warm whenever we want it in the evening. You can see part of the frieze of artists in this view and thru the mirror you can see the book cases that are on the other side. The winged victory is wonderful. She is an inspiration to have around. The windows are dark because this was taken at night with a flash light. In the daytime this room is very light.

In No 4 we tried to get a close up of the "victory" and the corner of Allene's couch. Then if you turn to the right you see No 5. Allene closed her eyes just as the flash went off. Turn a little more to the right for a close-up of the corner with the candlesticks and the beginning of the bookcase No 6. We are taking some more of the books.

In No 7 we have turned clear around to go out into the dining room again and in No 8 we have gone out the door at the farther end into the kitchen - taken from the sink.

No 9 is a glimpse into my room which leads off the hall.

I expect you are still in Citra but I shall address this there anyway. Miss Valentine might be interested. How I would like to be with you there!

I am reading H. G. Wells' new "Outline of History" It is a massive work of 1300 pages; but it is the most fascinating thing I have read in years! I had no idea that the narrative of the life and of mankind from the beginning of Time could be so interesting. I am almost half way through the first volume and have not yet gotten down to historic times - not even reached the ancient Egyptians. From the point of view of a spectator of the vast range of life development from 500 million years B.C. to date, how ghastly appears our modern state of civilization; our petty wars and murders - the individual struggles for financial supremacy. The bounteous philanthropy of Henry Ford who announces a $5 a day minimum - takes a salary of one million a year for himself and $125,000 for his son and then when sales fall off holds tenaciously on to his one millions and gives them ostentatiously to charity for advertising purposes and turns out of employment in one day 50,000 of the men who earned those millions for him - turns them out into an already overcrowded city of unemployed never once thinking of "human" values.

Edmund

881 Mass Ave Cambridge
Dec 31-1920
Dear Sister,

You are a genius at knowing just what people want. There is nothing I could have told you I wanted more than socks. They came just in the nick of time because my last pair that are good looking have come back from the laundry full of holes (the laundry is not at fault - they were full when they went but I thought when they came back the laundry might have exchanged them for a good pair) only silk is such a wonderful luxury! I fear I am being a real bourgeois.

Allene was equally delighted with her gown.

I am wondering if you are still in Evanston. I will address this there anyway.

I wish so much, Sister, that you could come to know Allene. I want it for her sake as much as yours. I have always thought that there is so much that each of you have that is so very fine I should want a perfect woman to have it all.

Perhaps you will be coming east or we will be going west before many moons and we will all get together.

I have not your St. Paul address. Will you send me a card from there giving it?

                                        Edmund

Jan 9 - 1921
Dear Mother,

Your very good letter telling all about the Citra trip was very much enjoyed and was sent on to Tom immediately. How I would love to have made that trip with you and seen the groves and hammock land and big live oaks and Spanish moss, and Miss Valentine, too. I was very glad that things are looking so hopeful and prosperous. Miss Valentine has had a long and, it must be, discouraging time of it, and for you too it has looked pretty black at times. I do hope that it will be very encouraging in actual returns. But even if it does turn out for the worst there is no need to worry. I am getting $30 a month from the government for you and I am expecting to send it to you as soon as everything gets cleared up here. And then as soon as I get to earning, my earning capacity will be easily more than my share of running the kind of home that Allene and I will always want, and we will expect as a matter of course and a privilege to make up to you (with Tom's help perhaps) the difference between whatever your income is and a decent living income of $1200 to $1500. Allene and I have talked this out carefully. Allene wants so much to have you feel that she is a daughter of yours, not merely a daughter-in-law, and as a daughter is entitled to participate in this. She feels strongly that in a decent state of society you would receive a pension and that in the present state we are privileged to tax ourselves for that pension as the state will tax us in years to come.

We both want you to be free of economic pressure. I am very strongly of the opinion that an individual does not need economic pressure or economic rewards to make him deliver his best work. I believe that very often we could deliver more in human values if we were relieved of this pressure. I would think you would react most strongly to the system which has so ruthlessly buffeted you during the past years. O I do hope you can visit us sometime, and perhaps live with us if you ever feel so inclined. But it must be of your own free choice, and we want you to know that if you do wish it, it will be in full economic independence. That is a rule of our household. You will receive your income and pay out of it your share of household expenses. Perhaps this may seem a very materialistic attitude. And yet I believe that it leaves us free of all oweing attitudes - free for the real human contacts.

Allene is pawing the air this year because she is out of a job in spite of the fact that her time is absolutely full. I believe you have many years of usefulness to the world and that as long as it is possible to work a job is almost necessary to one's self respect, if one has thought much about such things. You will remain young (I like your word elasticity) as long as you are producing - and it doesn't matter in the least whether you are receiving enough to live on from your work, or at least it would not matter if we had gone far enough not to need the encouragement of feeling that our work was producing something that society was willing to pay for.

But I am rambling on to no purpose. O yes - while we are discussing this we might as well do a good job. The above takes care of the situation if we all live happily ever afterward. If I die of sudden something-or-other Allene impressed me with the fact that she expects as a matter of course the privilege of taking my, or rather "our", place in the above arrangement, since her income when she is working will be more than enough for her living and since she has no one else dependent on her. If she leaves before I, her new will provides that everything of hers will come to me, and of course the arrangement will go forward as above. If we both depart simultaneously she has provided that the property will go in trust to you until your death and then the capital will be divided equally between her three unmarried sisters and Margaret. That will mean an income for you as long as you live and the money from my insurance, which is $13,000.00 and also $2000.00 which I have in the bank will go to you. I am keeping the $13,000 as a sort of old age insurance. I am putting it all in the form of "age 62 annuity" so that it will be a kind of savings account that will provide for our old age.

Now, I'm glad that is out of the way. Please don't take all this as any beneficience on the part of Allene. I know that would hurt her very much. It is entirely a matter of course. As soon as you really want to retire we will get together and talk about the amounts, etc. In the meantime we know we are doing our best to set to rights our little part of the universe.

I started out in this letter to talk about something else - about your letter.

Allene and I were trying to fit the brush and palmetto slab together - that's the reason we couldn't see where they went. It all seems so obvious now. Thank you for the hat brush and the dish protector! I'm glad you liked the pictures. I sent some more last week. The kitchen is a nice place. We eat breakfast there from choice.

This being Sunday Allene and I went for a walk. It could hardly be dignified with the name hike because it was around Harvard Campus and Brattle St. and Radcliffe College and the Washington Elm where

Washington in 1775 took first command of the American Army and the oak further on where General Stratten assembled his soldiers and after a word of prayer marched to Bunker Hill. And the historic telegraph pole that Paul Revere ran up to get away from the British on their way to Lexington.

The Charles River is the place I love most about Cambridge. It flows right in front of Tech. over 2000 ft. wide and has lots of gulls and wild ducks, and now ice floating around in it. I walk along it every day.

     Lots of love from Edmund.

Feb. 20 - 1921
Dearest Mother,

Your letter came some days ago and I have wanted ever since I received it to write you a long letter. But long letters don't get written. It must be wonderful to be able to run over to Daytona Beach whenever you want to have a picnic. Do you remember the all day and all night trip we used to make to New Smyrna and to Daytona?

Our apartment is still in our hands although I am waiting daily to be thrown out. We tenants have not yet recognized the man who has seized the building, as our landlord. He has told us the leases we hold are worthless, and has said if we wish to remain as tenants we will pay $50 a month instead of $30 a month as our lease says we are to pay for the next six months. We say, "Who are you? We don't know but what our old landlord will come back and make us pay all over again to him". The law seems to be all on the landlord's side. One of the tenants when asked for the rent said "First I wish to see about having my kitchen plumbing fixed". He said, "Well I have nothing to do with that. I'm not the landlord. I have seized the building and want to collect the rents". Eighteen of the twenty-eight tenants either have moved or are moving out before next month.

Allene has not been well. Evidently the nervous breakdown that was threatening when her mother died is more serious than we supposed. She is staying out at a sanatorium for a few weeks for a complete rest and I believe that she will be all right after that. She has been filling up her time too full - writing, and studying at the School of Expression, - and teaching at night school and lecturing at clubs. She doesn't know how to say no to people who ask her to do things with the result that every minute is full. So for these days I am alone in the apartment and I am enjoying it very much. I really like solitude.

I have been reading Matthew Arnold's Buried Life. I love especially the part that begins:

Only - but this is rare -
When a beloved hand is laid in ours
When, jaded with the rush and glare
Of the interminable hours,
Our eyes can in another's eyes read clear,
When our world-deafened ear
Is by the tones of a loved one's voice caress'd
A bolt is shot back somewhere in our breast,
And a lost pulse of feeling stirs again.
The eye sinks inward and the heart lies plain
And what we mean we say and what we would, we know.
. . . .

I am tempted to go on line after line. Is it possible, I wonder, for the thing the poet catches at such rare moments to become a fixed experience in one's life. If perhaps instead of waiting for the sacrament from the eyes of another or the voice or the hand, if we might be willing to offer that ourselves to our brothers, -- and yet Matthew Arnold did not see possibility of that. I have heard that there are those who look into the eyes of those they meet on the street and find a response in each one of very much the kind Matthew Arnold finds so rarely.

I think I will stay out of school for a year or so and teach. I would love to try teaching Physics or Math, or Chemistry. Two years of school is enough at a stretch. Teaching would be different. What do you think?
                                            Edmund

[March 7 or 14, 1921]

Dearest Mother,

It is Monday night before this letter gets written. Sunday was full. Allene and I got up early and took a train for Wellesley. When we reached Peggy's dormitory, she had just left on a picnic so we took a tramp around the lake. Wellesley lake is beautiful, large, blue, surrounded by heavily wooded hills. On a sunny hillside we had lunch and read and talked and bathed in the sunshine. Later we found Peggy and Mr. Ayres from Hamilton who is a Tech graduate student. We all went out to the aqueduct and the river which flows under it. I fear Mr. Ayres was not particularly pleased to be too much chaperoned but Peggy was insistent on our going along. It was bedtime when we returned. The trip to Wellesley is long and tedious. We are reading again William Morris' News from Nowhere, a delightfully written utopia. How impatient it does make one, though, with the outward conditions of our society that produce or permit poverty, misery, spiritual death. I don't know but what the spiritual death comes to those who cause the suffering rather than those who suffer. The way out for the individual is inward and I think it is for society too. "The kingdom of God is within you" is having a very special meaning for me. I have read again and again the first chapters of John. I love John all the way through.

I enjoyed your letter very much. Yes, Rachmaninoff was a wonderful inspiration. We are going to try to hear Fritz Kreistler this week. (Allene remembers Miss Gant very well). Boston is a wonderful place for art.

I like your suggestion of a cottage at Geneva. It sounds ideal. Your future sounds wonderfully interesting. I know you could teach effectively. And your children are all so interested in you. You have a definite place in their lives.

Edmund.

March 14 - 1921 We had Professor Warner around to dinner again and enjoyed a very pleasant evening. Allene likes him very much.

March 20 - 1921

Dear Mother,

The second term is finished and tomorrow we start the third term's work. I am certainly getting a technical education. My subjects for the coming term are Precision of Measurements, Physics of Light, Mechanism of Machines, Machine Design, differential calculus. I am certainly glad that Allene is not a physicist or I would go mad with technological hysteria.

It is all very interesting, though I was amused in thinking of the things I really want to do in Paradise to note that none of them are technological. I was thinking that if I could do exactly what I want I would divide my time between flying, tennis playing and playing the piano. The trouble is that this world is a place where we are bent out of our own instincts and desires into patterns which we do not like.

Allene and I went to see "When Knights were Bold" at the Copley Theatre this last week and enjoyed it very much. On Friday evening we had Mrs. Crane who was a classmate of Allene's at Vassar and Mr. Crane and their daughter Sally around here to dinner. Mr. Crane is Jake and Mrs. Crane is Tuffy. Jake began talking Philosophy immediately after dinner and an hour or so later it was frankly religion which we were discussing. At about ten-thirty we were each so interested in each other's point of view that we decided to continue next week.

Jake was an engineer for about three years - then he decided to change to landscape architecture. He is now studying at Harvard and Tuffy has a job. Saturday evening we went out to Wellesley to the big Vassar - Wellesley debate. The subject was the restriction of Immigration. Peggy invited us to come and furnished the best seats available. The cheering and college songs were good but I cannot say so much for the debate. In the first place it seemed to me that the delivery and argument was in a class with our High School work. I have certainly heard as good debating by Harry Singer or Irvin Fathschild. [High School classmates?]

I was much disappointed also in the type of argument. The negative (Vassar) opposed a further restriction on the grounds that (1) not so very many immigrants are coming in now and not many are likely to come (2) our present laws keep out all undesirables - (diseased, criminal and radical) (3) our plans of Americanization are able to assimilate all that get in (4) we need foreigners to do our work for us.

They ought to have lost and I'm glad they did. Not a mention of the great moral issues involved. Where is all the idealism of 1917? - Vanished into thin air. Where is our humanitarian-ism, for which we patted ourselves on the backs during the war? Was there a thought for the point of view of the immigrant? Was there any humility of mind which might have said to him "Perhaps you can give us something besides your labor"?

Even our revolutionary ancestors, even Thomas Jefferson, to whom democracy - a capitalistic democracy - was the highest good in government conceivable to the mind of man - surely even he would have objected violently to such prostitution of human beings, - of refugees to our shores - to whom we were to be an asylum. There was not much difference between this point of view and that of the Dean of the Harvard graduate school of Business Administration whose class I visited with Murry Espy. The Dean discussed for an hour the pros and cons of the question of the discharge of an employee who does faulty work - its effect on morale, the right of the foreman, shop discipline, efficient production, etc. and never one word indicating that he thought of the human being he was discharging as anything but a thing, a cog, a bad tool.

[He ran out of room on the page and continued across the top of his first page:]

Allene is teaching still at the Waltham High School. She is enjoying teaching French very much but Ivanhoe in 2nd year English is getting on her nerves. She wants me to go over and talk aviation to them. She says that as soon as her job is through she will write to you.   Edmund.

March 21 - 1921

Dear Sister,

I am sending you a letter of Thomas' which Mother wished me to send to you some time ago.

Mother tells me you are very happy this year but that the old high school appealed more to you than this one. I have often wondered if you would find teaching biology the thing you ultimately wanted to do.

I wish we could get together and talk over things. We are about due to do that soon, aren't we?

I have been wondering lately what kind of a person I really am. I think we change very little in the fundamental characteristics, and looking back over a period of yeas I think I see a certain type of development, not particularly lovely - but rather persistently working out through a variety of experiences. There is first of all a fundamental dissatisfaction with "the world". It seems to me that life at the level of the aims, methods and results of the world about us is entirely unsatisfactory. Of course I am a socialist (whatever that term means now) The social system of mankind (except in Russia, perhaps) seems to me fundamentally wrong. And even Russia where social idealism alone exists strongly enough to get people to act, even Russia will fail - as did the French Revolution when France was to be the Savior of the World.

It is not in the realm of Social or political changes that mankind is going to find its End or goal. And I find myself turning more and more away from a vain hope of making over the world into a beautiful instead of a sordid place, toward an inward change. It is said that the mystics knew where they were going and consciously realized that there was no other call but the attainment of their goal. But of this again

Edmund

May 1 - 1921

Dear Mother,

Allene has gone to New York to visit Lorinda Perry and I am batching in the apartment for the time. As neither of us will be here this summer we are taking advantage of the opportunity to sublet the apartment. When we do I will find a room out for the rest of the term.

Thank you for writing the way you did last week and for sending me Mrs. Healy's letter. I showed them to Allene and when I returned I found the note I am enclosing from Allene. Mother, you are admired by more people than anyone I know. I am reminded again and again of it by meeting or hearing from someone who knows you. "That wonderful mother of yours" is becoming a well known phrase.

It has been raining here for months it seems. Once in a while the sun comes out and it drys off enough so that we can play tennis. I went out for the Tech team this spring but was eliminated in my first match. I never played such poor tennis as I did that day, or rather in that match. I am learning that in a contest I usually do very poorly. My best work is never done when hard pressed. I practically always do poorly in an examination when pressed for time and I can easily trim Harold Harper at tennis although he can best the man to whom I lost. Strange, isn't it?

I am much interested in the way you have gone after your work in biology. You have such tremendous capacity for intensive work. I do want us to see each other for a while this summer. If you are in Chicago around the 15th of June we'll be together then. Your summer plans sound very attractive and I know you will enjoy Sister very much. (Do send me an orange blossom. The trees here haven't even become green yet.)

I just heard from Aunt Mabel that Aunt Ella died. This must be a great relief for all concerned. What a horrible thing is insanity.

           Edmund

2727 Woodbine Ave Evanston
Aug 6 - 1921

Dear Sister,

I enjoyed your letter more than you can believe from my delay in answering - but you and I have lived through so many delays in correspondence that an apology seems stupid. I am glad that you are enjoying so much your Michigan summer. Florence tells me that you have dropped enough courses so that you can appreciate life to some extent and that you are having some good times with some men there. That's fine! A fellow surely can have fun with men and I believe a girl can have just as much. I'm afraid I have a little lost faith in the capacity of girls to have comradeship together on as fine a basis as men can have together. This is, I know, unwise, and I hope that before long I will have an opportunity to renew this faith.

As you see from the headlines of this letter I am in Evanston. Tom wrote me that Florence was very happy at not having to carry Tommy up and down stairs because he now goes up and down himself. I decided I would have to witness this new development in the "Education of Tommy Allen" and so I came in for a couple of days to see him. And here I am in the mood to write to you and with pen and paper at hand.

I wish you might have the good time we had at Baraboo this last weekend. We had a real house-party - the first one I have had since Army life ended. We danced and sang and sat around an open grate telling wonderful stories and we put on plays and washed dishes. Lida would have been right in her element there. We read in parts Wilde's "The Importance of Being Earnest" and we put on about half of "Hamlet" in costume reading the parts, of course. Then Dick told, one evening when the lights were out, that most wonderful of the Arabian Nights stories - Prince Kawaralzaman, and then someone read "Alice in Wonderland." By the way, have you read "Alice" since you were a youngster? Well, if you don't enjoy it three times as much now as you did then then I miss my guess.

One delightful stunt was making up folk dances to some music such as Tschikofski's Rustic Wedding Symphony, and parts of the Prelude to Carmen Act 1 which we had on the Victrola. Then almost everyone played some instrument and everyone wiped dishes well. I do wish you could have been there. The hostess was one of the most gracious women I have ever met in whose house one felt perfectly at home and never bored. It was like one of those fine old English homes where as Galsworthy says "the host never knows if the guest has committed suicide after breakfast until next day" one is so free to be wherever one wants.

Dick was in his glory. I love to see him get away with a situation like this: entirely surrounded by people - the center of an admiring throng of girls - jollying them all along and setting everyone laughing. I wish you knew him. He is a real guy. With his black beard and heavy hair he looks like a distinguished French diplomat.

Do you remember about that loan we were negotiating with Miss Valentine? I wonder if Mother and you could handle the other $150.00. I sent $50 and now find that before I am going to be able to get started in school next year I will need every cent I have. If you can't make it I'll borrow on my bonds.

A letter from Allene last week sounds very encouraging for her recovery. I had hoped the summer's rest would enable her to get on her feet. This summer ought to be a wonderful means of giving us all renewed vitality and increased interest in our work.

       Edmund.

Delton, Wis.
August 22, 1921
Dear Sister,

Now that the summer is drawing to a close we all take stock to see just how worth while it has been. I expect that yours has meant much to you in the matter of a new grasp on your field of Biology and in your social relationships which I understand have been very pleasant. My own summer, the first real vacation I have ever had, has really been the happiest and finest summer I have ever had. Spent among congenial people (I have made a host of friends) the vacation has been a revelation to me that I have never had before, of the *possibilities of wholesome meeting ground between human beings. It has been considerably marred by the terrible interpretation put upon all my friendships by my mother. You are probably eminently familiar with her attitudes and actions which have sought to make me absolutely impossible socially wherever her influence has touched, and would make my friends everywhere shun me for the sake of their own social and economic positions. Mother apparently does not realize what a terrible thing she has done and what insult she has dared to put into a letter to me, if she assumes that I could write to her after receiving such a letter. My attitude is perfectly known to her. We differ not in values or standards. That is not the point. The important point is facts. She is putting out things that are lies and she has been told this. Since she denies this and cares not for my respect which she has lost, I fail to see how there can be any further converse between us, and I shall take steps to warn all my friends that they are assuming grave responsibilities in being known as friends of mine.*

I had wished that you might have been kept out of this, Sister, but this step seems to me necessary.

Edmund

[What a sad letter about a very sad situation. His letters in the first half of 1921 were full of Allene, then something happened that made him very angry with his mother, right after he had praised her in his 5-1-1921 letter to her with, "Mother, you are admired by more people than anyone I know. I am reminded again and again of it by meeting or hearing from someone who knows you. "That wonderful mother of yours" is becoming a well known phrase."

Then she wrote a letter to a Dr. Cobb that he referred to in the undated letter I gave the estimated date of 1-17-1922: "There is one thing I want to say to you, Mother, without having it bring up discussion, which is so futile. It is to correct a former statement I made to you and an impression I gave you this summer: Allene is not the "normal woman roping in a husband". Last summer when she was so ill and unresisting, those around her tried to make her out to be that. I have found that she did not do certain things for which I held her responsible. They were done by others. Allene is not the sort of person you pictured to Dr. Cobb. If that impression was given you by me it was wrong and I withdraw it."

I (Margaret Broussard) don't know who Dr. Cobb was. My imprecise memory of my mother's (his sister's) account of that rift is just that it was about Allene; that his mother was very concerned about the way Allene and her guru, Carl Rahn, were twisting her naïve son's mind. Mother said that her dear brother suffered from depression for a while as a result. If this is from a year later, he seems to have recovered and is on good terms with his mother again – and Allene is out of the picture, at least for a few years. So I presume that Abby wrote Dr. Cobb thinking he could do something about It, perhaps asking for some kind of intervention, and naturally, Edmund resented that very much, even if it was valid.] — MRB

881 Mass. Ave
Cambridge, Mass
Dec 2, 1921
Dear Mother,
I wonder if you have been getting my letters. You said Thomas was a busy man and did not write often and that Margaret wrote you every week. I thought that there was the implication that I was not writing often. If you have not received a letter a week, then you have not received all of mine. I do not write regularly, i.e. on the same day every week - but I get an opportunity always once a week to write you. You have not mentioned a letter from me for some time. Have you disapproved of them? I look forward to your letters from week to week as I expect you do to mine.

Thank you for mending that glove. I have not had gloves this winter yet and expect I will need them, for the streets are icy almost every morning when I go to school. We have the queerest samples of warm and cold weather here. Thanksgiving day was beautiful. Allene and I decided to forgo the usual orgy in which we always overeat, so we packed our packs and hiked off to the woods and hills. It was snowing heavily when we struck Stony Brook. The trees stood out in strong relief with each twig covered with white half an inch thick. We stopped on the top of a ridge overlooking two towns for a cup of hot soup from our Thermos bottle and then hiked out for the Blue Hills. For miles we did not see a house. At one place a tall cliff overlooked the trail we were following. As it looked inviting we climbed it and eventually came out on the top of the world, with beautiful snow covered wooded hills stretching in every direction finally lost in the blue haze of the snow filled air. And then it stopped snowing. As Allene and I stood there side by side a great wave of thanksgiving came over me and as I looked at her I knew that the same thing was happening in her. It was a queer kind of thankfulness - an inward kind - not for "things" of "opportunities", and yet I'm sure you know - you must have felt it often.

Dearest Mother, I so much want to be the sort of chap who can make his love for you a real force that you will feel all about you, warm and real; something you can rely on that will give you backing. I think that I have done this very little so far but I hope to grow that way.

I decided this summer (or was it last spring) to send you the proceeds of my summers work. It was coming to you. Thus far I have not done it; I am sending a check probably in the next mail. You see the gov't is paying my expenses.   E.

I was much interested in your account of DeLand and the walks and the trip to Daytona,. Yes I remember Mrs. Kruse [?] children and mirror lake & Mrs. Boydster. [?]

Please do not send this letter on. It may be misunderstood.

[This last note may be the reason we still have this letter. Others that were shared didn't get back to Abby to keep – for us!] — MRB

[Undated letter written during the gap between letters dated December 2, 1921 and January 22, 1922, a gap when other letters surely were written, but have been lost. This one reveals the emotional distress he was feeling over Allene, or over his mother's acceptance of Allene. She is not mentioned in the letters from January 22, 1922 on.

I will give this one the approximate date of January 17, 1922 because of his mention of a letter and present from his sister in this letter and the first two sentences in his letter of 1/22/22 to his sister: "Thank you very much for the Christmas and birthday present. It was much delayed in the mail and reached me but a few days ago."] — MRB

Dear Mother

No letter from you this week. Are you ill? I hope your school work is going along well and your papers are getting written.

I am designing a glider - as I told you, - a little beauty, with no struts and wires and wheels and propellers. It has just a body and a wing and tail like a bird - in fact, it looks much like a bird soaring high over Lake Gertie.

There is one thing I want to say to you, Mother, without having it bring up discussion, which is so futile. It is to correct a former statement I made to you and an impression I gave you this summer:

Allene is not the "normal woman roping in a husband". Last summer when she was so ill and unresisting [?], those around her tried to make her out to be that. I have found that she did not do certain things for which I held her responsible. They were done by others. Allene is not the sort of person you pictured to Dr. Cobb. If that impression was given you by me it was wrong and I withdraw it.

I have a letter from Sister who told me something of her work in St. Paul. I also received a present from her - a birthday-Christmas present.

Biology is only one side of History, Mother - It may be the most important side but at least it is but one side. If you can read Henry Caufield Osborn's book, "The Men of the Old Stone Age", you will see another side.   *Edmund.*

[* In the summer of 1921, his life was in the turmoil of his rift with his mother over Allene. His letters in the first half of 1921 were full of Allene, then something happened that made him very angry with his mother, right after he had praised her with, "Mother, you are admired by more people than anyone I know. I am reminded again and again of it by meeting or hearing from someone who knows you. "That wonderful mother of yours" is becoming a well known phrase," in his 5-1-1921 letter to her.

The 8-22-1921 letter to his sister is about that rift. My imprecise memory of her imprecise account of that rift is just that it was about Allene; that his mother was very concerned about the way Allene and her guru, Carl Rahn, were twisting her naïve son's mind. Mother said that her dear brother suffered from depression , then called " "Dementia Praecox," for a while as a result. If this is from a year later, he seems to have recovered and is on good terms with his mother again – and Allene is out of the picture, at least for a few years. So I presume that Abby wrote someone she thought could do something about it asking for some kind of intervention, and naturally, Edmund resented that very much, even if it was valid. But other letters from the first half of 1922 are full of news about his glider, so my guess could be wrong, only – no other year seems right, either. Perhaps it wasn't June at all.] —MRB

M.I.T.

January 22 - 1922

Dear Mother,

I have been skating on the Charles today. The ice is very thick and has been very good for skating until recently when a snow and hail covered it with two inches of soft material. Last night this too froze over and we skated on the surface of the hard snow.   *Just outside of my window, in the churchyard or campus of the Theological school the* grass is frozen over and boys of the neighborhood are playing a strenuous game of hockey. This is just such a Sunday after-noon in Cambridge as that one which was used so effectively by John [blank] on his trip to Priscilla's house. Every morning when I go to school I pass the house where the Village Smithy lived. I do not look in at the open door nor do I see sparks fly from the ringing anvil or hear the bellows roar but I hurry on to the garage nearby and watch the mechanic changing tires instead.

There are fine musical productions here constantly. I suppose Boston is a center of that sort of thing greater even than New York. The Boston Symphony Orchestra is very fine and the Popular concerts are also well attended.

I am afraid the aeroplane proposition is going to fall through again. There seems to be a scarcity of funds always connected with those who are interested in flying. The Aero Society, however, is planning to build a glider. This is an important undertaking in view of the revival of glider experimentation in France and Germany. it will be many weeks before it is completed but the design, development and construction work will be quite as interesting as the actual flying.

I see a good deal of Murry and Ruby these days. Murry is completing his work in the Harvard Graduate School of Business Administration and Ruby is private secretary to some business man down town. They are both old dears. I see Margaret Byard very little. The trip to Wellesley is a most difficult one and Peggy never seems to get in to town. When I do get out there she is usually so busy she can see me but little. She is a very popular girl. I would like to see more of Uncle George, but we fail to connect up. I am off very little from my work at Tech and Uncle George's time is limited in Boston.

I was interested in what you said of Leonardo de Vinci. I had always thought of him as a most versatile man - one who had touched almost every phase of knowledge and experience and had done work not only in painting but in medicine, in music, in anthropology, in literature, in philosophy and in engineering. I know engineers claim him as a great scientist, artists as a great artist, literateurs as a great writer and I believe the medical profession has some claim on him. Of *course, he was a genius, with an infinite capacity for labor and breadth of interest. What do you think!*   *Edmund.*

[1. No mention of Allene.

2. There are two comments written in the margin by his sister. One, alongside this about Leonardo de Vinci, says, "Definition of a genius! (M.A.R.)," and the other, alongside his lines about the Village Smithy, says, "dry humor!"

3. Here are some excerpts from Henry Wadsworth Longfellow's poem:]

The Village Blacksmith

Under a spreading chestnut-tree
The village smithy stands;…

Week in, week out, from morn till night,
You can hear his bellows blow;

You can hear him swing his heavy sledge,
With measured beat and slow,
Like a sexton ringing the village bell,
When the evening sun is low.…

And children coming home from school
Look in at the open door;
They love to see the flaming forge,
And hear the bellows roar,
And catch the burning sparks that fly
Like chaff from a threshing-floor.

[There are eight verses, 48 lines, in this poem written in about 1840.]—-MRB

[Cambridge]
January 22, 1922
Dear Sister,

Thank you very much for the Christmas and birthday present. It was much delayed in the mail and reached me but a few days ago. Did you know that I needed a pair of socks very much or did you have the larger bit of knowledge that men always need new socks?

I wish you could live in Cambridge for a year or two. It is very different from the west but I believe it has something to give to the young people who are growing up even in this age of intense modernism. There is here a very great love of beauty - old beauty. Shelly's line "the pain of old beauty" is very real here. Goldsmith's, "I love old books, old friends ____" explains much of the atmosphere surrounding Harvard. They may lack that spirit of progress, of pioneering, that has so wonderfully developed the west but in its place is something we westerners lack. The little old-fashioned Harvard professor can sit in his gas lit study and let the world pass by. Has he not seen the same thing happen in Egypt, in Greece, in Rome when wealth surpassed the dreams of avarice, and anything that was not new was scorned? Has he not looked on while a great renaissance [sic] turned over the social, economical and political machinery of France and the world hailed the newborn thing as "the Christ of Nations", while the industrial revolution started in England the movement toward modernism in industry that has now almost reached its summit in America? Has he not seen in the education of Greek youths hundreds of years before Christ an ideal of a perfect well rounded human being that cannot be equaled by our Oxford or our Technology? What does he care that New York and not Boston has the highest building in the world and has no painter that can produce the Mona Lisa before which he can worship eternal beauty? He is little interested that science has discovered or invented a machine that can make automobiles at a thousand a day or an explosive ten times more powerful than T.N.T. and has taken away the interest of the world from anything more lasting than styles. I have known such a man here.

    I am glad that teaching is being such a source of happiness to you. I want to teach myself some day. It may be that I can do it for a while. For the summer I am hesitating between Langley Field and Aeronautical Research and Italy. I shall probably do neither and so am making no plans.

    The group picture of our family turned up in my recent packing. I was rather surprised at the audacity of it. It was taken during the war, of course, and that accounts for some aspects of it. Still I cannot understand how it ever happened that you - who are by all odds the most interesting figure in the group got stuck back in the corner and T and E with their bright new uniforms occupy the foreground. It is really audacious. Do you get it? I am quite ashamed of myself for allowing it.

    You are probably having much more skating than I, but the Charles River is a great place for sport.

                        Edmund

The Allen Family: Eddie, Margaret, Abby, and Thomas

[M.I.T.]
Feb 1 - 1922
Dear Mother

Your letter came Monday. It seems to take a long time for mail to get from Florida to Boston. Perhaps when the aerial mail is running that far south and this far north it will be quicker.

I am glad you told me about those girls of yours[*]. I think they are fine! The splendid spirit in which they have gone ahead and finished their school speaks well for their training. It is your spirit - that is easy to see. That Hamilton School was a very fine institution.

Your meeting Mrs. Boydston and her boy. What happened to Charles. You told me he was married and then separated. Is he still living alone? Was he successful in business?

The grove sounds very good. I should like to live in the grove some time - working at agriculture and poultry, oranges and vegetables.

Did you know that Murry Espy finishes Harvard Graduate School of Business Administration this year. He is planning to get in with Henry Ford in the Muscle Shoals project. That will give him contact with the business side of agriculture - just the thing he wants. Murry is much changed - all his coarseness has smoothed out and he is quite loveable.

Our glider proposition is still up in the air until we hear from the administrative committee. We may have to agree to allow no one to glide in it unless they have the written agreement of the parents relieving the Institute of responsibility. Imagine the stupidity of such a conservative attitude! I suppose one might find enough justification for it, but at the one school in the country that specializes in aeronautics, this attitude seems to me singular. We have loads of enthusiasm. Eighty dollars are already subscribed.

I am interested in your attitude about Martin Luther. The opinions of history teachers are so different in regard to him that it is difficult to predict what they will say. There is surely an advantage in having a course on the Reformation under a man who is a sympathetic admirer of Luther. I am living in a building with several Theological Seminary Divinity Students[**]. Their attitude toward the Papal Election is interesting in view of the constant tendency of the Anglican Church toward Rome, and the fact that in increasing numbers Anglican clergymen are going over into the Roman hierarchy. These students are - as, I suppose, most all students are - strong conservatives. [Even history teachers at Tech. say they dare not say what they really think in historical ideas for fear of startling the conservative students and causing their dismissal - and so they keep mum - as Luther did not.]

But this is all very unimportant - only you seemed much interested in Luther.

Edmund

[* Just as a reminder, since I've quoted it before, from my "Mother's Book" by Margaret Allen Reynolds, about her mother, Abby Dyer Allen:

… her greatest pride was the work she did with and for "her girls" at the "Hamilton Training School." These girls had been through the courts for various reasons, from truancy to very distressing home conditions. As "Sup't. and Physician" for 6 - 8 years there she changed many lives. She was so proud of them. And we were proud of her!

**he is not living with Allene at this time.] — MRB

[M.I.T.]
February 8 - 1922
Dear Mother,

Your two letters came yesterday and today. That was the first news I heard of the tragedy in Evansville. I had thought Uncle Will had been through enough hardship of this kind so that this would affect him but little. I am so sorry for you, Mother. These things are all coming at once. I wonder if the tendency toward self destruction is hereditary. I have felt it in myself very strongly at times - especially before 1919 when it dropped away. But I suppose it is in every person to some extent. It was an honorable way of taking one's exit in Rome. Custom changes. I do not think it dishonorable in the least. Conditions surrounding it may be dishonorable. I do not think they were in Uncle Will's case. The act is one of a lost individual and it is a pitiful thing to be so lost and to believe that is the only way out, but it is not entirely unworthy.

I have spoken about this because of Aunt Margaret's letters which were distressed because of what your children would have to remember. The death of a brother remains the sad thing and you have all of my sympathies in your loss. [*]

The house proposition I like very much. Now is the time to buy in the business cycle. Of course I do not know the condition of the house. I would ask some man to look at the plumbing. That is important. Replacements cost hundreds of dollars. How is it lit? Gas?

But the real questions are: Do you want a home? Do you want a home in DeLand? Do you want a home now? I think you would answer them in the affirmative, and I should second you. There are many advantages. That is the thing that grandmother Dyer should have had in all those years after your father died. It is the mistake that is so often made for old people to give up their homes and be floating around. It weakens one psychologically and morally and people as they grow older need increased morale. Don't worry too much about the money. Remember the funds I have for you.   I do not understand the financial arrangements as you have given them, but Professor Smith, I believe, should be a sound advisor.

Go ahead with all of my backing, Mother.

Edmund

[* Edmund's mother, Abby (Abigale), his Aunt Margaret, Uncle Will, Aunt Mable and one to four other siblings were children of Elbridge Gerry Dyer and Margaret Teyrer Dyer. An excerpt from Mother's Book about Elbridge Dyer says:

Poor in money matters, Grandfather left school at 14 to learn a trade. He was apprenticed to a Mr. Nichols to learn to become a "machinist" or "artisan," as it was called then. He lived in the family (home) of his employer, a very fine man, and was treated as a son, with privileges of a fine library. He loved to read, so educated himself. By nature a "perfectionist," he became an excellent machinist. (two omitted paragraphs) He became very wealthy. He had a large home and a large family. Strongly "Anti-slavery," they harbored escaped slaves in their basement – an "Underground Railroad Station." During the Civil War, the government requisitioned the factory to make guns, cannon and other war machinery. When war ended, ruined financially, Grandfather committed suicide rather than face "dishonorable" bankruptcy. He left life insurance.

My mother thought Edmund's mechanical aptitude came from their Grandfather Dyer.]
— MRB

[M.I.T.]
February 19 - 1922
Dear Mother,

Your letters have been good during the last days. Thank you for writing so freely. Your latest action about the house may, after all, be the best thing at present. I am returning Aunt Margaret's letters.

Why should the University in DeLand grow? It is the popular thing for universities to grow I suppose - but if Tech gets much bigger I shall be tempted to go to an institution that is not so troubled with growing pains. Some of the old Universities of France have probably not grown a bit for a hundred years. It would be a shame if they did. Of course, I see your point, that if it doesn't grow it means that the quality of instruction and the character of the student body is diminishing.

Your discussion of Roman Catholicism and Protestantism is very interesting to me just now because I am living in a dormitory with young Anglican priests - an Episcopal Theological Seminary. The Oxford movement is not very popular among the younger men. It is the older and more "English" of the churchmen who are attracted by the Roman Church. The younger men at this institution are inclined toward the "low" church or the broad church as they call it. I am inclined to apply your judgement in its strictest terms right here. They "broaden" out to nothingness. I believe from a historical point of vantage a tendency toward the very high church can be discerned all through the protestant ranks. This is seen all the way from the introduction of Methodist prayer book and confessional to the change in name of the Episcopal Church to the Anglican Catholic Church and finally the transference over into the Church of Rome of Anglican clergymen every now and then. Of course the opposite tendency is also in evidence in the Free Thinkers and New Thought and the branch of the Baptist who emphasize social service and Sociological and Political pulpit discussions. Church history is a fascinating subject. I should like sometime to go thoroughly into it.

I wonder if you would enjoy Jean-Christophe by Romain Rolland - one of the really great novels of literature, I believe. Have I mentioned it before? I started in the second volume which is quite independent from the first, as is also the last. The analysis of France and the Comparison of French and German character is absorbing. Each character, one feels, represents a type. I have seen much of myself in Oliver. The book is wonderfully refreshing and is very easy to read.

Our glider program has almost reached the construction stage. My design was accepted to-day. It is a beauty still even though much of its looks were sacrificed for economy of construction. Aeroplane design is a constant compromise.

Edmund

Feb 23 - 1922
Dear Mother
Yesterday was Washington's birthday. It might have been spent under the old Washington Elm where he marshaled the first American Army, but it was snowing and the ground was covered with sleet and so I stayed in my steam heated rooms and studied Heat Engineering. "Heat" is taught by a very fine old white haired professor with clear blue eyes. He has taught many classes of Technology engineers and he knows just about what he can expect from them. Perhaps I am a bit

pessimistic in believing that our class is just a bit less interested than any before. When the bell rings, no matter how interesting a part of the subject is being covered, no matter how alert one of two have been to catch the whole essence of the lecture hour which comes at the last, no matter if we are finding out for the first time the answer to the awful riddle of waste in modern power houses, - when the bell rings the students in the back of the room are aroused from their little nap, those in the middle put up the watches they have been watching anxiously for the past half hour, and the shuffling and slamming of seat backs begins. The professor may look up almost imploringly to these young engineers of tomorrow who will not even let him finish his sentence; no, the bell has rung; It is dinner time and one has to stand in line if one is late at the commons and so the professor gives it up and nods dismissal, and one who is near can see the old tired look come into his eyes. I wonder what he thinks of us; - this old scholar, this engineer - of us - of the new generation, rushing to get our degrees and get a job with some corporation, with no time for research, no time even for professional reading, no time for fine things - for music or literature or religion. Professor Berry is a product of a generation that cared for quality.

 I was much interested in your account of the Kruse boys. They do sound fine. I do not remember them very distinctly.

 About the grove, I know so little that I could not advise you satisfactorily. I am glad it has at last started to bring in a return. If Miss Valentine would be interested in a scheme of packing family boxes, or if that would be possible with the facilities and labor available, I should think connection could easily be established with a commission man in some middle sized northern town.

      With love,
       Edmund _

99 Brattle St. Cambridge
March 4 - 1922
Dear Sister,

I was particularly glad to get your letter this week. The winter party sounds most interesting; I should like to have been along. The cold and the snow add a sporting element that is delightful. Do you do any skying - no, not skying; that is what I do sometimes. I mean skeeing.

Did I tell you about our glider? The Aero Society at Tech. decided they wanted to do something constructive in the field of aeronautics. They could not have an aeroplane; that was forbidden by the school authorities. The only other field that was open was gliding. The parallel in the case of Germany is striking. At the time of the armistice Germany was completely disarmed in the air. The interallied Air Commission refused to allow her to fly anything with a motor in it. And so, nothing daunted, she decided to fly without motors. The result was a tremendous revival of glider experimentation. The surprising thing about it is that before long such remarkable flights had been made that France became suspicious, decided that Germany was getting away with something, and was torn between the two alternatives of forbidding Germany to fly anything at all, or entering upon a program of glider experimentation herself. She decided on the latter course, and as a result we have international competition in soaring flights.

There is really quite a distinction between gliding and soaring which does not appear at first sight. Gliding is the operation performed by any aeroplane when the motor is shut off. Most aeroplanes have a gliding angle of about one to five, i.e. they can glide for five miles if they are one mile high. An efficient glider has a gliding angle much greater than this, so that if they get into the air from a hill top whose angle with the horizontal is 10 degrees, corresponding to 8 to 1, they can easily remain in the air as they glide down parallel to the surface of the hill. Glides of three or four miles have been made in this way. There is another factor which often enters into gliding of this character. The wind is normally horizontal, which strikes a hillside, is deflected up until it blows parallel to the surface. In this case there is a vertical component of the wind which gives a direct lift to the wings of the glider. There are also occasional vertical gusts called rising currents. These gusts of course take anything along with them which happens to be in the way. One time while flying at Love Field I came over a woods which had a very light color as seen from above. I felt myself suddenly lifted and as I watched the altimeter I rose 4500 feet almost vertically in about two minutes. It is these rising currents which explain to most scientists the apparently effort[less] flights of buzzards and cheel [?], flights in which the birds often attain great heights in very short times.

There are some investigators, however, who are not satisfied with this explanation of soaring flight. Motion pictures have been taken of bird flight, pictures at high speeds to detect very slight or very rapid movements which are indistinguishable to the naked eye. Instruments have been perfected to determine the vertical velocity of the rising currents. Dozens of theories have been offered to a credulous public to explain the phenomenon of soaring flight. The history of these investigations is very old. I have recently read the treatise of Leonardo de Vinci on the subject. About fifty years ago Lillienthal built the first successful glider in Germany. In this machine he rose to a height of fifty feet above his starting point. From that time until Wilbur Wright remained suspended in the air for ten minutes at Kitty Hawk in 1902 glider experimentation continued along the old lines. Soon after this time motors were put in the gliders and real soaring flight investigation was at an end.

Following the revival of this work in Germany a flight was made in a motorless soaring machine from level ground. I may now define soaring flight as distinguished from gliding, as the utilization of the energy of the wind for obtaining lift. You may have noticed that birds never "soar" when there is no wind blowing. The most plausible explanation of this fact is that the birds take advantage of horizontal gusts. When the soaring bird encounters a gust, a body of air moving at an increased velocity, he increases the angle of incidence of his wings and rises on the gust thereby reducing his own speed relative to the ground or relative to the main body of air. If now he continued on his course he would pass out of the gust and have to dive to recover his speed with reference to the body of air in which he is now flying. The net loss of height would be exactly the same as if there had been no gust and he had glided along at his normal gliding angle. Instead of doing this the bird turns when he has reached the peak of the gust and flies parallel with it until his velocity in the gust is reduced to normal. At this point hi turns again 90 degrees and flies out the front to the gust, thereby losing some height but not as much as he would have lost had he flown out the rear of the gust. He now immediately turns and flies in the front of the gust again. There is a net gain in height each time the entrance and exit from the gust is made. If you will notice soaring birds you will always see them constantly circling.

A soaring machine is one having a very efficient gliding angle and so built that it can be readily maneuvered quickly in gusts. We are planning to construct such a machine as soon as the present glider is off our hands. In the glider design contest we held, Otto Koppen and I pooled our ideas and as a result we won the award over contestants who have had several courses in design, something of which we are entirely innocent.

I am afraid I have bored you fearfully by all this technical explanation. I believe you are not teaching Physics this year. I am surprised that your Biology does not interest you in any sort of a historical background. It is quite plausible that in an engineering school such as this, men should leave without the least perspective on the place of their work in the work of mankind. But in the teaching of Biology, - it would seem to have so little meaning apart from the great story of man, not his battles and his dates, but the growth of his ideas. . .

Mother writes a very interesting letter about the grove and about the new ideas on the reformation that are being presented in her courses. We have a new nephew.

So you are dancing. I hardly know whether to congratulate you or commiserate with you. I have given it up. I am so completely fed up on modern dancing that I hope I will never be persuaded to go on the floor again until something more interesting happens to our dances. Do you really enjoy them? They are so absolutely meaningless, so barren, so devoid of any creative outlet that after one knows the steps that is all there is to it. But do not let me bring on the disillusion any *sooner than it will come anyway. Why don't your Sunday School class or your Hiking club take up something creative like dramatics. You know what wonderful fun that is.*

Brother Edmund

[No place or date indicated by ETA,
but In MAR's hand, simply "1922".
I give it the arbitrary date of March 13, 1922,
based on the next letter dated March 20, 1922.] — MRB

Dear Mother,

Exams are over and the new term well started. Our glider is also started. On Saturday afternoon we had a dozen men working on longerons, spars, ribs and skids. I am trying out a new system of shop management in exercising shop foremanship. I have omitted detailed plans and directions and give to each man only the general idea of the part of the work he is to do. He then plans it himself - his brains as well as hand work. The spirit of the job is wonderful. Everyone is interested in an inward way. We hope to have the fuselage completed next week and the wings will be coming along soon.

Your letter just came. I would like to have been on the trip to Lakeland with you. Some time I may come to Florida and really see it. I do not remember it as a garden country but as an undeveloped wild place - none the less wonderful perhaps. How I should love to see the groves in bloom. Has the Florida boom spread to Citra? And is Miss Valentine enthusiastic about prospects? I am very glad you gave her the $1200. I hope Miss Valentine comes out all right with things. I should like to see her.

The glider work as I said before is moving slowly. I have just finished an article for the Tech. Engineering News, - the monthly engineering magazine on glider experimentation. At the 1922 Competition to be held in Germany one hundred gliders are expected to be entered in a contest for long distance and length of time in the air. I wish Tech could send over their glider, but I am afraid that is impossible. Did you receive the paper I sent you?

The Flonzaley Quartet is an organization created by the will of a wealthy musician. The idea was to have a combination of four great artists who could play really great music without the necessity of having to watch public opinion and play up to public taste. They are financially independent of the box office. They are not in it to make money and so they play music for musicians as Browning wrote poetry for poets. Ask Miss Gaut about them. She probably knows much more than I do and then find someone who has a Victrola and get them to play a Flonzaley record - any one will do if you like chamber music. I think Glazenow's Interliudum in modo antieu [?] is especially wonderful.

                              Edmund

[in M.A. Reynolds' hand: Cambridge Mass
Boston Tech stricken out and
MIT inserted (M.R. Broussard)]

March 20 - 1922

Dear Mother,

Your letter came this afternoon and I am answering it immediately. Yes the orange blossoms did come but they were evidently much delayed and although for one day they gave the room a faint reminder of those wonderful spring days in DeLand. The white buds did not open. I wonder why. If they had the fragrance would have been sweeter. The green leaves withered very quickly. But they are all here still surrounded by the Spanish Moss of which I am very fond.

Last Saturday I had my last examination and since there was no symphony concert in the evening I went with a crowd of Tech fellows to a Bohemian restaurant down town. It is Greek a la Athens. We had Pilaph with Yaurti and Baklava and Dorpon and Svingi. It was some meal. Afterward we went to a theatre and saw three of Shaw's plays. "O'Flaherty V.C." is the account of a war hero returned to his home in Ireland for a rest and longing after two weeks for the peace and quiet of the trenches. "Press Cuttings" is an earlier play, a satire on the feminist movement. "The Dark Lady of the Sonnets" was in some ways the best. It was an intimate glimpse of Shakespeare and Queen Elizabeth - wonderfully done. Sometime when I can do exactly what I want, I am going to spend my mornings playing around with airplanes, designing, building and flying; my afternoons playing tennis and my evenings putting on amateur dramatic productions!

At the theatre I met Murry Espy and a girl from Urbana, Mrs. Kirkpatrick. I went home with them and had tea and toast at Mrs. Kirkpatricks - and discussed poetry, drama, -- science, law and theology. It was extremely amusing. Murry is so intensely serious this year in his investigations into religion and ethics. I didn't see Ruby. I guess she was in bed when we took Murry home.

On Sunday I went for another walk looking for gliding hills, with Koppen.[23] In the afternoon we went to vespers at the Church of St. John the Evangelist a very high church Episcopal service. In the evening we went to an Italian restaurant and had some real Spaghetti Marinara.

Last week I heard the Flonzaley Quartet which came up from New York for a concert. I wonder if there ever was such playing, since Orpheus tempted Providence with his harp celestial.

Edmund.

Tommy's picture is good. It is quite characteristic.

[Flozaley Quartet was organized in Manhatten, N.Y. in 1902. Members were devoted to quartet-playing. The quartet was highly praised for the perfection of its ensemble and its artistic finish. They played successfully until they disbanded in 1929.] — TKO

[23] Otto Koppen was one of the MIT participants at the French glider competitions with ETA. He went on to become a highly respected professor emeritus teaching at MIT until he retired in 1965] —TKO

April 15 - 1922

Dear Mother,

Tomorrow is Easter Sunday. I shall probably go to Trinity Church or else to one of the larger churches in the poor districts. It is a glorious time for Boston. There are many lilies, all hot house, of course, and many other flowers, too. The parks are beginning to show green buds on the trees and the grass, too is becoming green again. One evening it was very warm and I took a long walk about midnight out to Watertown and back. From where I live I can reach the country in about an hour's walking. One day I went to the hills near Arlington and a large grouse flew up near my feet. He was beautifully colored with metallic silver and red and purple and a long tail that streamed way behind him as he flew.

The glider is showing some signs of progress. The fuselage is almost entirely finished and the wings are nearly ready for assembly. We can only work on it about two afternoons a week and consequently it goes slowly. We tested a wing rib yesterday and found it excessively strong, but since we cannot well reduce its weight under its present 4 ½ ounces without affecting its stiffness we will leave it as it is. Professor Warner says it compares favorably with the best ribs that have ever been built.

Many of the boys here are musicians. In the evenings when I am studying they play the piano beautifully. They play so many of the things you used to play in DeLand, and I remember again the many nights when I would go to sleep with Hayden or Beethoven in my ears. I love those old things much more than any new pieces. The more one becomes familiar with old music the more one loves it. Symphony orchestras make the greatest impression when they play something we know. I wish we could know everything they are going to play.

                                                                     Edmund.

["where I live" is 99 Brattle Street, Cambridge.]
99 Brattle St. Cambridge, Mass
May 4 - 1922
Dear Mother,

I have been working night and day on our glider program for several weeks. Actually 21 hours a week outside of school work has gone into actual construction work, and evenings have been spent designing and planning. We have been rushing hard because we wanted to have the skeleton finished to exhibit at our annual banquet (The Glider Banquet this year) This is to be on Saturday, May 6th. The only thing that remains to be done is to put in the ribs in the wing. This is an intricate job but it can be done by three men in an afternoon. Almost the whole of this week has gone in on fittings. These small metal parts have been most difficult. Each one has been tested in the big testing machine before final acceptance. They are all done now and most of them are on the machine. Everything from tiny hinges to the big box spars of the ribs has been put in place. The only thing remaining is the covering, the fabric surface of the wings. For this we have the finest balloon clothe which weighs only 1.9 ounces to the square yard. When this is put on our glider will fly.

The glider banquet is a swell affair. The new president of the Aeronautical Engineering Society of M.I.T. is to preside as toastmaster. I have asked Professor E. B. Wilson the head of the physics department at M.I.T. to speak on "The Flight of Man". Professor Warner is to speak on "Soaring Flight" & Major Wooley speaks on "The Government and Flying". Major McDonnell on The University and Flying and several of the junior officers - on the research program of the Technology Aero Eng Society. In the afternoon we will go out to the flying field and have flights for everyone who wants to go up. I have asked Peggy to go with me. As yet I have not heard. It is most difficult to reach Peggy even on the telephone.

(N.B. "Peggy" is our cousin who was at Vassar M.A.R.) [" Margaret Byard" was written above "cousin" and "Vassar" stricken out and "Wellesley" added by MRB at some forgotten time in the past]

Last Saturday I went to the last concert of the Boston Symphony Orchestra. They played the "Schaherazade" [sic] Suite of Rimsky-Korsakov. Get Miss Gast to play it for you. I got a copy of the suite in book form before the concert and had looked it over and I was surprised to find that I could easily follow the whole thing as it was played. That sort of thing may be highly artificial and very academic but it does enlarge one's enjoyment. You may be able to hear parts of this on the Victrola.

        Edmund

Excerpts from "Boys' Life" - March 1931   **Nearly ten years after ETA's glider experience in France and Germany, he recalls the preparations and the events.**

"During my last winter at MIT, I helped organize a glider club. The Aeronautical Engineering Society of the Institute offered a prize for the best glider design. Otto Koppen, a member of my class who is now on the faculty, worked with me on the plan for a glider that won the prize.

The next thing was to build it. It took us several weeks putting in every minute of spare time and cost about $100. It had a twenty-four foot span, a 54 inch cord, and a fuselage 20 foot long. The total weight was about 75 pounds. When we had it finished we got a crowd together and tried to get it into the air by towing it into the wind with ropes. But we weren't able to get it off the ground.

That was a pretty discouraging day for the Glider Society.

But we wouldn't give up. We borrowed a truck and hauled our glider about thirty miles to a small island near Ipswich, Massachusetts. It was called Hog Island and was about a mile long and and half mile wide, with a hill on it about 200 feet high. There were no trees on it, but a good many rocks that stuck through the grass. We had to wade across a shallow channel to get to it, carrying the glider. It was a dull rainy day with a light wind. One of the MIT faculty, Edward P. Warner, who is now editor of "Aviation", went along with us and worked as hard as anybody. From a point part way up the hill we made a short, successful flight, that lasted perhaps ten seconds, with the glider never getting more than a foot or two off the ground. I found that the controls handled perfectly. It gave me a great thrill but as I landed the skid struck a rock and broke.

Professor Warner said: "Thats the first successful glider flight made in America since the Wright Brothers!".

The next day was clear, with a much stronger wind. We repaired the skid and waited for the wind to die down. We had left the glider overnight in a dilapidated old hay barn, the only building on the island. When the wind dropped enough to make it safe, I made a number of flights. On the longest I was in the air for about 35 seconds. That one was from the top of the hill and at one time I must have been 50 or 60 feet above the ground.

The worst trouble was the skid; it broke nearly every time I hit a rock and then we had to go to work and repair it again.

It was almost the end of the college term and we had to hurry back to Boston and plug for final examinations. About commencement time we got an invitation from France to compete in an International Gilder Meet for a 10,000 franc prize. We decided to accept it, provided we could raise enough money to make the trip.

The members of the Glider Club contributed what they could, and several alumni promised further help. We understood we could get special rates on steamship tickets and for the shipment of our glider. Altogether we raised about a 1,000 dollars.

We had about three weeks to get ready. We took the skid off the glider and replaced it with a pair of light wheels from a boy's bicycle. Then we made a second glider that we thought would be an improvement on the first. It weighed about five pounds more and cost us nearly $150.

When the time came for sailing we were in an awful tangle; we didn't have nearly money enough. Some of the people who had promised money were unable to make good on their contributions, and the steamship company insisted on charging full price for everything.

But I was crazy to go, and persuaded the rest to start, even with the handicap of not having money enough, and trust that we would be able to get through some way. If we could only win the prize, we'd be

One of Allen's successful glider flights during the early part of the International meet

all right! Four of us made the trip: Koppen and I, and two others who paid their own expenses.

It cost $300 to get our two gliders to France, and another $150 to get them down to Camp Mouillard at Clermont-Ferrand, where the meet was to be held. What with our own fares and other expenses we didn't even have enough money to buy a rubber-shock-cord for taking off, but had to depend on the little ordinary rope that we'd started with.

The meet was held in a country where several small craters of extinct volcanoes rose out of a fairly flat surrounding valley. The hill selected for the contest was perhaps 500 feet high, with sides so steep that it took nearly an hour's hard work to get the glider up to the top of it.

There were about forty entries. Although the meet was "International" they were all French except ours and one crude little bi-plane, with a strap suspended from the lower wing for the pilot to sit on, entered by a Swiss bank clerk. At that some of the gliders looked much better than ours. The best of all was a beautiful Farman, of perfect workmanship

Our hopes faded. We felt we had little chance of winning against anything as fine as that.

That shows how little you can tell about gliding contests in advance. On the first day we made test flights from about a third of the way up the hill. Chardon, the Swiss, made a short, successful glide that lasted only about five seconds, controlling his glider by swaying his body back and forth as you would in riding a bicycle. My own first flight lasted about thirty-five seconds. When the beautiful Farman took off, propelled by its fine shock-cord, it shot about twenty feet into the air, nearly went into a stall, and then, as the pilot pushed his stick too far forward in a desperate effort to save himself went into a dive and came down on its nose, a mass of wreckage.

At the end of a couple days of trials, it began to look as though we would win first prize, after all, and get out of our money difficulties, for we found ourselves leading all the other contestants. Then along toward the evening of the second day, one of the French boys who was helping to pull us into the air at the takeoff, stopped to let the glider go over his head. He was pretty tired, after helping pull flyers into the air hour after hour, and didn't even bother to turn his head to make sure that the glider would go over him.

It didn't. I yelled to him, but it was too late. The wing hit him right in the back of the neck. Down we crashed, glider and all. I got a good shaking up, the French boy wasn't hurt at all, and the glider was smashed.

Our other glider wasn't even assembled. By the time we got it together and ready to compete, the meet was over. The winning glide was only about three minutes long. After the meet was over, the winners took their plane to the top of a much bigger crater some miles away - the Puy de Dome, about 3,000 feet high, and made a flight that lasted more than nine minutes.

While this French meet was going on, another was being held in Germany. There, while we were gliding for minutes, they were

E. T. Allen with one of the two gliders that he took to France in 1922

breaking world records and staying up for hours. Two flights were made that summer of nearly three hours each.

Our money was all gone, but I decided to draw on a fund of my own that I had saved up from my pay during the preceding years and take our second glider to Germany. I wanted to find out whether the German records were being made because of more favorable terrain, or because they had better gliders.

They had, as I learned, both.

The Wasser-Kuppe, where the German records were made, is a great hill with a gently descending slope that drops away for nearly 1,800 feet. With a twenty-five mile wind blowing from the west, there is a vertical lift of as much as 3 feet a second, which is enough to keep almost any sort of glider in the air. With a 10 mile wind the lift is almost two feet a second; that enables only skillful pilots with good gliders, to maintain their altitude. The best glider pilots are those who can get back into rising air currents quickest and oftenest, and stay in them longest.

Although the German meet was over by the time I reached the Wasser-Kuppe, a good many of the competing gliders were still on hand. On one day I counted eight gliders in the air at the same time, two of them more than 1,000 feet above the take-off. One of them was piloted by Klemperer, the world's record-holder. But by the time I had our MIT glider ready to take off, all other activity had stopped.

Still I wanted to try a flight. For two weeks the weather was bad and the wind unfavorable, and the best I could do was to make a couple of short test flight on the south slope of the hill, where conditions weren't nearly so good. Day after day there would be only rain and fog.

Then the weather cleared suddenly, with a good wind springing up from the west. I hurried to the top of the hill. By the time the glider was ready to fly, half a gale was blowing, with a gusty wind of perhaps thirty-five miles an hour, that made any gliding dangerous. The German pilots who were still there looked a little anxious about it, but didn't try to advise me: I was supposed to be an experienced flyer, and they assumed that I knew my own business. They even went further and loaned me a rubber shock-cord to take off with.

I remember sitting in the glider and hollering "Los" - which is the equivalent of "Let's go".

The next thing I remember is opening my eyes and looking up into a tree. I hurt all over. I was lying on my back on the ground and high in the branches above was what was left of the glider. The Germans came running up to see how badly I was hurt. They told me over and over again that I had made a wonderful flight, and gave me all kinds of praise, but I knew perfectly well that I had done no such a thing. I guess they thought I was going to die.

Later I saw motion pictures that were taken of my flight. They showed the glider took off at about forty miles an hour, and was immediately carried up by a gust of wind to a height of 70 or 80 feet, almost directly over the starting point. It was evidently out of control from the very first, with the ailerons jammed under the unusual strain.

ETA at the Wasser-Kuppe

Presently it turned over onto its back, while I hung on as well as I could, and finally came crashing down onto a tree at the edge of a grove some distance to one side of the starting point.

The next three months I spent in a German hospital. For a time it looked as though I would have to lose a leg. Finally I was all right again. I had to borrow money from my relatives to get home. On my way back to America I stopped for a couple of days in England to see still another "International" gliding competition held on the South Downs, near Brighton. The flights averaged better than those in the French contests, but on the whole did not nearly equal the Germans. At that the contest was won by a Frenchman, Mauney-holle, who stayed up for three hours on the last day of the meet in a glider from which nothing was expected. It just happened that at the particular time of his flight gliding conditions were more perfect than at any other time during the contest. In that particular wind almost any sort of glider could have stayed up for a long time. The next year Maney-holle was killed when a wing of his glider broke off in the air." — ETA

[This letter and its enclosure are carbon copies of the letter to the addressee at the Chicago Tribune. The carbons were sent to ETA's brother, Dr. Thomas D. Allen, who sent it on to their mother and sister. TDA wrote a note at the top: "(This is a personal letter and is not to be published.)" Sister Margaret wrote, many years later: "I do not think this restriction holds any longer." That was probably when she included it with letters she provided Thomas Collison, who wrote several articles about ETA after his death.]

Landkrankenhaus, Fulda, Germany.
September 26, 1922.
_Floyd Gibbons,
Chicago Tribune, Paris.

My dear Mr. Gibbons: [24]

Now that I am again able to dictate letters, it occurs to me that you may be interested in having an account of my flight and crash. The memory of that day, (which was temporarily lost, as is often the case in crashes) has been coming back to me. My own recollection is supplemented by the accounts of German pilots who saw the crash and have since visited me in the hospital. One of these, Herr Hans Gronau, has written his account, a copy of which I enclose. I may add that nothing could have exceeded the kindness and comradeliness of Herr Gronau and the other German pilots. If I had been one of their own people they could not have done more for my comfort.

I had waited ten days for clear weather and a west wind. Although on the morning of September 20th it was still raining and the clouds were hanging low, I decided to go up to the Wasser-Kuppe on the chance of its clearing up later in the day. On the way up the clouds hung so low that we walked through them practically all the way and could see but a few feet ahead. Several times I lost my way in the dense mists, and kept to the road with difficulty. On arriving at the camp on the top of the hill we found a strong west wind, but the mist had not yet lifted. But about noon the clouds broke away, giving a clear view of the valley. The wind was a little north of west, not quite the best wind for soaring. However, it seemed so favorable that the opportunity was not to be lost.

The German pilots were much interested in the American attempt for this day. They advised a preliminary flight from a lower slope. I replied that as the machine had already shown satisfactory performance in trial flights from the lower slopes, I was not interested in taking the machine from the hangar again except for the main flight from the west slope, for which I was waiting.

---

[24] Floyd Gibbons was the editor of the Paris edition of the Chicago Tribune; was an aviation enthusiast and possibly a financial supporter of the MIT glider activity.

In comparison to the solidly built German machines, the light American glider seemed very fragile, and the German pilots seemed dubious about its ability to stand the severe blasts on the west side. But I know that it was only in a strong wind that record flights could be made, and although the wind that day was more gusty and irregular than I liked, I feared the season was drawing to a close and no better day might be found.

We took the machine from the hangar to the west slope. There is a saddle at this point between two peaks. The slope is gradual for about a hundred yards and then comes an abrupt drop to a forest in the valley below. Beyond the forest the valley widens. The gusts were blowing up the saddle at from 20 to 50 miles per hour.

After assuring myself that the machine was alright, I strapped myself in and gave the signal for launching. In a moment I was fifty feet above the starting point. Caught in a rising gust of wind, it seemed that the machine would be carried entirely over the Wasser-Kuppe. I dove steeply to gain control, but on account of the rapidly rising air current I did not approach the ground. At this point as everything seemed satisfactory I leveled the machine out with the purpose in mind of gaining height from the rising current. The wind was so strong that the machine was practically stationary over the ground for some time. I intended to steer a course over the Felskuppe and continue flights between the two peaks. On the first time over the Felskuppe I found myself forced to the edge of the "air fountain", to where the descending current begins. With one wing in the descending current, it cost me several moments of struggle before I recovered from the violent sideslip. I finally forced the machine back into the air fountain just in time to keep from striking the ground. The rising current was so strong that I almost looped before I could recover equilibrium. When I finally leveled out I was several hundred feet high again. From this point I flew straight out over the valley towards what I hoped would be a region of steady air. But I again struck a descending current. This time it turned the machine completely around. The fragile controls of my light glider were quite inadequate to combat the gust, and with jammed ailerons I was blown before the wind directly towards the woods. I attempted a dive and turn with the rudder, but before I had gotten half way around I reached a tall pine on the outskirts of the wood. The fall through the tree to the ground is missing from my memory. The next thing I remember was waking up the next morning in Gersfeld. I was taken that day to Fulda where under an anesthetic my nose was sewed up. I am now receiving excellent care in the Privatklinik of the Landskrakenhaus. Within a few weeks I shall be about again; I hope, in time to attend the English gliding meet.

Although this flight ended in a crash, its experimental purpose has certainly been admirably fulfilled. The American glider has been flown now in three countries, under the most varied conditions, and in comparison with almost every known type of machine. Its last flight afforded a test under as severe conditions as any machine is ever likely to meet. In that we are ready to make a material contribution towards putting gliding on a basis of accurate knowledge.

We know something more inclusive than how to build a machine. We know now definite facts about soaring flight, its limitations and possibilities, and the direction of its development; facts which were not known in America before the American team sailed for Europe. I intend to place these facts as speedily as possible before the public, in order that the progress of soaring flight in America may be saved the needless waste of time, materials, and perhaps lives. It is an activity in which thousands of young men will wish to engage. All the information which can be gained by experiment and the experience of other countries should be placed at their disposal. My efforts in this direction will be the best expression of my thanks to those Americans who so generously backed the French and German expeditions of the American Glider Team.

                    Very sincerely yours,
                    (signed) Edmund T. Allen

Translation of letter from Herr Hans Gronau

On the morning of September 20th, I accompanied Mr. Edmund T. Allen of Boston to Wasser-Kuppe on the Rhon. The summit was hidden in clouds. About noon the weather became clearer, so that Mr. Allen decided to fly. This surprised the pilots who were present, because the wind was very strong and Mr. Allen's machine frail. We tried to persuade him to make first a trial in a small declivity of the mountains, but Mr. Allen argued that in Clermont-Ferrand he had already flown in spite of similar strong wind, and so he decided to start on the abrupt west slope.

When Mr. Allen was fastened to the machine, he tried the starting gear, and started to fly. Immediately the machine got perpendicular during 10 to 20 meters, and then erect. It seemed that Mr. Allen steered very deeply and the machine flew always higher over the valley, but then was taken into a gust and fell on the left wing toward a group of rocks. Just before reaching the ground the machine rose again with the right wing upward, but in a manner threatening it would be turned on the back. I saw Mr. Allen squeezed in the machine vehemently trying to work the gear. The machine obeyed and flew again over the valley. After a very similar current of wind , the machine did not come back but turned on the right wing and shot at a furious speed toward the wood. It seemed that Mr. Allen would succeed in bringing the machine over the forest, and turn against the wind and soar high over the valley, as he had enough room under him. But unfortunately the machine was taken into a violent gust, and Mr. Allen flew at all speed down on the top of a fir tree. The machine was crushed on the top of the tree and Mr. Allen was somewhat 10 meters from the ground, and as soon as he succeeded to unfasten himself from the machine and had cut some branches, he fell on the ground. His bleeding nose was broken, and he complained about a pain on the hip. He made the impression of a man who was suffering much, although with his strong willpower he was trying to silence his pain. Afterwards he asked for the length of the flight, later he lost his memory. "The length of the flight I cannot estimate as I ran under the machine" where there was much excitement.

The flight was an achievement in its kind. According to a pilot the wind blew at 18 to 25 meters a second. In a spirit of comradeship all the flyers hurried to help Mr. Allen until he was safely handed to the Doctor in Gersfeld. Later he was transported to the hospital at Fulda. —— Herr Hans Gronau

[Notes: My mother, Margaret Allen, had saved money for several years of teaching, since WW I ended, to take a trip to Europe in particular to see the famous Passion Play at Oberammergau in Bavaria, which had been postponed from its normal date, which would have been 1920, because the war had disrupted everything for a few years. She was excited to be able to meet with her brother at the port where he was awaiting his glider, which had been dismantled and boxed as freight. He then went to France for the meet referred to in the above and she went on to Bavaria and then Italy. I think she didn't know until she got home to America that her brother had been injured. – Margaret R. Broussard

"Krank" = "sick." "Krankenhaus" = "hospital."]

[The glider pilot competing that day was Wolfgang von Gronau who was about 29 years of age. Herr Hans von Gronau was Wolfgang's father who was 72 years at that time. He may also have been in attendance at his son's competition.] — TKO

[Also competing that day was the glider designed by Dr. Georg Hans Madelung. Madelung was the designer of the Vampyr glider which became the dominant design for future gliders. Madelung became the Director of the Aviation Research Institute and would meet with Allen in future visits] —TKO

[The pilot of a competing glider was Wolfgang Klemperer who set the world record. Klemperer would emigrate to the US in 1924 and would become a highly respected aerospace scientist at Douglas Aircraft and would collaborate with ETA in the 1930's.] — TKO

Motion pictures in 1922 at the Wasser-kuppe

Wolfgang Klemperer's prize winning flight 1922

Vampyr Glider Designed by Dr. Georg Hans Madelung

Landkrankenhaus, Fulda Germany
Dec 1, 1922
Dear Thomas,

I have just had an interesting operation for appendicitis. There was a stomach pain Thursday noon after dinner. It increased until it was rather severe about seven o'clock when it suddenly disappeared. The Director of the Hospital said it was appendicitis and proposed to operate immediately. I objected, but was finally persuaded that delay was dangerous and so at ten that evening I had it out. The nurses tell me it was very bad and almost ready to burst (I take this as somewhat of an exaggeration). The incision was less than two inches long and on the fifth day I dressed. (But the first day I had very bad pain.) I heard that another young man was operated on an hour before me for the same cause. He had waited too long and the appendix burst. He was given the last rites by the priest yesterday but today he is improving.

I thought these cases though common might interest you in German practice. The hospital here is up to date in most respects although it is at present bankrupt as is everything in Germany except Berlin.

Professor Warner has just written me to buy him a lot of German engines and instruments. Next week I will feel more like venturing out on an expedition of buying engines.

Don't know when I'll be back. Perhaps next Spring. No word from you about the money.

E.T.A.

Landkrankenhaus, Fulda, Germany
[undated - perhaps about 12-6-1922]
Dear Thomas,

Your letter has recently been forwarded to me from Berlin. I was surprised at the news of the check. I thought the check cost $1000. The $200 cabled I received, but I never knew from whom it came, and I credited it to the Aero. Eng. Society thinking it was a contribution to the fund. I received $600 by cable, in all, and I was never able to find out from whom any of it came. So - at any rate, the loss is not so great and I can make the change on the books and charge the Aero. Eng. Society with $200. which I spent on glider expenses. Then we can settle when I get back and find how much you got for the check. As for converting it into marks that would have been extremely unwise because the mark is always dropping. I keep as few marks on hand as possible. Never more than 30 or 40,000. One example of loss incurred this way is my glider bond. The customs office required a deposit of the value of the machine that would be returned, they said, when the machine was taken out of Germany or broken so as to be unsalable.

I named a figure as low as I dared at 25,000 M. when the dollar cost 1000M. I told them that the labor had cost nothing, being voluntary help and that the materials were very cheap. They accepted the 25,000M giving me a deposit guarantee. When the crash occurred the matter was reported to the "Solhaupt" with the prospect of much red tape. After weeks of waiting I again asked for the return of my 25,000 marks which was then worth but $12.50. There were certain matters which must be referred to the "Hauptsolampt" and another wait ensued. When I returned from England I again asked for the money and was informed that all the papers had been forwarded to Berlin. It is now worth $3 and I have lost $22. Further delay doesn't matter now because I can't lose more than two or two and a half dollars more in the next year unless the revolution comes and it all goes. If I had been wise I would have deposited $25. in U.S. bills and then I would have gotten it back.

The condition is so serious that people are afraid to keep money for over a week. Everyone rushes out to spend his money because goods are worth so much more than the paper money. This causes prices to mount enormously. Every day they go up 20% or so. Wages do not keep up in spite of the strength of Socialism and everyone who hasn't a government job is in need, except, of course, shopkeepers.

I am feeling much better since my appendicitis operation than I did before. The blood clot has not been removed but no sciatica has appeared. I am having salt baths and mercury vapor light (called

Hohensonne) baths, a daily hypodermic and medicine that tastes like salts.

The note about an account in the Scientific American Literary Digest etc. is news to me. I hope someone is saving them. I should like to see them. I have also seen very few of the Tribunes - and no movies. I shall be very glad to see Mr. Singer. He sounds most interesting. I hope to enter the Tribune Competition next Spring near Chicago with a glider although where I'll get the glider I don't know yet. There is another meet here on this side in January. After I have seen that, I will know just what kind of a machine I want. It is down in Africa near Algiers. I expect to leave Deutschland about Christmas and go down to Genoa and across by ship to Algiers. I hope it is warm in Africa. I'll get either the Chicago Trib. or the Associated Press to pay my way for reporting the meet.

I have just received pictures of the Curtiss Army racer which now holds the world's speed record of 230 - 240 miles per hour. Have you seen it? I can't stop looking at these pictures. It is the most beautiful airplane I have ever seen. I would give my eyeteeth for a chance to fly it!

Love to Florence and remember me to Bob and Ruth and everybody else.   Edmund

Leipzig, Dec. 31, 1922

Dear Mother:

Your letters were forwarded to me all in a bunch from the American Express Company, arriving just at Christmas time with the cable from Evanston. They were quite a Christmas greeting. I was much interested in Margaret's letters enclosed and in your accounts of the new De Land.

My nasal operation I postponed, perhaps indefinitely. I had my appendix removed, however, and am now well and busy in Leipzig writing. The meet in Algiers was postponed until summer and I look forward to returning to the States for the meets there. I do not think much of the Daytona meet proposed for January because it is backed by the Aero Science Club of America (which is nothing). The big meet will be early in the summer.

I welcome a return of our correspondence and friendship. Such a return, however, cannot ignore the past, or the mistakes which we all made. We can forgive, and I think we have all of us forgiven, but that alone does not wipe out the harm and suffering caused by our mistakes.

I am assuming that the basis of our renewed correspondence is that you too feel the desire to do everything in your power to restore good will and understanding all around, -- to put the past definitely in the past by undoing your share in the mistakes which we all made so plentifully.

In this connection there is one thing I want to ask of you now. The renewal of the cordial understanding between myself and my wife which I am now enjoying, had, as one of its conditions, the return of all letters which she wrote during the period of our mutual misunderstanding, that she may destroy them herself. She feels this as very necessary to her own peace of mind. I have already returned to her all the letters which she wrote to me, keeping no copies. If you will do the same it will be a great relief to us all and will relieve the shadow of the past from my marriage. Indeed, I think it would be a good thing if you would take this one step further and send back to me all the letters I sent to you in the year 1921. I, too, should enjoy making a bonfire of a mistaken correspondence.

Affectionately

Edmund

Washington DC /Dayton, Ohio (1923-1925)

In 1923 Edmund continued some test pilot relationship with the NACA. Based upon letters of inquiry, both NACA and Langley Memorial Aeronautical Laboratory were seeking to have Eddie join their staffs. Whether he did or not has not been confirmed. NACA did grant Eddie's request that he be "given the privilege of flying one or two of our [their] fast airplanes, to become acquainted with their characteristics". They also granted him the opportunity of discussing some of "the aerodynamic characteristics of the present Berliner helicopter" with "Mr. Bacon" of the NACA.

Eddie became acquainted with Henry Adler Berliner and got the job testing Berliner's helicopters. Richard Smith's research concluded Allen made about 36 flights. Berliner, however, failed to solve the basic instability of his helicopters[25]. Fortunately for Eddie, the flights were mostly "tethered", saving disastrous consequences. Following the crash and total destruction of the Berliner helicopter, ETA built two different light planes with the view of commercial potential. The first built with Berliner was a four passenger 80 horsepower craft. It did not survive its initial flight and ended up in the top of an apple tree near College Park. The second craft built independently of Berliner was a light weight single passenger craft powered by a Harley Davidson motorcycle engine. This craft was successfully flown many times but failed to be commercially successful. These ventures are described in more detail in following writings.

Concurrently in 1924, Allen started some association with the American Security and Trust Co. and writing for "Aviation" magazine. This magazine became "Aviation Week and Space Technology" which remains today a dominant aerospace publication. Eddie wrote a weekly column, mostly about light planes and gliders. He did this for almost a year. Interestingly, Edward P. Warner had editorial responsibilities for the magazine for the next ten years and published ETA's articles throughout that time.

Eddie traveled to Germany again in the summer of 1924 although the purpose was not clear. It may have related to his editorial responsibilities for his weekly column and his continued interest in gliders and light aircraft. He traveled to Austria and possibly Paris during this trip.

In 1924-1925 Eddie worked as a civilian test pilot for the Army Air Service at McCook Field, Ohio.

---

[25] Henry Adler Berliner became President of Engineering Research Corporation (ERCO), makers of the Ercoupe light aircraft.

This is also the last time "Allene" is mentioned in correspondence and it is assumed the divorce is part of the "situation" that racks the family. Allene enrolled as a student at Carnegie Tech in Pittsburgh in the drama department. The two maintained some level of correspondence. She reported that in 1926 that she and Eddie had been divorced for "some time". Allene would eventually move to New York, remarry and become editor of the <u>New York American's</u> women's page, also writing a "homemaker's column" under the pseudonym of "Prudence Penney". Allene Gregory passed away in 1948 at the age of 59.

Berliner Helicopter with Triple Set of Wings

Excerpts from "Boys' Life" - March 1931 Years later ETA summarized his experiences with the Berliner helicopter and the design, build, and flight of his two "homebuilt" aircraft.

"Coming back to America early in 1923, after a few desultory months of investigating and writing about light planes and gliders, I went to work with Henry Berliner as the pilot of a helicopter.

Airplanes, as nearly everyone knows, are sustained in the air mainly by the pressure against the under side of their wings or planes, as they move rapidly forward. Whenever they slow down below the speed necessary to keep them in the air, they stall and fall toward the ground. Helicopters are designed to go straight up and down so that falling they fall as a bee or humming bird can.

Up to the present time no helicopter has been entirely successful. The Berliner machine was no exception. But it is not generally realized how close we have been for many years to the solution of the vertical flight problem. There have been many helicopters that have flown, and flown greater distances than the Wright brothers in their memorable flight at Kitty Hawk, N.C., on the seventeenth of December in 1903. At the time I was working with Henry Berliner, a prize of 50,000 pounds was offered in England for a successful helicopter. It was this prize - a trifling quarter of a million dollars! - that Berliner hoped to capture.

The machine was a curious contraption that scared me nearly to death. The fuselage was that of an old Nieuport war plane with two horizontal propellers above the pilot's seat and a small third horizontal propeller mounted above the tail. The two main propellers were to lift the machine into the air while the third one was used to assist in controlling the flight by raising or lowering the tail. As I sat in the pilot's seat, the tips of the main propeller blades whirled about only a few inches above my head.

In all, I made thirty-four different flights in that machine, without counting innumerable trips back and forth across the field when it failed to get off the ground. On dry clear mornings it would often fail to rise; in moist heavy air it would do better and go up three or four feet. Then, I would sail it cautiously across the field, afraid that at any moment something would go wrong and those twin propellers would chop me into hamburger.

One of the main troubles with all helicopters, so far, has been that they have to have adequate system of control. We do not yet know how to fly them. Chained to the ground they will rise a foot or so into the air and strain contentedly at their cables for hours. Once off the ground in free flight, however, it is almost impossible to tell what they will do next.

I began flying the Berliner helicopter after a good deal of preparatory work in May. Toward the end of July [1923] the machine came off the ground one morning with surprising speed. Conditions, I suppose were exactly right. The engines probably developing a little more power than usual. Almost before I realized what was happening I was twelve or fifteen feet in the air. The machine seemed to be trying to get its tail under itself, as if it were going to turn over on its back. I shifted the controls as far as I could to make the nose go down and the tail come up, and so avoid the crash that seemed imminent. I was able to stop the tipping over-backwards business, but in the process the machine went into a sort of whip-stall and plunged to the ground on its nose. It was smashed completely; entirely beyond repair. By good luck I escaped the propeller blades and got only a shaking up.

That was the end of my helicopter flying." - ETA

"After the helicopter crashed without cutting my head off, Henry Berliner and I worked on a small plane (among other things) that we hoped to enter in a St. Louis economy contest. In designing it I had a lot of my glider experience in mind; the wings were long and narrow, planned for low speed with maximum lift and a minimum of what is known as "induced drag". With an eighty horse-power motor, it was to carry four passengers. Berliner designed special wheels for it that he wanted to try out, and helped me with the rest of the plane, but for the most part it embodied my own ideas. Perhaps that wasn't so good.

We built it at a small field at College Park, near Washington. It took us two months and cost nearly a thousand dollars. .....

Then, on one of the short flights across the field, I let the plane get too far above the ground. I realized it was perhaps ten feet up, and that the small field did not give me room enough to get down without risking a crack-up at the end of the field. Straight in front of me were the hangers.....there was nothing to do but pull up .... over the buildings.... It was the first time anyone had ever banked that plane, and it was the last time anyone ever did, too. It went into the bank all right, but it would't come out again. The design of the aileron control was wrong. The ship just flew along in a slow curve, gradually losing altitude, and I couldn't do anything about it. Presently, about half mile from the field, still turning on that slow circle, we hit an apple tree in somebody's backyard. The plane took most of the shock. It was traveling slowly, anyway; only about thirty-five miles an hour. It simply disintegrated. All I had to do was pick myself up, step out of what was left, and walk back to the field." —ETA

"My next plane was begun while I was still with Berliner, but was not finished until several months after our connection was ended.....I finally finished my plane in the middle of the summer of 1924, at Bolling Field. It had a seven and a half horsepower Harley Davidson motorcycle engine. Plane and engine together weighed only about two hundred pounds......Finally, I changed the propeller, taking off the one I had designed and made myself and putting on one designed by my glider companion at Massachusetts Tech, Otto Koppen.....After all the discouragement it certainly felt great to be circling smoothly around over Bolling Field and looking down at the Potomac River and the city of Washington beyond.

I flew that plane many times without ever having a bit more trouble with it, except once when I overshot the field and broke a wheel. Once I kept it up more than an hour testing fuel consumption. I took it up three thousand feet or more and even stunted it a little, rather gingerly. But I was never able to do anything with it commercially..... When I left Washington the plane was still in one of the hangers at Bolling Field. When the space in the hangar was needed, it was pulled out and put on the dump heap. One of my friends saw it there and towed it around to his garage. I had advertised it for sale and received an offer from some fellow in Canada. Two hundred and fifty dollars for the whole works, f.o.b. Toronto. When it got there my purchaser backed out; said it was too much money. The shipping and customs charges were about seventy five dollars.....as far as I was concerned it had turned into a sort of flying white elephant.....That lad who wanted it was able to buy it in by paying the seventy-five dollars shipping charges."

[Bolling Field /Anacostia was a joint Air Force/Navy facility deactivated as a flight center in 1962 due to its flight interference with air traffic from Washington National Airport. Originally established in 1918 as a Naval Air Station, the field was shared with the Army Air Corps which named their side of the field Bolling Field] — TKO

American Security & Trust Co.
Washington D.C.
May 2, 1924.
Dear Uncle D.R.,

Washington is wonderful at this time of year. The cherry trees are blossoming and the parks are showing as beautiful settings for the Washington monument, Lincoln Memorial and some of the newer buildings. If you or the family come east, plan a day or two here. I may be able to show you a new "light plane" as well as a beautiful city.
Edmund

American Exp. Co, Berlin
July 1, 1924
Dear Mother:

I'm back in Germany again after a trip to Austria. Your letter from Daytona reached me today. I am glad you are there and are enjoying the ocean with Margaret. And that you look forward happily to a more comfortable winter in the sanatarium. I hope you will not worry about the money. I am sure that I will have enough for a salary when I return to take care of a share always.

Germany is very uncomfortable to live in just now. Everywhere is the greatest poverty except in the large cities where foreigners display their money. In the hotels one has no hot water, no baths except at enormous cost. No one can afford to live as we live in America.

I hope to go to Paris next month for another visit to the airport there. It may be necessary to omit part of my program of study unless I stay longer in Europe than I had planned.

With love,
E.

[No place or date]
Dear Mother,

I have been thinking of the shallow philosopher in Voltaire's Candide who insisted that "All is for the best in this best of all possible worlds". That attitude is so characteristic of the people who go in for

"Christian Science" and Coné and the "glad" cult. Those who "see life steadily and see it whole" do not merely look for the "good". No, life is terrible, stark, cruel, lustful, sinister, as well as happy, truthful, self-sacrificing, altruistic, lovable. And the strong man who desires to see only the latter qualities has not gone very deeply into his own nature nor does he have a passion for "truth" as differentiated from "what makes him feel good"

Some people by nature are chiefly interested in themselves. They are not what we call "selfish" but they let their minds be chiefly occupied with the aspects of experience which expand their natures. They may be very altruistic, with "service" for their mottoes but the right hand always knows what the left hand does. Some people are born with an impulse always to champion the oppressed. It is quixotic, of course, but it is the essence of the Anglo-Saxon tradition of freedom – not freedom only for ourselves but for all those whom our lesser impulses tend to "lord it over". Imperialism does not grow from these.

The spirit of Nordic struggles for liberty has given us Goethe's statement that the liberty which is worth anything must be re-achieved, "re fought for" every day. It is not something our grandfathers established so that we can now sit down to enjoy, or that we fought for in a "war to end all wars" so that our children's children can have "safe democracy" served them on silver platters. The spirit which looks only for the good in life is the greatest enemy of "freedom" because it is neither for or against; it is the dead weight that those who stand as eternal watchers must drag along in combatting eternal encroachments.

Personally, I prefer in my friends the attitude which champions the oppressed to that which says "There are no oppressed".

Remember the Lady of the Lake?
"Whether joy danced in her dark eyes
Or woe or pity claimed a sigh
Or filial live was glowing there
Or meek devotion poured a prayer
Or tale of injury called forth
The indignant spirit of the North."

This is the spirit which so many of us have lost, -- which cries through the Eddas, the sagas of our forebears.

Take an illustration:

Shall I interest myself in the floggings, hangings, tar-and-featherings, burnings at the stake conducted by the Ku Klux Klan or shall I say "No - all is for the best in this best of all possible worlds" and try to find something good in the Klan (very easy to do if you read their literature. They are the only patriotic, protestant, law-abiding body in the country!!) Those who take the latter attitude permit and actually assist in further floggings, hangings, tar & featherings and burnings at the stake. This is implicit in the attitude of the man who prefers to believe what makes him feel good and get on well in life, to having an indignant spirit of the North called forth. The former attitude is the one which rights wrongs!

St Paul may have talked about thinking on whatsoever is good and just etc. but he hardly practiced it. His writings are full of maledictions against evil doers. You have taken one verse and, out of its context it may appear to say what you wish it to say. As you know one can prove anything by Bible quotations if properly selected and removed from their contexts. Even if it were demonstrable that Paul did wish Humanity to think only of the good this would hardly convince anyone whose moral fibre is built on solider foundation. Jesus was a living example of one who did just the opposite!

Don't you see that happiness built on such a basis of looking only for the good in life is of the shallowest kind? Then when misfortune comes, the basis of life is swept away and one sinks, floats, or starts all over again. One who sees all of life with, perhaps, an emphasis of sadness or indignation at the evil and folly, -- such an emphasis as is evidenced in Jesus' life, -- does not build upon a false structure which can be overthrown by misfortune. Some characters who have achieved depth within themselves who see life steadily and see it whole are quite independent of prosperity or adversity, success or failure of their plans, the greatest woe of human life, the defection of a friend - only such a one can say, I am the captain of my soul. My head is bloody but unbowed by any adversity.

But I am preaching, and that will never do. You, however, invited it so I relieve myself of blame. Isn't that easy? If all the preachers could get out of their responsibilities that easily we would have more of them. But fortunately their number is decreasing. We have enough to do to live our own lives sanely and nobly without telling someone else how to do it, haven't we?

Affectionately,

E.

Air Mail Cheyenne/Salt Lake City 1925-1927

Edmund makes the move to fly the air mail mostly as a financial decision. He initially inquires about a position at Monmouth, Illinois, which he believes is an experimental field where new Air Mail aircraft types are tested and developed. The $2000 per annum salary is well below his salary at McCook Field but he is interested if "adequate compensation" is possible. The Postal Service flies rebuilt wartime de Havilland DH-4 bi-planes and pays on the basis of 6 cents per mile (day time) and 12 cents per mile (night time) which Eddie eventually accepts and agrees to fly the challenging Cheyenne to Rock Springs to Salt Lake City route. This is 417 miles of flight over the most demanding terrain flown by the Air Mail. The flights are made without the aid of radio over an airway that has only an occasional beacon to guide the way. The flight is scheduled for 5.5 hours each way. He is paid typically $30 each way in supplement to his $2000 annual compensation. He flies for the postal service from July 1925 until June 1927 and is involved in eight forced landings during that time. He is obviously a very good and lucky pilot! His flight pay logs show little time for other activities during that two year period.

Eddie takes courses in electrical engineering and philosophy at the University of Utah in 1927. His letters to his mother during this time are almost all oriented to discussion of his readings of philosophy and in fact he questions that philosophy might be his real calling as he stands at the top of his class. In electrical engineering, he struggles. This is when he repeatedly recommends to his mother the "greatest novel in the English language- Jean Christophe by Romain Rolland".

During his time in Cheyenne, ETA also registered a new "home-built airplane; this one a biplane powered by a Wright-Morehouse 2 cylinder 80 cu. inch 28 hp engine. The aircraft was identified as an "Allen" and the date of manufacture was May 15, 1928. He later sold the airplane in 1934 to Ellery Bennett of Cheyenne (a Boeing Air Transport mechanic) who built the airplane to Eddie's specifications. Letters from Ellery Bennett's family also note that Eddie Allen was the god-father to one of Ellery's sons and that only one of the sons flew the "home-built" with Eddie. Ellery often flew the airplane but never with one of his sons. The plane was called the "Guillotine" by the Bennet family since piano wire directly behind the pilot's shoulders was used to hold the fuselage together - thus the potential for beheading in a crash. The airplane was eventually dismantled and destroyed in Cheyenne. Many of the family letters which follow address very little on aviation and much more on some personal problem he is having with his brother and on his readings in philosophy and religion; specifically, his focus on philosophy from classes at the University of Utah. The letters by Allen to the Air Mail management show why he was valued much more than just a pilot.

de Havilland DH-4 Postal Service

Boeing Model 40A Mail /Passenger Plane

McCook Field, Dayton, Ohio,   May 17, 1925

Dear Mother:

Will you please send me my New York Life Insurance Company policies which I turned over to you for safe keeping?

Faithfully yours,   Edmund

McCook Field, Dayton, O,
June 24, 1925

Dear Mother:

Thank you for sending the policies. They were received all right.

I am just now in the bustle and excitement of making a move. Everything must be sold or packed and shipped. The place is not very definite as yet but it will probably be out west Cheyenne or Salt Lake City or somewhere out in that direction. I am going with the Mail Service.

Thanks for the pictures of DeLand. I do not remember any of the places, try as I will. I remember that home of ours and the wonderful oaks and the great pines across the way.

How is Miss Valentine? Have you been to Citra lately?

It must be pleasant in Seabreaze. I remember Daytona Beach with much enjoyment. The air and the see seemed so intimate. I shall never forget it.

I am glad you are with Sister this summer/ you will have a great relief and rest and will feel young again. Glad you enjoyed Osborn's book. He is now planning a new trip through Asia exploring North Tibet and Siberia for traces of earlier men than the Neanderthal variety. You might be interested in a new book of Jensen V Jensen called The Cimbrians that I have recently read. It is a record of psychological evolution in novel form.

I called on Uncle, and Aunt Margaret some time ago and had a pleasant evening there with Bob Ellis. I hope to be able to call again before I leave Dayton. Sorry I missed Margaret Byard Kellar's wedding. I should have enjoyed most kissing the bride.

Your son,   Edmund.

**Excerpts from "Boys' Life" - May 1931   Flying the Air Mail
Cheyenne to Salt Lake**

"While I was still testing planes for the army at McCook Field in Dayton, in 1924, I made various efforts to get closer to the job of air-mail pilot, that I wanted most of all. I wrote letters to different pilots that I knew, who were already flying mail, telling them that I wanted to join up with the air mail whenever opportunity afforded, and asking them to recommend me if they happened to have a chance. Among the friends and acquaintances to whom I wrote were Slim Lewis, who had been one of the civilian instructors at Scott Field, where I received my first flying lessons after enlisting in the army in 1917, Frank Yaeger, another veteran mail flyer, and others whom I had come in contact with at one time or another.

Presently, to my great delight, I got a letter from the superintendent of the air-mail service, Mr. Egge, offering me a job as a mechanic and reserve pilot. It was the only position that was open at the time. Frank Yaeger had said to him, I found out later: "Whenever you have to take on another pilot, I know where you can get a good one."

In spite of the waiting list that contained more than 500 applications for air-mail jobs, good pilots always were in demand, just as they are to-day. The problem is to determine which ones have the particular characteristics that make them valuable to the service. That is where personal knowledge and a personal recommendation backed on that knowledge come in. ....

On the day after I reported for work at Cheyenne, I was called at half-past three in the morning. The mail truck took me out to the air field. There I found the mail plane already out on the apron in front of the hangar, waiting for the arrival of the west bound mail that left Omaha at midnight. The engine was being warmed up. It was still dark. Just at the break of dawn the Omaha plane came in. It took about ten minutes to transfer the mail sacks. I climbed aboard, took off, and turned to the west where everything was still dark.

Gradually the ground turned gray, as the light increased; I was flying over empty miles of sagebrush, with distant mountains ahead. By the time it was full daylight I was passing just to the north of

Laramie, Wyoming. A thousand-foot ridge loomed ahead of me, and I had to climb steeply to clear it. As I topped the edge of the plateau I leveled off and went flying across the sagebrush within a few feet of the ground. An antelope, startled from his sleep, jumped up and ran to one side. I was so close I could see the sand thrown up by his tiny hoofs.

Next came another butte, a cliff, ragged rocks, and canyons that descended gradually toward a desert plain. Far out ahead of me I could see a tower with a searchlight beacon on it. As I passed above it I could see the small landing field smoothed off for emergency landings. Then sagebrush again, and after that, miles of desolate desert without even a sagebrush. Here and there, I passed over flat, dried lake-bottoms.

At one place I passed a group of lakes, some of them a mile or so long, but all without any sign of vegetation. Then another plateau, roughly cut by ravines and dry gullies. Miles and miles of this rough country without a sign of life - human, animal, or even vegetable.

Suddenly I saw railroad tracks, flashing out of a tunnel. The Union Pacific. They told me I was exactly on my course. Ahead of me was the smoke from a long passenger train. I came down low as I caught up with the train and passed it, and waved at the engineer. A road, that I saw on my map was the Lincoln Highway, came down toward the railroad, paralleled it for a while, and then went skirting off toward mountains to the north.

Presently the railroad turned north also. Ahead of me was another high mountain-range. I had been flying nearly two hours, and had travelled, against a fairly stiff wind, a hundred and sixty miles. I crossed two ranges of peaks at 9,000 feet altitude and ran into sudden curious wind-currents that almost turned the plane over. Then I saw Rawlins - a tiny town with a landing field beside it. I maneuvered for a landing, remembering that I was still nearly 7,000 feet above sea level, and had to touch the ground nearly ten miles an hour faster than I would have had to come in at Dayton.

After a few minutes I took off again. For the next forty miles I followed the railroad. Just before I left the tracks I caught sight of a couple of coyotes running through the sagebrush. I passed Cherokee emergency field, and then Red Desert emergency field, with the ground brick-red because of some strange geological formation.

Above all these miles of desolation I began to wonder what would happen to me if I had a forced landing, and managed to get lost without water.

Crossing a high, level plateau I saw from my map that I was on the Continental Divide. Presently it broke off into a ragged canyon marked as Point of Rocks, Wyoming. A little farther along I passed Rock Springs - the present intermediate stop for the air mail between Cheyenne and Salt Lake City - at the base of a cliff nearly 1,500 feet high. It seemed strange to be flying down into a valley with land on each side towering far above me.

Another landing at the Rock Springs field, and a take-off under the brow of the cliff, over which I found the wind blowing in strong gusts. Climbing was nearly impossible. By circling about the little town I managed to climb to the 8,200 foot altitude necessary to clear the cliff. Behind it I crossed a great level pasture where cattle and horses were grazing. Next came a series of high, parallel ridges, round buttes with vertical sides and flat tops, all getting more and more rugged until I was flying over "bad lands" so broken that any forced landing would have meant a bad crack-up.

From this broken country I climbed gradually into a range of high wooded mountains with snow-covered peaks. The range loomed so high in front of me that it seemed almost impossible to cross it. As I flew closer, various passes opened up, and I found I could slip through a great canyon without climbing above ten thousand feet. Here and there were possible landing places, but a good deal of the time was necessary just to trust the old de Havilland and the mechanics who checked the motor.

Skimming through the great canyon, two miles above sea level, and with magnificent peaks towering north and south of me, I suddenly came in sight of Great Salt Lake spread far below. A few moments more, and I could see the city roofs, and a few minutes later I taxied up to the hanger of the airport to deliver my mail. The end of my first day's run."

Return Flight   Salt Lake to Cheyenne

"On my return flight from Salt Lake City I had a real adventure. I climbed from the Salt Lake field toward the great Wasatch Range, and cleared the 10,000 foot saddle between the peaks with altitude to spare. For an hour I flew eastward, leaving the wooded range behind me, and flying out over the bad lands that had impressed me so much on my first west-bound flight. The country all looked the same.

But presently I began to worry because I was not arriving at the other side of the broken country. There were no landmarks that I could recognize. Not a sign of any sort. I became convinced that I was off my course. For some reason, some trick of the subconscious mind that so often guides us even when we are unaware it. I decided that I was south of my course. I turned until the compass told me I was headed northeast. I flew for two solid hours without seeing any railroad track, emergency landing-field, or any sign of human habitation. With the wind on the tail of the plane, that meant more than two hundred miles across the great waste stretches of that mountainous country, without anything in the world to tell me where I was! Nothing but a wilderness of rocks and sand and sagebrush.

I turned again, following a compass-course still further north, determined to find the railroad that had been so much in evidence on the west-bound trip.

I began to wonder what I would do when I ran out of gas, or had to make a forced landing for any other reason. I had brought no food with me. I would be able to get water from the radiator of the old Liberty motor, but it might well be poisoned by the pipes and metal and concentration of impurities through long heating. It would take days to get to the edge of the desert on foot, even if I had food and water, and knew the way. The country was all rough, with no level ground that I could see—merely peaks, piles of boulders, ravines, cliffs, dried river beds.

Another hour's flying and I began to wonder if I had passed the railroad tracks without seeing them and had reached the desert to the north. From my map I tried to recognize the courses of the dried river beds, and make out whereabouts in all this labyrinthine watershed might be. The sun was at this time so nearly overhead that I could not check up on my compass with it.

Finally I came to a river. No amount of searching on the map gave me even a hint of its location. I decided to follow it downstream. That took me still further north, by the compass — another forty miles.

At last I saw a small town, the first sign of human life I had seen in hours. I came and circled it, looking for a landing, but could find no field big enough, or smooth enough to make a safe attempt. I knew that my gas was nearly gone.

For nearly half an hour I circled the town, trying to decide on the spot I would use to come down on when the motor quit. Then I saw, on a narrow strip of ground against a nearby hillside, a man waving a white cloth. On closer examination, this strip of ground looked pretty good and I decided to try it. I got down safely, without even nosing over.

Where was I? The town proved to be Boggs, Wyoming. I was still forty miles south of the Union Pacific tracks. I had been down in the Colorado desert for nearly three hours. My compass had gone wrong. On checking it up I found that it was nearly fifty degrees off.

I called up the field at Rawlins, and told them where I was. Then I got a new supply of gas and oil, took off safely from the hillside I had landed on, and flew on to Cheyenne. I found that the people at the airport there had been on the point of starting a plane out to search for me when they got my message, relayed from Rawlins." —ETA

July 18, 1925
Dear Margaret:
Your letter reached me in Cheyenne day before yesterday. Today I flew over the Rockies to Salt Lake. The Red Desert, 200 miles of it, is very near the limit for barrenness and death. I was lost for a while because I could see no landmarks for so long. The Rockies are as wonderful as ever. Snow-topped peaks are a relief to see these days even in Salt Lake City.

Salt Lake is a very beautiful town. It is clean and healthy and prosperous. There are a great many other churches here besides the

Mormon's but, of course, the Mormon Temple is by far the most beautiful.

The annual roundup is being held in Cheyenne. The town is wide open for these days and cowboys and girls dash through the streets day and night.

Glad you are having a rest. I hope you and Mother both have a wonderful time at the beach this summer.

Edmund.

U.S. Air Mail Field   Cheyenne Wyo.
August 19, 1925
Dear Mother:

How you would enjoy this country! The weather is delightful, the scenery grand and expansive, the people friendly and liberal and broad-minded and progressive and the cities modern and improved. The schools out here would satisfy even Mr. Wells if he were not too grouchy on the day of his visit.

As far as the Air Mail is concerned, the citizens of Cheyenne regard it as their own particular pet. They come out en masse every evening to superintend the arrival of the east-bound mail plane, the transfer of the pouches to the outgoing ship bound for Omaha and due to arrive there at 1 a.m., and the departure as the pilot takes off into the inky blackness. They do not, as a rule, get up early enough to see the west-bound arrival and departure which is due to take place about 4:15 a.m. But nevertheless Cheyenne shows more interest and sends more Air Mail than any town of its size in the country, and in fact even out-distances Omaha, ten times its size. Cheyenne is booming and consequently its prices are very high.

Margaret wrote of the wonderful time you are having at the Beach, your hopes for a trip abroad and the interesting new developments in Florida. I was very much interested in the pictures of DeLand you sent, which arrived just at the time of my departure from Dayton. I do not recognize many of the places. How St. Petersburg has changed! Do you remember the night we spent there with the mosquitoes? The desolate landscape! The God-forsaken citizens and the dreary bay that had to be crossed to Tampa!

How well I remember Miss Valentine and our recitals of Idylls of the King. I believe I started with "The Lady of Shallott" and then we read Elaine. Not long ago I reread Guenivere. It has the same appeal that Tennyson always has. A lot of bunkum has been written about the alleged sentimentalism of Tennyson. Service is ten times more sentimental with all his pretense of he-man roughness.

I have recently read Mr. H.G. Wells' latest: "A Year of Prophesying" which I regard as better than "Men like Gods" that Sister liked. It is tremendously stimulating and it enlarges one's horizon and sympathies but it tends to make me intolerant of human inefficiency and lack of purpose and desire for the highest good. One of the articles in the book is called "The Beauty of Flying". It is very good.

      Your Son,
       Edmund.

Air Mail Field, Cheyenne Wyo.
August 30, 1925
Dear Mother and Margaret:

Your letters were so interesting and contiguous I shall answer them together. Thanks for the views of Florida. No, I cannot say that Florida eclipses California out here. We are too far west. But in Chicago it seems to. I saw a mention of it in RHL's Live [Line?] some time ago. Do you ever see the Live? A friend in the east sends it to me and I will forward some of them to you because I believe you used to enjoy B.L.T. the predecessor of R.H.L.

Did I tell you about Well's new history? I have found it quite a new experience. It is quite short and one can get the entire thing - from the earliest spiral nebulae to today almost in a continuous panorama. It gives balance to the picture in a way a longer, more detailed history could not do. And yet there is enough essential detail to make it as fascinating as a detective story. I stayed up all night one night reading it.

The atmosphere of the west is very refreshing both internally and externally. The people are liberal-minded. (We even have a woman Governor, which is going a bit strong. But she is quite efficient and certainly beyond the taint of graft.) It is curious to contrast this country with such a hotbed of Mrs. Grundyism as Urbana, Illinois. People here think in terms of generosity and cleanness. They champion sportsmanship and help the under dog. One finds people acting on the basis of the "Nation" and the "New Republic" while at the U. of I. these were to be ruled off the library shelves as "revolutionary". But, of course, it is not fair to compare any place with the U. of Illinois. Some day I am going to put out a book on "Why Universities Die" with the Pres and Deans of the U of I as Exhibit A.

People out here are apt to be physically cruel to animals, perhaps not any more so than in the east. I notice they recoil from flagrant vivisection. But roping steers and throwing wild horses, shooting coyotes and deer makes men firm and sound with a look in their eyes that one can trust. I like this about the West.

So Margaret may become a realtress! Well, I don't know why she shouldn't. I notice all the real estate people out here ride around in Cadillacs. She will have the advantage of knowing the country and then she may be able to sell your land. It might be safer to go into a real estate office at first to get on to the hang of things at someone else's expense.

If you are interested in fiction I would suggest Jean Christophé by Romaine Rolland. Rolland is a "great" writer. He has some remarkable biographies. A recent one on Mahatma Gandhi is excellent. The Nation has some good book reviews but I presume books mentioned would not be in any but the largest libraries. I enjoy the articles on current events in the N. - our relations with Nicaragua - England and Opium - LaFollette. If you are in a cynical mood the American Mercury is in order. There is an excellent article of an analytical nature on American Politics, by a Chinaman in the last issue.

In the mass of reading matter now available one must pick carefully or fill up with trash. I have discarded the Sat. Eve. Post. It is too difficult to find something good in it among all the rest. And the Lit. Digest has gone the same way because one gets nauseated with the mass of meaningless arguments on meaningless issues. The press that is "digested" is all on one side anyway so why digest it. It is like the grand old parties which raise a lot of dust making you think they are *fighting for something worth while, when there isn't enough difference in their stands on important issues to make it worth listening to. There are, however, a few analytical papers. One of these is the Century. One, I think, the Mercury. With the Nation and New Republic as weeklies.*

But this isn't philosophy. It is criticism although I believe that may in some forms be considered a part of philosophy. A book on Philosophy I liked was "The Philosophy of As If". It was awfully technical and deep and somewhat dry.

I'm afraid this has been an awful bore, this letter. Next time I will tell you some adventures.

[no signature (no space)]

Do you know the Philosophy of Loyalty by William James? Or The Theory [?] of the Leisure Class by a [illegible - see below] of Chi. Prof - Veblan.

[No place named, but probably Cheyenne]
October 15, 1925
Dear Mother:

I am taking steps to go into manufacturing in the east and at the same time some interesting editorial work is being offered. My work on the staff of our professional magazine has borne unforeseen fruit, and I think I shall like the new work, and this is something which in a few years may lead to big business.

I shall be going to Washington where I must consult the lawyer looking toward a further clearing up of the outrageous situation in which I have been placed. To that end I wish to know if, by chance, you were led into unwise correspondence or conversation involving me with a certain Mr. Bassoe of Chicago. What do you know of any correspondence that other persons may have had with him, and what is the nature of the personal relationship between him and Tom. This person has been going about doing what is commonly called pimp's work for Tom Allen. Deans and doctors have curious ways of sidestepping issues and covering each other's misdemeanors.

Please write me in that matter before October 25th if possible in care of the American Security and Trust Co., Washington D.C.

Have you seen in the papers last week how Kent of Yale again whitewashed our scapegoat Clark? This is the same Kent that I referred to in my open letter. He certainly had a thick skin. The magazines are full of this sort of thing now. Kent is great in religious writing, but from what we know of the way he meets his own moral problems it is obvious that he hasn't much to offer to a world that is apparently honestly looking for guidance!

Edmund

[In a handwriting that I do not recognize is written, "please return" below Edmund's signature. Possibly Abby sent the letter to Tom to inquire about Mr. Bassoe and this case, whatever it is.] — MRB

faded letterhead says:
"EDMUND ALLEN
AMERICAN SECURITY TRUST COMPANY
WASHINGTON, D.C."]
Nov., 4, 1925
Dear Mother:

Have not received the word regarding the Bassoe matter. I need this very much to clean up the situation, which has now dragged out so long and in which I have had no cooperation. If you are waiting to communicate with Tom, I hope the wait will be productive of results. If you have not written him, it seems to me it would be well for you to send him my letter and get what you can out of him, sending it along to me immediately together with the information regarding your own concourse with the man. This sort of nastiness has got to stop in its relation to me. If Tom's concern over his precious career is so great that he must strike as a madman at other people, he had better be put in an asylum where he can get well, or be prevented from broadcasting the seeds of his dirt. The sickness of everyone in the crowd is so manifest that all my friends are incredulous and feel it necessary to withdraw from me when I inform them of the situation, or to take precautions against being hit by such irresponsible muck slinging. I must guard my friends from any connection whatsoever with my family. During the past few years help was extended to me in an extremity by friends. The events of the past indicate that their very helpfulness would be interpreted by the deans and doctors and medical advisers, with their pathological slant on life, as of evil origin.

But that is neither here nor there. I have a definite program for cleaning up the mess that, I am sorry to say, my own people created for me.

The magazine and clothing reached me OK. Thank you.

Ed.

November 6, 1925

Dear Mother:

Your letter disclaiming connection with Bassoe is a surprise, for it shows up the Chicago matter in a new light. The lawyer was incredulous, but I insisted that it must be so.

I am looking for connections in N. Y. State. A career is opening up and I am bent on getting on at all costs. In regard to the Florida proposition, even if I had time, I am not awfully good at selling things. Sister is much better than I am at that. - Why not have Tom go down to manage the grove while he arranges to sell it? His affairs are now in a very much better state than mine, and it seems to me he would be much better able to spare the time. Whatever prejudice I may have, it does not blind me to the fact that I think him a pretty good business man.

I do not, however, wish to have communication regarding me go to Tom. I want to work out my career & my life without interference from such connections.

Your difficulty with young people seems unreal. The restraints of your generation have been taken away for them by the sudden unveiling of the hypocrisy of their elders in the war and afterward. They have seized their freedom and have not, on the whole done badly with it. There is little abuse and human nature is mighty decent. We are coming out all right, with a fundamental cleanness and honesty. Young people are hopeful; they have their eyes peeled - that's all. Toward older people, especially toward one's mother, one takes an *attitude of respect and for that very reason it seems incredible to me that you so lightly waive aside all moral responsibility in the matter you refer to as "our trouble". I am still dealing with that "trouble", and what is left for me to deal with in particular is just the incredible mistakes of evil intent which you attempt to assume is a closed book.*

One would have to go a long way in history to find an adequate parallel of the situation. The Medicis as I remember used the stiletto and poison vial, but even Catherine felt inwardly that confession and repentance were necessary before she could again take the mass. Even she did not attempt the injured innocence rôle.

The things you sent to Cheyenne have reached me at last. Thank you. Eddie.

Field Manager December 9, 1925
Air Mail Service
Salt Lake City, Utah

On December 7th, 1925, when I arrived over Salt Lake City, the entire valley was covered with a solid blanket of fog. I flew west to the lake and south about ten miles trying to locate a hole through which to come down. Then I flew back into the mountains where It was clear. I landed in a large field near the highway about 27 miles from Salt Lake. While still rolling about 25 miles per hour, I saw a hay wagon drawn by a team of horses pull over in front of me. I was unable to steer out of the way in time to avoid it and collided with the left wing on the rear of the wagon. No damage was done to the wagon, horses, or driver.

If I had had some means knowing when over the Salt Lake Field how much ceiling there was under the fog I could have come down through it. I understand that such fogs are common at this season. Could not a system of rocket signals be devised for communication?[26]

E.T. Allen
Pilot, Mountain Division Air Mail Service

---

[26] In 3 short years he will help develop a radio communication system with Thorpe Hiscock of Boeing Air Transport which will solve this specific problem.

Air Mail Service, Cheyenne, Wyo.
February 26, 1926

Mr. S.A. Cisler
General Superintendent
Air Mail Service, Omaha, Neb.

Dear Sir:

In view of the crash and fire of Art Smith resulting in his death and the destruction of the mail and the airplane, I wish to submit an outline of the recent work of the Engineering Division of the Air Service in their attempt to arrive at the cause of fires in crashes and to eliminate this hazard.

A 500 foot inclined runway was built at Wright Field with a concrete wall at the base into which the airplanes on test were crashed at speeds of 50, 100, and 150 miles per hour. Slow motion pictures were taken of the impact, and in all the cases where fire resulted the source of ignition was determined. Over fifty airplanes of all types and with varied equipment, crash-proof gas tanks, special exhaust manifolds, etc., were put through this test, by far the greater number, however, being standard DH4B's. The most frequent cause of fire was found to be the exhaust manifold which was white hot at the instant of the crash when everything was flooded with gas. In no cases, however, did a fire result with an airplane equipped with crash-proof gas tanks where the impact was made at the ordinary speeds in which the pilot might be expected to come out alive.

An investigation was made of the design of the exhaust manifold with a view to so cooling it that fire could not start there even though it was covered with gas spray at the instant of crash. A finned aluminum stack was designed and tested which seemed to solve the difficulty. It is now to be standard equipment on the new 02 airplanes of the Air Service. I am enclosing blueprints of this stack. It has been put through extensive service tests and has been crashed on ten DH4B airplanes without a single fire. If this equipment is satisfactory for night flying, eliminating the glow from the exhaust, it might be used on Air Mail planes with a saving in weight and a great decrease in fire hazard. If it is not satisfactory for night flying it might still be used on those planes which are not flown at night.

The crash-proof gas tanks are now used on many service airplanes. The Air Service specifies in contracts either droppable tanks or crash-proof tanks, leaving it up to the manufacturer to provide either one. In some cases the manufacturer prefers to supply crash-proof tanks because of the difficulty of eliminating structural members under the tank in the design. Crash-proof tanks have certain other advantages which are coming to be recognized in the service. There are many occasions when the pilot does not have time to drop his tank before a crash. Art Smith's crash was a case in point, as was also Frank Yager's and more recently, Jimmy Murray's. The crash-proof tank is good in every case that the droppable tank could be used, and it is there as a sure safety device in many cases when the droppable tank is useless. This equipment cannot, of course be used on our present DH4's, but with the new planes of greater carrying capacity, the crash-proof tank could easily be provided.

There is considerable experimental work being done at McCook Field which has a direct bearing upon our problems. One instance is the design, construction and test of a DH4 airplane with Clark Y wings which was completed at McCook Field almost three years ago. The report on this test is available to anyone writing to McCook Field. The airplane was found to be not sufficient improvement over the R.A.F.15 winged DH4 to be noticeable in the performance tests.

Yours truly,
Edmund T. Allen
Pilot, Mountain Division
Air Mail Service

Air Mail Service, Cheyenne, Wyo.,
March 13, 1926.

Mr. V. E. Clark,
Consolidated Aircraft Company,
Buffalo, New York.

Dear Col. Clark:

I am building a light plane which I designed at McCook for the Morehouse engine while I was test pilot there. Facilities out here on the frontier are not so good for technical engineering requirements and I am having my difficulties getting materials and special work done. One of the things I am unable to get anyone to do is to bend and heat-treat my axle. I have a trick landing gear consisting of the bent axle shown in the print, inclosed, passing through the fuselage above the lower longerons, and connected to them by shock-absorber cord. I understand you have facilities for bending axles like the PT1 and I would like to ask if you can bend my axle for me.

The Pipe and Tube Bending Corp. of America offered to bend and heat-treat the axle according to my print, and I sent them the specifications and a draft for their charges last December 23. Since then I have been unable to get a word out of them. They must have had bad luck with the tube and been too sore to write about it. I doubt if they ever tackled high tensile steel before.

My axle is to be 1½" O.D. x either 1/8" or 3/32" wall depending on whether I can count on 120,000 or 160,000/sq.in. yield point on the tubing. If it will not be too brittle if drawn at only 600 deg., I would, of course prefer the thinner wall. Otherwise it will be better to draw at 1000 deg. and use the heavier tube. My idea of the material was chrome-molybdenum steel such as the Ohio Seamless Tube Company makes.

If you can help me out of this difficulty, or know where else I can get my axle bent, I will certainly put in another cheer for the Consolidated Aircraft Company.

Yours cordially,

Edmund T. Allen

Post Office Department   *June 26, 1926*
Air Mail Service

While flying mail plane No. 611 on the morning of June 25, 1926, the motor stopped suddenly at approximately 6:40 AM. I was west bound and had passed over Bitter Creek Emergency Field a few minutes previous. Thinking the motor failure might be due to gas trouble I switched to the gravity tank and turned the airplane toward what looked like a smooth hillside. The motor did not start and I hit the ground in front of a sage brush hillock which pushed the landing gear up through the lower wings. Upon examination it appeared that the only damage was as follows:

    Right wing stub broken
    Right lower aileron damaged by ground looping
    One wheel bent
    One landing gear brace strut bent
    One center section strut bent
    Horizontal compression struts in fuselage buckled

The pilot's cockpit was not damaged, nor was the mail pit. There was no tendency whatever to nose over. The plane ground looped when landing gear bent sideways but this did not occur for 200 yards after first touching ground.

                                        Edmund T. Allen
                                        Pilot, Mountain Division

c/o Mr. F. B. Hughes
615 H street N.W. Washington, D.C.
[No date, but context suggests about November, 1926]
Dear Mother:
The first rush of winter has abated and left a wonderful Indian Summer over the land. River and woods and farm all remind me of that fall, which now seems so very long ago, when I had my first company under me during the war. It was cold but we officers had to pretend at least that we did not feel it. Drilling, drilling, drilling –

never asking questions – we were unified in a simplified world. The struggle to make a living now, to get established among the ranks of those who are already getting along is far different. The enemy is not all "evil" and ours is not an "all holy" cause.

I have been reading some of the Russian novelists, Turgeniev, and Andriev and Dostoievski in my spare time in the evening and I find them looking at life from a wonderfully detached attitude. Andriev's "Life of Man" was presented as a play at Harvard while I was in Cambridge six years ago.

I have been working tremendously hard for the last year and have been sick part of the time. It seems at times as if I could almost see the light, and perhaps, if I "make good" at my present undertaking I may be out of debt in a year and a half. As you know Allene pushed me pretty hard after she had the backing of our people in Chicago. But one must expect that sort of thing if one allows one's self to get into the clutches of such people.

At any rate one must not be downhearted. There are lots of bright spots and one finds comfort in hard work. What do you think of Mr. Glenn Frank's philosophy of modern life? He writes for several newspaper syndicates since he has become president of the University of Wisconsin, that the modern engineer can solve all the problems of the modern industrial life. He idealizes Henry Ford.

I have wondered many times during the last year what your studies in philosophy led you to. If I had more time I should like to read enough to get a good background of philosophic training. Perhaps some day more leisure will be available. Mr. Frank's thesis is that people should work hard in their youth to get economically independent not worrying too much about ideas. These he can leave until he achieves leisure. One wonders, however, if people will be able to have any interest in the "higher things of life" after a life of drudgery. Aspirations are usually lost in such an environment.

I would like to have a statement of your symptoms from you personally. I presume Tom is arranging for the very best consultation in the country with a view to the removal of the cause of the paralysis. It seems to me that this is not at all difficult. There were many such removals and complete cures in the war. And the will to health is one of the greatest factors as you know.

Affectionately
Edmund.

[In Mother's book, she wrote that Abby had a "stroke" causing partial paralysis Nov 11, 1926. Unfortunately, there was no cure, and she lived five years with that paralysis, using her left hand and a typewriter to write.] —MRB

[No place named. Tom's note to Margaret tells us the letter was written late December 1926.]

Dear Mother,

Our Christmas rush of business is somewhat abating and we are getting back to the normal activity of between times.

I was very glad to have your Christmas letter together with Margaret's and to hear of your happy Christmas together. Your progress toward health is excellent but I was hoping you would be able to use your right hand soon. The clot theory sounds very shaky in this case especially in view of the recovery symptoms. "The regaining control of arm but not yet the hand." To one who understands the nervous system in the light of some of the more recent developments in physiology, this does not sound like a clot. I have been studying the physiology of the nervous system since you have been ill, and I have found it illuminating. There are many things being cleared up nowadays. Dr. Herrick of the U. of C. has been doing some very interesting work in the theory of nervous functions. Some of the scientific reviews of recent months have been fascinating in this regard. General motor paralysis or paresis is a very different sort of thing even in its partial manifestations. It is apparent that the first diagnoses were of this condition. Even then I doubted it and now I am quite confirmed in the opinion.

Have you ever read Romain Rolland's Jean Christophé? I read it a second time not long ago and again I pronounce it the greatest novel in the English language. Of course it wasn't written in English but was translated from Rolland's French. But my enthusiasm is none the less great. I believe you would enjoy the novel very much now that you are at rest, mentally, physically and spiritually, and have time to read and aesthetically enjoy fine points of feeling and insights into human character that Rolland presents. There are very few authors who can so move one in the sophisticated world of 1927!

I have just read Count Keyserling's Travel Diary of a Philosopher which is certainly written with a view to pleasing an American audience. For that very reason it is worthless as philosophy although some of the analyses are keen and all the balderdash of the professional philosopher is there to fool the casual reader. I was interested in the account of the difficulties of a youngest son in getting himself accepted in the family as a human being. As we all tend to become habituated in our attitudes as well as our acts and thoughts, the members of a family tend always to attitudinize toward the youngest as toward "the baby". His opinion is not sought; his views, even though they are proclaimed as marvelous in one so young, are never taken seriously; his judgment is questioned.

It often takes, as was pointed out in the story, all the energy of the man to attain the position of human being. This often happens also with an older person who is bed-ridden or financially dependent. There is a tendency to put such a one under-spiritually. Keyserling is good observer. His discussions of oriental religions and the constant comparison of them with Christianity is significant. Christianity, according to this philosopher, is superior to all because it proclaims the godliness of success. The Hindu Yogi's inferiority is on the basis of his unproductiveness. I enjoy Keyserling's descriptions. He visits India, Burma, China, Japan and America and philosophizes in each. It is refreshing and flat by turns.

I wonder if I mentioned reading Watson's Behaviorism in my last letter. I finished it last week. This is a very readable book and gives one a new slant on life - perhaps a one-sided view - but nevertheless a view we would hardly ever get without Watson. The author is charming personally. His love of little children is very fine and his accounts of experiments on infants, form a day old to several years old - is fascinating. He manages to be very convincing in his elimination of the entire concept of instinct.

I am hoping that my work with Van Gregory will presently be productive of results. I will then have full scope for my engineering training. Those years with Allene were a total loss. The awful drain of energy which can be caused to all around her by a leisure class woman who can afford to give way to her sicknesses is a curious phenomenon. The pretense of such a world makes great demands for a similar pretense on the part of all who allow it to touch them, and then it takes years of work to recover one's normal environment where one can begin to build a career. It isn't a matter of money; those debts will take care of themselves. Professional or business success can be attained in spite of financial status but not without connections that take time to build up. I enjoy designing as I do flying. Of course flying pays better but that is not likely to last long for pilots. Engineering is a very interesting life work. The engineer surveys his problem, finds out what materials he has, what operations he must perform and goes ahead adapting means to ends till it is done most efficatiously.
Affectionately Edmund

2/14/27
Dear Mother,
 I am not at all pleased over your "progress" I think that too often one gets into the attitude of being pleased with "progress" which gradually becomes purely imaginary. The point is that when you accept conceptually your "condition" and gradual improvement you thereby erect the greatest barrier to real reconstitution. After all the Christian Scientists achieve results by their very refusal to admit that anything is wrong. They "kid" their nerves and muscles into continuance of function just as one can often kid one's employees along by refusing to allow their grievances to come to a head. Old man Gregory was stricken with paralysis, and the specialists had told him he would never arise from his bed again. With a tremendous effort of his entire will bent upon one object, he rolled out of bed and got to his feet and walked, streaming with perspiration, into the room where the doctors were consulting with his wife. He learned to ride a bicycle after this event.

But then if one pins one's faith on doctors, one must let them make one well.

No, I have never read "The Way of All Flesh". I have seen some reviews of it and talked to people who thought it a great book but there was something about the comments that made me feel that the author had no light upon the vital problems of life.

I never felt about my parents as you suggest in your letter. After I was of age I decided to live my own life as we would want all young people to do unless they are mentally deficient, but until that time I think I was, on the whole, obedient and respectful. Conditions are changing rapidly now and the youngest generation takes a somewhat different attitude toward their elders. I think it is hopeful and may lead to fruitful character. It is important for young people to keep their eyes peeled. Only when they have seen through all the bunk of their elders' world may they start on a search that may discover for them a way out.

Have you seen Dorsey's book on "Why We Behave Like Human Beings"? Personally I can't handle these new popularizing books much. If people are unable or unwilling to put the small amount of effort into their reading required to grasp a really scientific presentation then the little learning of the book which comes down to their level is a dangerous thing. And Dorsey is worse and more vulgar than Dean Clark. Br r r r

E.T.A.

Dear Mother:

I am glad you are so comfortably situated in Florida. As I remember the autumns in DeLand, they were delightful. And now for your complete recovery, rest is vital - a rest which means freedom from irritating stimuli, freedom from cold and also from overheated houses and from exigent persons.

I am sorry the Golden Book does not intrigue your interest now. There is a great variety of literary things there, everything from O. Henry's delightful fascinating short stories to the most high brow biographies. What are you interested in now? Perhaps I would be able to find just the things that appeal to you. How about asking someone to bring you from the library a play of Maeterlink. Let me suggest <u>Aglavaine and Selysette.</u> If you like that I can suggest others which you would probably like also. Maeterlink has the most exquisite mastery of his art. Consummate skill is shown especially in his earlier plays and some of his essays. If you happen to be in the mood for lilting satire, try George Bernard Shaw's <u>The Doctors' Dilemma</u>. If you like this I can name a dozen others you will like also. If you would like to read about Japan and China from a very appreciative author who combines an unusual knowledge of oriental art and life with a very beautiful and easy flowing style of writing, try some of Lofcadio Hearne's books - any of them.

Maybe you can find in the University library, if not in the town library, some of L Adams Bek' books. They are about India and China and Philosophy and age - old things of beauty, or perhaps you may be in a mood of great hope for the near future - the outlook in a mechanical utopia about to dawn upon the Western World. Then try Well's <u>Food of the Gods</u> and his <u>Modern Utopia</u>. These are better than his later books.

Have you ever read any of the novels of Arthur Machen? You would enjoy <u>The Hill of Dreams</u> I believe.

I have been reading some of Ibsen's plays when I could get a few moments - when traveling or in an off hour in the evening. Study keeps me very busy. I hope you will continue to write to me of things you are interested in and upon which you have put thought and mature deliberation. There is often a tendency to forego analysis and sink back into sentimental repetition of prevailing opinion around one. It is of the essence of mental awareness and life and health to resist this tendency, don't you think so? For that reason I have felt you would always want me to discuss frankly all questions which come up - not to say what I thought you would like to hear. To meet human beings as intelligent, reasonable, responsible men and women we must do this. It is not always possible in the world in which we do business, for one is betrayed often. One must be very circumspect when one walks among those one does not trust. And so it is a mark to confidence to disagree.

Perhaps the thing I disliked about Durant's <u>Story of Philosophy</u> was that it closed more doors than it opened. Almost every one of my friends who has read it has taken the position that now they know the philosophers and they don't have to read anymore philosophy. One would expect just the opposite result from such a book. Now Schwegler's <u>History of Philosophy</u> while not "easy reading" does give one a passionate longing to get hold of Plato's <u>Symposium</u> of Zeitzche's <u>Thus Spoke Zasthustra</u> or Royce's <u>Philosophy of Loyalty</u>.

Durant has recently written an autobiography - a sort of confession of his "inmost heart". It certainly is a revelation for him who looks upon life as "something to be understood". It appears that during the writing of his <u>Story of Philosophy</u> his mind was filled with very unphilosophic thoughts! "Shades of Plato! where live the great ones of earth? In their bellies?"

Are Blue Lakes still as beautiful as ever? I remember this wonderful clear bodies of crystal beauty with their live oaks and their "floating islands" and the great cumulus clouds overhead. And the whole thing so utterly remote from the finger of civilization. I suppose now the real estate people have made the can dumps, if not factory sites, bloom there and Lake Geri where we used to go in swimming with the alligators!

And I suppose DeLand has its aviation field where you can get a ride for $5. Have you been up? Isn't progress wonderful. Twenty years ago Father was telling me about the Wright Brothers experiments and now flying is so common that nobody gives a second thought to taking the Transcontinental airplane in New York and getting off in Los Angeles 26 hours later. Of course, from another point of view, such progress is of the least interest for racial progress. It is hard to imagine him who said so subtly "Give unto Caesar's the things that are Caesar's but ……," getting excited about the latest inventions or scientific discoveries. Human nature, he knew only too well, does not change much with the years. And progress is not measured in dividends or airplanes.

                                        Your son,
                                            Edmund

[No place or date]

Dear Mother:

How the weeks do fly around! Soon it will be summer again. If time is only relative I wish we could enter the fourth dimension and live timelessly for a while. I am sitting at my window watching the setting sun and I see it move down so rapidly that even while I write the words it is gone. How, I wonder, might I look at it relatively so that it would not move for me? There is a story of Dumas in this month's Golden Book in which the Compté Cogliostro, who has discovered the secret of perpetual life, is one of the fascinating characters. The story is called "The Queen's Necklace." I enjoyed it immensely.

Our psychology course goes on as gaily as ever. I enjoy talking over its possibilities with our professor. He is a mechanist as distinct from the vitalist school and believes that all behavior, human as well as animals, is reducible to physio-chemical changes in the organism. I object that this removes all subtlety from human life. All those exquisite and unusual modes of behavior where subtleties of character and personality give a pattern quite distinct from the usual purely animal drives of most men - are hardly to be explained entirely by chemical changes of electrical changes in the body. He disagrees with me there pointing out that such subtleties of character are extremely rare, that on the average, humans are animals with animal desires and reaction - tendencies and that society as it is organized holds in check the cruder expressions of animalism. When, he says, here and there, a man emerges who is quite different in his reactions, we all hate him for being different, or dislike his ways - and eventually, if possible, we make him an outcast. Didn't we do this, he will ask, to all the philosophers?

I am enjoying the Golden Book this year more than last. I never realized what priceless gems of literature there were in the past of the English Tradition. Lanier [?], the editor of the Golden Book, is a very unusual person.

The Electrical Engineering is getting ahead of me this quarter. I have been tutoring in it to try to keep up with all the material given.

The course is very difficult. I am hoping to get a course in radio-frequency next year. I feel that I have a good grasp on the fundamentals of electrical engineering up to the point of alternating circuits.

Are you taking sun baths now? And have you regained complete control of your right hand? What sort of physical culture are you doing to build up?

I have been traveling quite a bit locally here and flying some in between classes. Lindbergh's flight to Mexico City and back from Havana were quite remarkable, weren't they.

Affectionately,
Edmund.

[Reference to her paralysis in the next-to-last paragraph put this letter after November 1926. The first paragraph, second sentence indicates it is perhaps late Spring, probably 1927. I'll arbitrarily assign the date of May 1, 1927 to put it in order in my record.] —MRB

[ETA was enrolled in two courses at the University of Utah autumn quarter 1927] — TKO

[Cheyenne
No date - probably July or August, 1927]

Dear Mother,
My Washington trip is delayed, perhaps for the summer and I am in Cheyenne, enroute to Seattle again. From there I expect I will go to San Francisco or Oakland for a short time.

I was very happy to receive your last two letters and I am quite anxious to be able to see you this summer. Daytona Beach must be quite nice to live in now. I hope you have a comfortable house. I am earning more money this summer than I need and I can send you whatever you need. Please do not hesitate to let me know and don't stint yourself in the little comforts which make such a difference when one is not very active.

I am getting Jean-Christophé for you and I will send it soon. I know you will find in it a mine [mine?] of inspiration and enjoyment. I am glad Sister is with you. She will enjoy Jean-Christophé too, if she has not already read it. Rolland has a new cycle called The Soul Enchanted which I like.

I wonder if you are enjoying Hegwood Brown's page in The Nation. I always look at that first because I have faith in Brown's reactions to situations. They are always right and high-minded.

I will write again in a day or two from Seattle, probably.

Affectionately, Edmund

[References to Daytona Beach and sister being with her places this in summer of 1927, but obviously before the anger and distrust stemming from that legal situation involving brother Tom and lawyer Bassoe. Filed with arbitrary date of 8-1-1927.]

[no place or date]
Dear Mother:
I have read the book The Christ of the Indian Road with much interest especially in view of your letter about it. It is in many ways a typical missionary book as it reveals the sort of "world" used in the philosophical sense that church people live in. In this case it is a world made up of that small group of serious-minded Christians who are "cultured" but are somewhat outside the reaches of modern criticism. The existence of such "groups" - isolated among themselves - in the midst of a cosmopolitan culture speaks for our gregariousness. It explains how it can be that in these United States we can have "cultured" people who cannot believe that there are millions of their neighbors who believe in the Volstead act, - or do not believe in it; who believe the world is rapidly becoming Christianized, or that the "second coming" is about to happen, or that the world is becoming daily less religious or less concerned about religion, - people for whom Anatole France is the spokesman and idol, and those who never heard of him and think Babe Ruth the greatest man on earth.

What I mean is that one so seldom manages to get above the scene enough to see it all in perspective that we are not in any sense cosmopolitan.

This book is far more so than any other church book I ever read. I remember the books in Aunt Melissa's library. There was an eight volume work on "Why I am a Baptist" which was typical of the rest. The Rev. Mr. Jones has "progressed" "beyond" that and feels that he is entirely emancipated and one cannot judge merely from his milieu. I like his supreme faith - or his profession of it - for in places it sounds a bit like "protesting too much", and his supreme egotism like the place where he says he asked Mahatma Gandhi "what, in his opinion, was the reason for the collapse of his movement while he was in jail". Gandhi replied by asking Mr. Jones what he thought was the cause, whereupon Mr. Jones devotes three pages to telling what he had said on this occasion and there the matter drops as if we were interested in what he thinks when we might have heard what Gandhi himself had to say! No, Gandhi was merely a peg on which to hang his own opinions.

But it is easy to find fault. It is not so easy to discover the good in the book. I am glad you sent it to me and I thank you very much. It is stimulating to read literature of this kind even if one's impulse is to pick out the author's obvious faults of character. The most interesting thing to be noted is the desire of the missionary to make over other people. He is not satisfied to help them realize themselves in their own way. He does not see their essential human beauty. No, he must give them a new religion. Now it is becoming increasingly clearer that it never pays for anyone to change his religion. Buddhism - stripped of its tomfoolery leads to God just as Christianity stripped of its tomfoolery does. Only it happens that Buddhism is especially adapted to the Eastern temperament as Christianity is adapted to us. It was interesting that Jones reports the great interest of Hindus in Christ apart from the church. So also he might find a great interest in Buddha in this country. It merely means that in both places there is an interest among intelligent people who are spiritually - (as distinct from doctrinally -) minded in high manifestations of spirituality. It doesn't mean that the Hindus are becoming Christianized or that we are becoming Buddhized. Here and there an intelligent Hindu will commit apostasy and go over to the missionaries just as the opposite is true. I felt that Jones' "converted" Hindus smacked just a bit of the renegade. Character there is in India - but not in every man.

I presume you wanted me to tell you frankly what criticism I could find. It is only so that there can be honest human intercourse. I have read a great deal about India and have had correspondence with a Hindu student over there. Jones is by all odds the most honest and intelligent of the missionaries but in my judgment he falls short in universal spirit, in charity and in insight. My judgment is that in places he does not know whereof he speaks. I may be all wrong, of course, but that is as far as I can say.

Again I repeat I am glad you sent the book and I hope I have not hurt your feelings by giving an opinion at variance with your own.

Did you read all of "Our Times"? It is a massive work which I could not manage to wade through. I got the impression of immense scholarship and a rather "engineering" ideal of material wealth as representing progress and advancement. Quite right, of course, but so also said the Romans at the height of the Empire. Most of us do manage to get a great deal of "spiritual" élan out of a contemplation of scientific progress. I believe The Decline of the West is a truer picture even if it is less agreeable to contemplate.

I am reading a History of Philosophy now - the only one I could get out of the library. I want to get a better background for world-view and I find it in addition quite interesting reading. I hope Sister does not swallow all that Wiggam writes in The Fruit of the Family Tree. A good antidote for him is Watson in his Behaviorism. Wiggam becomes very nasty. He has an idea and is riding it to death without careful criticism. I think a subsequent generation will consign all his works to the waste basket. It is difficult to see how one manages to reconcile the teachings of the Christ (of the Indian Road) with those of Wiggam. If the Hindus read this book they would have nothing to do with us as a Christian nation. -

Affectionately

E

[The last letter is interesting, not only for its content, but also for its illustration of what intelligent folks did before television and other distractions of our world today. They read.

They thought more deeply, perhaps, than we do. They expressed their thoughts to others who were important to them, whether they agreed or not.

"Mr. Jones" is E. Stanley Jones, widely read and respected by many Christians. It is well to remember that Edmund's father, Dr. Edmund Turney Allen, Sr., was born in Burma, where his father, the Rev. Thomas Allen, was a "Missionary to the Heathen," as it says on his tombstone. That history was a matter of pride to many of the Allens, and still is to many of their descendants. ETA Sr., newly married to Abby, spent time as a medical missionary to American Indians in the Oklahoma Territory before their (surviving) children were born. (One son died there.)

There is little in this letter to fix it in place or time; however it was in the stack of letters immediately after the other undated one that I arbitrarily assigned the date 8-1-1927. That letter said he would write "in a day or two, possibly from Seattle" so perhaps this is that planned letter. If it is, you can see why he wanted to wait until he had time to write and perhaps also time to think about it first. I will therefore assign it the arbitrary date of 8-3-1927 in my file] — MRB

[Albert Edward Wiggam - "Intelligence appears to be the thing that enables a man to get along without education. Education enables a man to get along without the use of his intelligence"] — TKO

[The National Prohibition Act known informally as the Volstead Act, implemented the intent of the 18th Amendment] — TKO

[No place or date]

Dear Mother:

I have been about to write you half a dozen times when interruptions have come and I was compelled to postpone it. I have been sent traveling about from one city to another with barely time to get a breath, and so one rushes about all one's life for a pittance eventually achieving what one calls a position and a "competence". But to look at all life from the point of view of a competition for a pocketful of gold or honor is about as one-sided an existence as to look at it as a chase after the most beautiful wife. Both are products of the materialism of a rather limited and fleeting civilization.

Your last letter was very interesting. You said something at the end about the author presenting Christ to the Hindus without dogma as a character they could admire. I think they have always admired the character of Jesus - those who are educated and intelligent. But to admire is one thing and to find light is quite another. We may admire

Gautama Buddha but unless we approach him in a different way we do not find the "water of life". There is a series of articles in the Atlantic Monthly on "Christ and the Buddha" which is really enlightening. Another article in this month's Atlantic tells of the simultaneous conversion of a Christian missionary to Buddhism and of his Brahmin friend to Christianity. It lifts one a little above the battle to read things like this.

Theosophy is a study of comparative religions with a point of view that from time to time great religious leaders come to earth in accordance with the needs of men. Christ was one of these, Buddha another. Their followers add to and mix up their teachings till the original can scarcely be discovered.

You spoke of Hindus being bereft of their national prejudice. What did you mean by that. It seems to me that few peoples are so free of national prejudice as they are. Americans certainly are not.

Do you ever read the American Mercury. It is a critical magazine published by Menken. There was an excellent article in the Feb. issue on the Parsons and the War giving extracts from famous preachers in this country during the war showing how far they went in the opposite direction from Christ's teachings.

I'm glad Tom is home now. The important thing is now that he gets the very best medical advice for your paralysis. I think that can be cured in a short time if treated properly.

            Edmund

[Her stroke was in November of 1926. This could have been written in 1927 or 1928.]

[no place or date]
Dear Mother:

I am going to send you Will Durant's Story of Philosophy if I can find it soon in the book stores. Then I have been trying to get some of the collection of Mosher's little magazine, The Bibelot. Did you ever hear of it? Mosher was a retired sea captain with a great love of great literature. He started a collection of rare "forgotten masterpieces" and published them in a monthly magazine something like The Golden Book only more carefully selected, usually more beautiful and applicable to life's real problems. Mosher had a point of view which it is quite worth while to understand and his interests went in that direction. But the collections of his masterpieces are exquisite.

The Golden Book is a much more comprehensive proposition. Stuart Pratt Sherman of Illinois was on the Editorial staff before his recent death. Now they have a staff whose chief figure is Henry W. Lanier. I liked his editorial about his recent "discovery" of Nietzche in the February issue and his quotation from Also Sprach Zarathustra which I have always much admired. The curious intellectual dishonesty or density of those who put the entire cult of Militarism on to Nietzche was one of the curiosities of the war period.

Do you remember Murry Espy? I had an interesting letter from him about his struggle to make a living and his desire for mental and "spiritual" "progress". His reading in philosophy is prodigious. He has apparently scoured library shelves. He mentions as recent reading, This Believing World by Browne, "The Ordeal of Civilization" by James Harvey Robinson, Virgin Spring by Waldo Frank, as well as literature from Brahman, Janistic, Buddhist and Confucian philosophy, Lao Tse, Zoroaster, the Hebrew prophets, Paul and the early Church fathers and Mohammed. He says, "From these books I could visualize the birth, growth and senility of our world's established religions. Now I cannot stop with any one in as much as I am sold on an evolution of humanity."

I answered that I hoped things were looking up financially for him by this time. "In order to get ahead it seems as if one had to abandon too philosophic an outlook upon life and too high a standard of individual conduct. At least most of those who have piled up fortunes have been of this type. The pity seems to be that those who have so deep an interest in finding the End of Life and a way to that End must need spend so much energy in the struggle to keep or to get leisure. Fortunately life itself as it goes along in the marketplace as well as in the retreat is an opportunity for growth. Our Western Civilization puts us of necessity into the struggle and it seems to be our job to work out our "solution" within that life. I think that here is the fundamental fallacy of those who speak with authority among us, such as Henry Ford, Glenn Frank, et al. in that their programs prescribe first the *achievement of economic independence so that one may devote the latter part of life to the so-called higher "pursuits of the spirit" and "high adventures of the mind". The point is that life goes on and he who cannot live finely and "spiritually" in the factory cannot do so in the hermitage. Modern economic life fixes certain attitudes and when a man reaches the age when retirement is possible financially it is usually impossible mentally and spiritually.*

[The following two paragraphs are an extended quote, as explained after them.]

"This is, of course, the very point you mention, viz, 'a relation of our world of elections and the world of spirit'. It is after all the point all philosophies have dealt with since Aristotle (and Marcus Aurelius). For some, Einsteinian relativity might offer something if they had eyes open for inner implications, but as a 'way out' such as you suggest, it seems to me lacking. I have a friend, a mathematician who has found much in relativity for the spiritual life in the sense of an explanation of the universe. But a mere explanation of the universe is "cold comfort" for some and inadequate for those who don't mind the cold.

"You say 'Buddha points the way for Hindu life but it does not fit my life if it is to be full" I should say - if Buddha limits life at all, in any way detracting from its fullness, it would not fit any life, least of all the Hindus who are surely not less than we in life's fullness. I believe the same for Jesus. If he would detract from the fullness of life, he is to be discarded. On the other hand one might reach an extremity where life would be too full and one's only desire would be to empty it - to empty it completely. In such an extremity either Buddha or Christ might offer a way. I was interested in a biography of a man who tried all religions one after another - really tried them in the sense of going the whole way without compromise - and the result was reported to be that the End was identical in each".

I quoted my letter to Murry at length because I thought you would be interested. Murry has lots of brain-stuff and what is more important, his attitude is right. For seven years now he has pursued his quest. I shall be interested in the result seven years hence.

You are certainly more "Tolerant" than I if you find a common way with the Christian Scientist. They are always "awfully nice people", but my experience has been that I never found one who would not crucify Christ today. Their claws are nicely hidden under lots of pretty fur!

I hope Tom is consulting with other specialists than Dr. Keller about the paralysis. That overworked clot theory is an easy out for inadequate mental energy.

One of the Rentschler boys, I think it is Gordon [27] - has gone in with Pratt and Whitney in the production of airplane engines. The first type they manufactured is a great success.

You have to educate your legs all over again just as you will the right hand. A strong will and patience and keeping at it together with healthful living and the elimination of tense feelings will do the thing. The affirmations of the Christian Scientists are useful. One doesn't have to be like them in order to use their tools.

Edmund.

[No place or date in, apparently, 1927.
The printing looks like mine, but I have no recollection of reading this letter before – MRB]

Dear Mother:

I was just sitting down to write you a letter when the postman brought yours. It was very good to get it and to hear that you are well and happy. Don't worry about my working too hard. It's good for me. And it is not the kind of work I was driven to at Wheaton. I have found, however, that most people one works for are just like the man at Wheaton in driving employees. And the business of life becomes to learn how to resist being driven.

I am enjoying my social life very much, especially among those friends who are interested in Philosophy. Philosophy is the history of man's thought about Life. It is a rich heritage we have. The finest minds of the ages have left us their ripest wisdom. The sort of people

---

[27] Frederick and Gordon Rentschler were founders of Pratt & Whitney which today provides the engines for more than 25% of the worlds commercial passenger aircraft. The first engine was the 425 hp "Wasp", an air-cooled light weight engine which would dominate the radial engine field for 20 years.

who are interested in these things are usually wonderfully alert mentally and fine in every other way.

I wonder if you have read Jean-Christophe. If the library in DeLand has it do get someone to get it for you. It is an experience of a lifetime to read this. Romain Rolland wrote it. You will enjoy it very much, I am sure.

How has the winter been in Florida? How pleasant it must be not to have ice and snow and sleet and extreme cold. I think Florida will be more and more popular as the years pass. I enjoyed California very much when I was there two years ago. San Diego is a very delightful city. I think I would like the Pacific coast down farther south than Los Angeles - way down on the Peninsula of Lower California. I have never been below the Mexican border but I think it is very beautiful there.

You say DeLand is built up south as far as to coalesce with Orange City. How strange it must be. If we still owned the grove there it would be very valuable now, I suppose.

Do you have difficulty in reading my handwriting? Some people do. If you wish I will write you on the typewriter. I write just as rapidly on the machine as by hand.

I have been having quite a correspondence with Van Gregory about an airplane design. Van is Allene's nephew. He thinks Allene is unbalanced mentally. I told him I was sure of it in 1922 and had it confirmed by her actions in 1925. Van is flying and selling and building airplanes.

You are right about giving one's self time to think. I do not need any more sleep than I am getting but I feel the need of composure, of quiet, of meditation. After all, the most interesting world of all is right inside our own minds, if we only had time to visit it once in a while.

Have you read "Aglavain and Silysette" by Maeterlink? It is a beautiful little short play. You would enjoy it.

My course in Electrical Engineering is very difficult. We are working on alternating current machinery now and have touched on radio. Theoretical radio-frequency problems are very difficult for me. I find psychology much easier. Perhaps I am not an engineer after all - but a psychologist. Or perhaps there is a field in between the two. I seem to have a special knack for psychology. I stand at the head of my classes with no effort at all.

The course in comparative psychology is quite fascinating. I have the head of the Department teaching me. The study of the "mental" behavior of all the lower forms of life and the perfectly uninterrupted chain of development makes one wonder where consciousness begins. Some think even the amoeba has consciousness and others go to the other extreme of doubting if any animals (except man) have consciousness. Of course it depends on what they mean by "consciousness". The two schools of thought are called vitalists and mechanists. I suppose I am a vitalist by temperament but I am fascinated by the mechanist point of view.

Wouldn't it be interesting if man could begin life with all the equipment and wisdom he acquires in the normal lifetime? Suppose you were now just at the threshold of life - about to enter the stage. How would you go about it? How would you draw the ground-plans for your house of life? When one is immersed in activities one has little inclination to wonder what it is all about and why all the hurry and what is the end and whether that end is really worth struggling for, does one?

         Affectionately,
          Edmund

c/o Mr. F. B. Hughes
615 H Street NW, Washington, D.C.
Oct. 15, 1927
Dear Mother:

I have been working pretty hard during the last few months with studies. And in addition I have had some outside duties which took lots of time. But now I feel that things are going smoothly. One of the biggest problems in the school work is Electrical Engineering. I have forgotten all of my electricity and now I must dig it up for the advanced course I am now taking.

I wrote you immediately upon receipt of your check thanking you for it. Uncle Sam's mails seem to be somewhat less reliable than they were once upon a time. Or is it unpatriotic and disloyal to hint at such a thing?

No, I shall not neglect liberal subjects. I am at present taking a course in Psychology for which I was well prepared by the reading on the nervous system I did during your illness. I hope to take an advanced course in Philosophy next semester.

The tragedy of the Sacco - Vanzetti execution is not in the death of a poor shoemaker and a fish peddler, as Vanzetti, himself, said in a wonderfully objective statement. It is rather in the easy way in which we 110,000,000 people (human beings with the attributes of mercy and a sense of adequate justice) can shrug our shoulders and pass it off with the mere remark, "It seems they had plenty of time to prove their innocence." There lies the tragedy. In the Dreyfus case in France the parallel is quite distinct - up to the end: - there after two juries and two courts had pronounced "guilty" - a third tribunal reversed the decision in the nick of time and Dreyfus was later proven entirely innocent. And now even the best of people dismiss out Dreyfus case with such phrases as "men of the caliber of Lowell" "Judge Brandeis would have intervened if" etc. France had its Lowells and Brandeis's.

The change in world-spirit is very evident in the classroom of 1927. Young engineers are preparing to be spotters for the boss. (Of course I am probably entirely mistaken. I have probably misconstrued something the professor said.)

On what basis do human beings meet now? is it only sentimentality, or mutual prejudice, or sex indulgence? Is there no detachment possible wherein might grow an appreciation of subtle human values?

"Ah! Beware young man of the study of Philosophy. It doth unfit thee for thy daily tasks. It maketh thee to think too much. Analysis, methinks, tends to inhibition. Better act - not think. Get married so as to escape from thyself into action. Find thy niche in the system of society in which thou findest thyself and do not subject that system to inspection. Act as those do around thee. Discover their aims and desires and appropriate like ones for thyself. Do not endeavor to find anything back of life. It will only make thee unhappy. Happiness is thy aim. Close thy mind to overmuch questioning. Sleep - sleep is good for thee. It maketh thee to forget".   Edmund

First Air Express with Boeing Air Transport

The first air express with pilots Winslow and Allen. The package they hold is a side of bacon sent to President Coolidge via Boeing Air Transport at Black Hills in the Dakota Country

*[University in the D.C. area
in March, 1928]*

Dear Mother:

It is raining to-day, a slow drizzle, which soaks up the ground and makes muddy pools on the campus and roads. I suppose it is good for the farmers, however, and it is not without its aesthetic aspect also. I have been studying as usual and have not been out to try my shoes in the mud. Our second quarter final exams are directly ahead and I am quite anxious to get A in everything because I might want to ask the authorities for exemption from rules next quarter and these requests are usually granted A students.

I have been struggling with the mathematical functions in electrical engineering until my head is in a whirl.

In reading over some of your letters I noticed again your account of Dr. West's address in Founders' Day. International love, inter-racial fealty, and religious tolerance are to be the watchwords of the New Education. It strikes me that these things ought to go without saying and that the New Education might profitably spend some time training the young in a technique which educators seem to lack entirely, - namely the ability to discriminate falsifications in the New World which is being born. What, indeed does interracial love mean when we all lack the discrimination to see that our national professions of interracial love are the cloak to cover financial spoliation of small peoples. Do educators accept the newspaper accounts of our good-will in Central America? Do they teach youth how news is manufactured and manipulated? Are they aware of our preparations for the next war? Do they analyze situations in the world today to discover the forces operating under the surface? Are even their hearts responsive to the cries of suffering humanity? Not where their prejudices are concerned. Are we keenly sensitive to injustice? No, we utter platitudes about the New Education and sign the death warrant for two humble alien workingmen guilty of trying to make the world a better place to live in and charged by all of us with murder.

The point is that when anyone starts talking platitudes from a pulpit or lecture platform, I at once look under the table and usually find the joker. Sentimentality usually covers an utter lack of sentiment.

The human breast is a strange mixture of striving and deceit, love and subtle hates, charity and strange prejudice, lofty idealism and credulity, capacity at once for acts of great beauty and bestial iniquity. And I feel that him to whom one can bind oneself with hoops of steel must indeed be, in addition to strong, courageous and trustworthy, also analytical and capable of seeing through the bunk both in his own breast and in his fellow men. Only then can he really appreciate beauty, truth and righteousness.

        Your humble son
        E.

[envelope says May 2, 1928
Return address gone]

Dear Mother:

I have been reading Nietzsche this week and finding him very refreshing. Did you ever read "Beyond Good and Evil"? I had hardly realized what a relief it is for the mind to get down to rock bottom and off the very unstable foundation of Hebrew mythology and folk-lore. To realize that the Hebrew bible can be destroyed and the human race be no worse off morally than before - to discover that the dignity of the human intellect far surpasses the limited conceptions on an anthropomorphic deity - to find that Greek, Oriental, Arabian, ancient Egyptian philosophies were just as truly honest, idealistic, moral and "spiritual" if not more so than that of a wandering Semitic tribe of nomads which has come down to us as the "only" "true" philosophy.

Knowing that these conceptions are not unfamiliar to you and that you agree with most of them although you call yourself a "Baptist", whatever that may mean, I write freely to you about them. There are always in every age of the world a small number who have insight into the real significance of the age in which they live, who are not lost in particulars but see things in the large. Such I suppose are amused at those who are caught up in the political, social, economic systems and are pretending, or really thinking, that these are the best the world ever knew. Voltaire was one such, Swift in England, Nietzsche, Emerson,

Shaw: these are a few. Their spirits are unalloyed with baser or extraneous considerations. In my class in Philosophy I am discovering how subtle are these forces which cause us to make false premises and draw false conclusions. Objective thinking is one of the most difficult things to achieve because all our previous education, all the needs for adjusting to the uncritical society about us, all our interests as biological beings, subtly influence us, as well as those inward urgings which make us think the things which make us feel good rather than the things that are objectively true. If humanity for one generation could think objectively wars would cease, crime vanish, superstition and most religion be done away with and society would be based on cooperation and love and tolerance.

    I guess it is the desire for such a society which made me enjoy The Nation and which made me desire for you to see in it as I do beauty, truth, love of humanity and justice, hatred of injustice, cruelty, lust for killing. I hope you do see these things in it.

Affectionately
E.

    Are you using your right hand in writing now? How are the sun baths? If you need money write me and I will send some. I am working all the time at school and I have some money left from the 500 dollars you sent me last fall. I have to hold on to my job since it has a future and I can work part time at school. So far I am in the top 1% in scholarship.

ETA and the Boeing 40-A Passenger/Mail Plane- 1928

Boeing Monomail

Independent Flight Test and Consulting (1928-1929)

Following his completion of service with the Post Office Air Mail Service in 1927, Eddie joined the newly established Boeing Air Transport, dividing his time between flying Boeing 40-A mail-passenger planes and consulting/flight test work for Boeing Airplane Company in Seattle. His route for the mail-passenger flights was San Francisco to Chicago. Boeing Air Transport would become United Airlines.

In 1928 he tested the Boeing Model 83, the model 95 mail cargo plane, some of the tri-motored Model 80s, the Model 203 trainer, and the "Monomail", the first of the all-metal low-wing transports with retractable landing gear. The Model 83 was a pursuit ship and the forerunner of the Navy F4B and the Army P-12.

Eddie worked with Thorpe Hiscock of the Boeing Air Transport to improve voice radio communications with aircraft. This work was conducted in the fall of 1928 and winter of 1929. This consisted of working out techniques for shielding radio transmissions from the radiations given off the airplanes's engine ignition systems, i.e. magnetos, spark plugs etc. Eddie flew around the Oakland area in a Boeing 40-A bi-plane while trying to understand what Hiscock would say and visa-versa. Gradually Allen flew at wider and wider circles around Oakland until finally they could talk as far away as Sacramento, almost a 100 miles away. Most of the ground work was done at the Boeing School of Aeronautics at Oakland, California/ This was probably the reason that Eddie took the electrical engineering course at the University of Utah in 1927. Also this would be why many of the letters are addressed from Alameda, California, nearby Oakland. Hiscock would become recognized as the developer of the first reliable system of aircraft radio communication which Boeing patented and implemented on Boeing Air Transport routes.

In February of 1929, ETA piloted a rescue mission to drop emergency supplies to a snow bound truck caravan in Wyoming. The caravan was part of a natural gas pipeline construction crew which became trapped by a series of blizzards. This may be one of the earliest examples of air drop rescue missions.

In the summer of 1929, Philip G. Johnson, president of the Boeing Air Transport Company (BAT) and the Boeing Airplane Company (BAC), sent Allen to Europe to study European airline operations. His passport says he planned to visit England, France, Germany, Poland, Russia, Italy, Czechoslovakia, Austria, and Romania. Allen was in Europe three months performing the inspection of airline operations and taking advantage of contacts he had make in 1922 and 1924. He apparently turned in a comprehensive report to Boeing.

U.S. Grant Hotel　　　　　　　　　　　　　　July 2nd, 1928
San Diego, California

Dear Johnson:[28]

We were just about ready to send a tracer after the airplane when we received word that it was arriving on the 1:15 PM train today. The train (once a week) to North Island had already pulled out and we got a price of $100. plus freight for a special train so we rushed out to the N.A.S. to claim the truck they had already promised us in case we had need of it. True to military traditions 1:30 PM is too late to get a truck that day although the entire trip would take but two hours. The result is that at 8:00 AM tomorrow we are to have a spiffy truck and trailer and a crew to get the plane over to the station. We will be flying tomorrow afternoon.

Commander Wilson talked as if this plane had already been purchased by the Navy but he did not have any orders to do any performance tests on it. He has already assigned it to a squadron for flight tests and pilots observations which will be carried out immediately. He seems worried about the engine's ability to stand much flying and he spoke of the airplane going into the Santa Monica races in September. He expects the other model (89) to have performance tests made at Anacostia very soon. What's the dope on this? When will it be ready to go? Apparently they are going to take our figures on weights and performance here on 83. Wilson said the only thing to determine was the consensus of opinion of the pilots as to changes to be made (presumably on the production order they expect).

Wilson got a long wire from Rentschler [29] of the Pratt & Whitney Co about the performance of 83. Rentschler seemed extremely enthusiastic about this performance. Wilson also has a letter from Flagg about the airplane. The officers to whom we talked here seemed particularly interested in the improvement in the landing hooks. We talked to Lt.

---

[28] [Addressee is Philip Johnson, the President of the Boeing Airplane Co.] —TKO

[29] Probably Frederick Rentschler - the founder of Pratt &Whitney and the developer of the Wasp engine

Com. Faight aboard the Langley and learned that plans had been made for landing the XF4B on the Langley.

The pilots here are much disturbed by the gas tank in the F3B leaking. They say the design is at fault and they showed us one tank which has come apart - the baffle plates pulling away from the flat side walls. They say the edges of the baffle plates which stick through the side walls were not soldered before the patches were put on. In the high accelerations of stunts the walls bulge and pull away from the baffles and start leaking around the patches. Some of the patches came off with a little tapping. They suggest riveting the baffles to the sides as well as to the ends. I don't quite see how this could be done unless we rivet a man inside, but something certainly ought to be done on the production jobs at once to remedy this fault.

Wilson wanted to know when the first twelve could be ready and whether they could be taken aboard at Seattle. Kneip has written down here that the F3B's are "never going to be done" or something like that.

Wilson told us that Champion had been up to 39,000 feet in the Apache and would still go higher if they could get everything (including the pilot) supercharged. He said the pilot "exuded" oxygen from his skin faster than he could breathe it in! and that the magneto, crank-case-breather, fuel pump, oil pump, etc. etc. had already been supercharged. This sounds like exaggeration to me, but the Apache must be some little airplane! 3500 ft/min climb at the ground! Visited the Douglas factory at Los Angeles and saw some beautiful fast riveting going on. Douglas has relocated the ignition switch under and in rear of throttle. Seems good idea. Douglas is using anodic oxide process of coating....

U.S. Grant Hotel (July 3, 1928)
San Diego, California

Dear Johnson:

We all hoped to get the ship flying today and spend the fourth down in Ensenada, but we found that it needed to be re-rigged and that took us till 7:30 PM with the cowling left to be put on in the morning and Ensenada a long way off. It almost broke Mike's heart, but we decided to work even on July 4th and so tomorrow we will be flying, I hope. Our only trouble now is to find some gas.

You should have heard of the Oh's and Ah's when we lifted the motor cover. There were about 50 Naval officers in the hanger and they were making all sorts of remarks from extravagant praise to criticism of the "enormous nose" on the front of the small fuselage. Then we raised the motor cover and got a real kick out to the unguarded expressions of surprise and pleasure. The mechanics were beside themselves and all wanted to hold something or hand us tools as we set it up. We had the wings on in about an hour and then had to take time off to demonstrate all the gadgets. Nobody seems to like the ratchet retracting mechanism for the hook. They like the former Boeing device better.

We had wonderful cooperation from Commander Wilson, who arranged for a working party to lift the ship out of the car. It was in perfect condition in the express car, and we had it over to the hangers before noon.

We heard some wild stories about the fabric ripping off in strips from the wings of the F2B's. It starts at the doors and rips to the trailing edge.

Since only one will be at the post tomorrow we will have a good time checking our new rigging of the ship so as to get it balanced just as the pilot's here want it. So far we have not seen any letter of authority which would justify our turning it over to them, but they may have this by the 5th.   Eddie

U.S. Grant Hotel
San Diego, California

July 4th, 1928

Dear Johnson:

We flew today (July 4th) after spending most of the day rigging and getting everything fixed up. The few pilots who were at the station were wildly enthusiastic over its take off (4 seconds), its slow speed when I flew across the field about fifty feet high at 1050 rpm, and 60 knots, and its stability in inverted flight when I did some gentle turns and banks in inverted flight.

Boeing Model 83

Tomorrow we will see Wilson and get some definite dope on when they will take it over and just what demonstrations they want. We will also get "U.S. Navy" on the ship for publicity.

When will 89 be finished and shipped? I was wondering if I am to go to Washington with it and if so whether I will be likely to go directly from here or via Seattle. I suppose, however, you will not know about this until we get some results here.

Here's to a first order for 200!   Eddie

We claim the streamlined cotter pins for fast rigging. Three of us completely re-rigged 83 in 30 man hours. It took the plant 63 man hours.

U.S. Grant Hotel (about July 5th, 1928
San Diego, California

Dear Johnson:
Great demonstrations all around! Eighty three now officially turned over with many prayers to the tender landings aboard the Langley. They found lots of things they thought should be changed. They want the cowling flared near the cockpit to keep the wind from their tender shoulders, they insist that the stabilizer control be put over on the left, some want a rudder bar instead of pedals, a new retrieving mechanism, etc. etc. Egvedt took a list of it all for future reference. Lieutenant Jeter flew it and reported to Commander Wilson that it was "a good airplane". Wilson asks if it can be supplied with the standard Wasp and what performance it will have.

We go aboard the Langley in the morning and I am going to fly the F2B and F3B for comparison with 83.

U.S. Grant Hotel July 5th, 1928
San Diego, California

Dear Johnson:
Force of habit, I suppose, impels me to drop you this line, although I know Egvedt[30] plans to see you before this will reach Seattle.

We spent the day aboard the Langley and witnessed the XF4B do its stuff. Lt Jeter flies it beautifully and he landed with great precision. The hook works to perfection and pleases all concerned with its non-

---

[30] [Claire Egvedt was the chief engineer at Boeing in 1928 and will later become Chairman of the Board of the Boeing Company in 1935] — TKO

rebounding quality. It seems that the diagonal post in front of the tail skid is going to get lots of abuse from the fiddle bridges. It got dented in the landings today. That deck rigging certainly handles airplanes brought than I ever saw them knocked about before. I wonder how the 12 passenger plane would stand it?

Tomorrow I fly the F2B, F3B, and Vought Corsair for comparison with 83. Also we will see slow motion pictures of 83's landings on the Langley.

E.T. Allen
U.S. Grant Hotel
San Diego, California

Seattle, Washington
July 6th, 1928

Upon completion your work at San Diego please report back to Salt Lake.

P.G. Johnson

Dear Mother,

It is tremendously hot today. It seems that the whole valley is boiling and sizzling with no breeze to relieve us. I have been trying to read, to write letters, to work at my books, but little success has attended me so far. I wonder how it is in Daytona Beach. If the breeze ever stopped there I suppose it would be hard to bear the heat. Does the breeze ever stop?

I have recently had a very pleasing experience. The president of our company told me personally how very highly pleased he was with the job I did up in Seattle last month. I was quite gratified for I had endeavored to have everything go just right up there. Apparently my work was satisfactory from a technical point of view and also resulted in a large order for the company.

 I made quite a few friends and enjoyed them on a visit but it seems that the friends we really hold to most are very few. The old ones who know our hearts and on whom we can rely absolutely. I have had a very difficult problem recently. In fact it is one of the most trying circumstances of my entire life. It seemed to me very important that I solve it finely and wisely. I had no one to consult with and so was thrown on my own resources entirely. Such situations try one and they also show one's weaknesses and strengths.

 I hope you received the book I sent from Salt Lake. I have not received Jean Christophe as yet but will send it when it comes.

Your son
Edmund

Western Union   Roseburg, Oregon
Boeing Airplane Co   September 8th, 1928

Staying overnight here   F3Bs waiting parts

Allen

Alameda, California
January 17, 1929

Dear Mother:
I've been away on a spring field trip and so busy trying to manage all the odds and ends I have not had a minute to myself until now. I wonder if you could let me have part of that $500 you set aside now. I can pay you back very soon after I graduate. I have been trying to avail myself of all the opportunities that have opened up here this year for improving myself and it has meant that I have spent all the available money I had saved up.

There are several very excellent opportunities for next year, one with the old company where I am offered a splendid advancement. I shall not decide just now but wait until summer.

Have you been well this winter? Florida is probably no more balmy than California is now

[The rest of this letter is lost. This first page is torn and ends in the middle of a sentence. Only one letter, estimated to have been in March of 1928, has been found since 10-15-27. However, the next letter was written only four days later and there are several more from 1929.] — MRB

[no place named -likely Alameda, California]
Jan 21, 1929
Dear Mother:
I have thought the enclosed clipping from the New Republic would appeal to you and so I am sending it. I have wanted so to travel before I became "set" in my ways and incapable of "humility". I've wanted to stay young in the sense of being able to learn, of taking attitudes characteristic of children toward life, of even being a bit "impractical" because one who is too "practical" in his personal life becomes closed to new possibilities of worth in human living. I've wanted to take piano lessons, even at 33, to study new foreign languages so as to be able to understand and love people quite different from ourselves, to fly, to

keep my mind agile enough to consider appreciatively unconventional attitudes. It takes effort to achieve this even at 33 in any considerable measure and I have wanted to put this effort into life.

You have been afraid that with these desires in my head and heart I would make a failure of my life, but I do not feel that I have missed out thus far on much that is really worth going after even in the way of a career. I believe one may be able to have a brilliant career without hardening into the groove which is waiting for the young man maturing into career hunting. The enclosed article expresses something like this.

I am writing this on the train, which will explain its wobbliness. This is a wobbly train. It is now passing very beautiful mountains with snow sparkling on their tree-tops. Snow covers everything and the sun is blinding on the white surface. I am on a trip for the department. It seems that routine study is a small part of my work this year. They think I will learn more by sending me around to lecture at little out-of-the-way colleges and even high school popular science courses. I don't mind; I find lecturing very easy and I enjoy meeting people. The only difficulty is that it breaks up my regular work so that I can't earn as much money as I did before.

I have not decided upon academic work for next year. The possibilities in commercial developments are far more attractive. I am fond of the social life among the faculty here but that is not so important at present as getting ahead in the industry. And although university salaries are far higher than even a few years ago, outside salaries are higher still.

How are you feeling in this cold weather? Or is it cold there in DeLand? Are you in need of anything? I can't send you money just now unless I borrow but I can do that if necessary because of my splendid offers for next summer. Are you feeling strong?

I will try to get you a copy of The Long Journey. I know you will find it quite fascinating and instructive.

Affectionately,   E.

Alameda, California
January 30, 1929
Dear Mother:

Your letter containing the $500 check was received by me at the same time as the post card written three days later according to the postmark. Perhaps the letter was delayed by storms which have tied up air mail for days. The check would be safe even if the letter were lost for you could stop payment on it. I would send you a telegram to tell you it is safe except that there is no danger of loss anyway and it would cost a dollar or two, which I think we might as well save. Thank you for sending the money now. It will relieve a lot of worry for the next four months. I will not use all of it and can at any time send you some. Wouldn't it be well for you to have some cash for use in a hurry if you should wish it?

I am very much disappointed in one sense about my work and school at present. My advisor tells me that I am in a very peculiarly advantageous position and that I should not worry. But I see it from quite another angle. If I were an ordinary senior of 21 years, it would, of course, be advantageous to be doing consulting work for a large outside corporation while "attending" school, even though it did not pay very much. and also it would be very nice to be going around delivering lectures on aeronautics at schools throughout the state - if I were a 21 year old without experience. But actually it is quite different and I do not feel with my experience and age that these things are anything to get excited about. Nor do they seem sufficient excuse for me to give up certain class work I would like to have.

What if the company has offered to send me to Europe this summer to study aeronautical affairs there? I don't know that I want to go to Europe just now, and I don't just see what sort of a position I would be in when I returned. It might be that by being away I might lose out on a really important position during the busy summer season. Summer is the time things expand in aeronautics and I feel that I ought to be here at that time.

During the Fall and Winter I did an important piece of work for one of the big west coast companies.[31] At least I thought it was important. It was a survey of radio telephone communication on the air mail line. It was necessary to determine what frequency was best suited to this job and to outline a set-up and method of procedure, with detailed estimates of personnel, apparatus, regulations, etc. the entire program, practically as I outlined it, is now being adopted and installed. It will certainly go a long way toward increasing safety of operation on the air mail and the new air passenger lines. This was written up by the newspapers and some erroneous statements were made which the company did not like such as that I had an official connection with the company. Apparently they thought that I had given interviews to the papers and stated that I had such a connection, whereas the papers got hold of all their information from some outside and, so far as I know, irresponsible, source. This has not helped my standing with the company although they have not as yet withdrawn the offer of sending me abroad.

The airlines of the country are in about the same position that the railroads were just before the big consolidations and expansions which made it possible for them to earn profits. The air lines are not earning profits now but with new improvements in equipment and savings which will accompany consolidation, they will get on a sound economic basis. The improvements in radio telephone communication and direction will also assist in making operation regular and efficient as well as safe. I have a hunch that we are already ahead of Europe in most aspects of air travel. Some things they have better than ours but on the whole we are about even. Neither side can keep a new development secret long enough to get very far ahead and each side is watching the other with jealous eyes and is quick to copy anything new and better.

I suppose I should rush this letter off to you with the reassurance that the 500 dollars reached me safely. But I will add something about books and things you mentioned in your letter.

---

[31] This is the Thorpe Hiscock/ETA research on aircraft communications conducted in 1928/29 and is summarized in following text

 Jean-Christophe is not "understandable but better toward the end". How strange to hear you say that? Is it difficult to understand a nature like Christophe's because of his honest living of life fully, - free from inhibitions? Do you not feel Beethoven's symphonies throught [sic - for throughout] that life. The deep calm passion of the Fourth Symphony, or the majesty of the Fifth or the clear high inspiration of the Eighth. For Christophe you know, was Beethoven. How different from us was he both in his wisdom and his unwisdom! And yet how beautiful for his difference. Was it his setting at naught of our accepted social values that you didn't understand and thought got better toward the end? Or was it that you objected artistically to writing so frankly of aspects of life we usually cover up. Or did you object to Christophe having those aspects of his life at all? If you felt it lacking in a certain aristocracy of attitudes, I think I know what you mean. But we have all failed miserably at times to achieve this aristocracy. What did you think displeasing about the book.

 Have you read Woodward's "G. Washington: The Image and the Man"? you would like it, I believe.

 I read Bojer's "The Power of a Lie". It is a strange Norse tale of a man who becomes deeply involved because he told a little fib, and because he dared not back out of it he sent an innocent man to Prison. It is very moral - or immoral. I read also Black Majesty, a story of the Negro emperor of Haiti. It is a rare tribute to the Negro character in all its noble as well as petty qualities.

 There is snow 3½ feet deep outside the car windows on my trip back to Alameda   E.

Aircraft Radio Communications Study[32]

Research at that time (1928) with Boeing, meant Thorpe Hiscock, W. E. Boeing's brother-in-law, an inventive genius who was determined to perfect radio telephony for aircraft—and did. While it is true the Army and the Bureau of Standards had been fussing around with the problem for several years, the aircraft radio was not perfected until Hiscock and Eddie Allen were brought together as a team. Hiscock was a big bear of a man whose appetite for work and play knew no bounds. A genius, he would work almost without sleep or food until he was so fagged he would fall on his face. Then somebody would shovel him into a sanitarium where, upon recovery, Hiscock would go back and repeat the process.

Hiscock was a practical joker whose idea of fun ran to the destructive side. When receiving men for a conference in his hotel room of elsewhere, he was likely to whip out a pair of scissors and cut off the neckties of all those within reach. Or, he would receive his guests with great solemnity, take the hats and trim the brims down to the general shape of a fireman's helmet. Other times he would fill the hats of his guests with water, placing the hats in the bathtub. Jack Knight, veteran Mountain Division pilot, once curbed Hiscock's enthusiasm by cutting the heels off his shoes. It was on a Sunday and Hiscock walked all over Cheyenne without heels on his shoes, seeking a cobbler!

Hiscock had never found a man who could work as hard as he until he met Eddie Allen. Together they flew back and forth over the Division in search of mountain storms of intensified fury. They wanted to try their radio out under extreme conditions of lightning and static electricity.

Some idea of the scope of Hiscock's investigations during this period, is suggested by this list of devices, the fundamentals of which Hiscock discussed on a single evening of record, with an associate: electric clocks, chronometric instruments, ignition harness improvements, voltage regulation , control box test facilities, gasoline flowmeters, flexible drives for generators, automatic mixture controls for engines, remote control fuel valves, air relieve valves to vent air from the line when the pilot switched fuel tanks, horsepower hour meter, accessory power plant drives, ammunition rounds counter, auxiliary jackshaft to be driven by a flex shaft from the engine and to mount such items as the generator and vacuum pump (the forerunner of the auxiliary gear box to drive the accessory group) and radio filters for contact points. Most of these gadgets have found common acceptance

---

[32] Material summarized by Thomas Collison ~ 1944

since Hiscock's untimely death.[33] He was a man who worked on inspiration. On a freezing day at Chicago airport he observed a frozen flag whipped up suddenly by a gust of wind; he saw the ice shake off. He spent the next five days developing the first practical set of de-icers for plane wings! In his laboratory improvements were made on automatic pilots, radio transmitters and receivers, quartz crystal cutting, generators, voltage regulators, dynamotors, power disconnect switches. Hiscock laid the groundwork for instrument flying.

It was with such a hard-driven genius, Eddie worked the hardest years of his life, interspersing this work with scheduled airmail flying.

Eddie with Radio Head Gear

Popular Mechanics July 1929.
"Laying the world's longest day-and-night air line out in signal control blocks like a railroad will be started soon, when the Boeing company completes twelve radiophone stations between Chicago and San Francisco. With the beginning of regular passenger

---

[33] Thorpe Hiscock died September 28, 1934, at the age of 41

service between in eighteen-passenger air liners this summer, each passenger and mail plane will be in constant communication with the ground over the entire route.

From the moment he takes off until he lands, the pilot will get reports at twenty-minute intervals on weather conditions ahead, and will be able at any time by throwing a switch from receiving to sending, to communicate with the nearest ground station.

Months of experimenting, both with radio apparatus and ships, to eliminate static and other interference, has culminated in the development of a combination transmitter and receiver which weighs only 100 pounds. For transmission, a 1,000 volt generator winding has been superimposed over the regular winding on the plane's engine-driven generator, which supplies the current for the lighting and starting systems.

The radio apparatus itself requires no tuning or other attention from the pilot, being operated from a remote-control switch on the instrument board. Tiny earphones, imbedded in soft rubber plugs molded from a cast of each pilots ears, and a transmitter, swung like a telephone girl's phone in front of his lips, provide channels of communication without interfering with the operation of the ship, or distracting his attention from that duty.

Before the radio tests were successfully concluded the experimental plane had to be almost entirely rebuilt. The shielding or the motor to eliminate interference from the high tension ignition current was only a part of the problem. Every loose part that rubbed, and every moving part of controls, produced static which, while not interfering with code transmission and reception, made talking virtually impossible.

With the radio system in operation, the movements of the fleet of passenger and mail planes will be as completely controlled as those of railroad trains traversing electric-block-protected rails. The twelve ground stations, spaced at about 200 mile intervals, will enable a plane to always be in touch with at least one, for the equipment in the ships has a 200-mile range. With that range, ships also can talk with each other, while to reach a brother pilot at a more distant point, the flyer in one ship can have his message relayed through the ground chain.

The twelve stations, some of which have already been completed, are to be at Oakland and Sacramento, Calif., Reno and Elko, Nev., Salt Lake City, Utah, Rock Springs and Cheyenne, Wyo., North Platte and Omaha, Nebr., Des Moines and Iowa City, Iowa, and at the Chicago municipal airport.

Describing the operation of the system, E.T. Allen one of the pilots who aided in the experimental work, describes an experience , when, flying above a deep fog bank, he heard the voice of the superintendent at the next stop, reporting 1,100 foot ceiling over the airport, with five-mile visibility and no indication of change in the next thirty minutes. The pilot reported back that he was thirty miles away, making 115 miles an hour. With a directional loop, the ground operator located the position of the plane and reported back that it was five miles south of its course, gave the course correction to get back in line to the airport, and announced a yellow signal rocket would be sent up when the plane arrived within three miles of the airport.

Allen reported back to the ground that the top of the cloud bank that he was flying over had an altitude of 8,800 feet, which would give a cloud thickness of 1,500 feet resting on the mountains above which he was flying. The ground station replied that at the Summit, a station in the mountain pass, the fog began to rise as the ground dropped away to the valley.

Back from the plane came the message that it would arrive at 9:13 p.m. over the airport, and that the signal rocket should be fired vertically at 9:11 to mark the field. The switch was thrown to receive, and the ground reported that the second section of the eastbound mail was seventy minutes behind Allen's plane, but that the westbound first section was due about the same time as Allen. Orders, however, had been phoned to the westbound pilot to delay his arrival until after Allen had landed."

Plane Saves Snowbound Truck Party
Fifty persons see needed relief come falling from the sky - San Francisco papers February 22, 1929

"The margin for nearly 50 men, women, and children between perishing from starvation and sufficient food to tide them over the time required to dig them out of massive snowdrifts where they had been held since Monday at Table Rock 55 miles east of Rock Springs, Wyo., was an airplane. The margin was enough. Piloted by Edmund T. Allen, air mail ace of the Boeing Air Transport Inc., one of the company's fast "95s" (all mail ships), flew over the beleaguered transport truck outfit, consisting of 23 of the big motor carriers, and dropped into the snow along side the enforced campers, 1600 pounds of foodstuffs late Thursday afternoon. More food will be carried to them in the same manner, in case of need, it was announced by D. B. Collyer, Vice President in charge of operations, after Allen's successful flight._ _ _ ."

Hotel Alameda
California
Dear Mother,

Another of these special missions which apparently have nothing to do with school has been completed satisfactorily and, it seems, it has added considerable prestige to my position. I am hoping it will enable me to prosecute successfully my European trip this summer.

I have been working very strenuously lately on a new airplane design which was submitted by one of the western companies for consultation work in the Department. No time for reading or play although I picked up a new book on China in the library. It is called "China's Millions" by Anna Louise Strong.

Expect to be in Sacramento next week for a visit.

Yours affectionately   E

[Hotel Alameda
Alameda, California]
February 24, 1929.

Dear Mother:

I'm afraid the last letter I wrote you was lost. I gave it to one of the fellows who flies out of here and I suspect that, although he goes to Salt Lake, he forgot to mail it, because when I asked him he did not know what he had done with it. I will mail them myself after this.

I am doing some work at present on stability, an investigation of a problem that is only too little understood. There may be some folks who really understand it but if so they are keeping it to themselves. There is nothing in the books about it.

Did I tell you about a new book I read recently by a Norwegian. It was "The Power of a Lie" by Bojer. It is not as it sounds, a conventional sermon on morality but a very subtle exposition of the psychological complexity of the forces of social approval in society. We see in it how the opinion of a man in a community is built up on a host of little items and may be built up on a lit. it shows also how a man's own opinion of himself may be falsified so the he will come to believe in a lie he has told.

To day I had a beautiful trip in the warm balmy sunshine over hillsides green with spring vegetation. Mt. Diablo alone stands out brown in the landscape. The beautiful golden gate gleamed in the setting sun across the bay. I hope Florida is as lovely at this time of year.

Have you plenty of ready money so that you are comfortable? It is so inconvenient not to have cash even though one may have money in the bank one does not want to spend.

Your son

E.

[Alameda
4-25-1929]

Dear Mother:

Things are nearing their winding up stage here and maybe I will get to see you this summer on my way to Europe. I'm not yet definitely assured of the European trip but I have it in mind still. I was told that I would lose out in my position with the company if I did not go right into the factory as soon as school lets up. But I have had in mind your admonition about working too strenuously and continuously and I am sure I need a rest.

We take physical examinations every month now for some queer tests and my score has been lowering as the year progressed. The doctor says there are signs of fatigue and so I have decided to have a vacation even if it means the loss of a good position which I would get by rushing ahead with a heavy schedule of work.[34]

It is quite cold here early in the morning and late in the evening and some days it seems to stay cold all day. This part of California is quite different from the Los Angeles region where it gets very hot. In many ways it is pleasant but sometimes I wish for the sunshine. Last Sunday I went to the beach and lay in the bright sun for three hours, turning over and over to get well tanned. It makes one feel so much better to get the sun all baked into one. I want to do it every Sunday if possible.

I hope you are well and are enjoying reading and the prospects of very pleasant weather as summer approaches.

               Affectionately   E.

---

[34] The work with Thorpe Hiscock the previous winter may be the source of this fatigue. Also the exaggerated ETA contribution to the study described in the newspaper article and failure to credit the primary contributor, Thorpe Hiscock, may be the source of the "erroneous statements" he referred to in his letter of January 30, 1929. It may also have contributed to his inability to secure a job with Boeing when he returned from the European trip.

Airmail Pilot Going Abroad for New Study
Boeing Company Sending Aviator for Inquiry in European Lands

Airport - Edmund T. Allen, air mail pilot on the transcontinental line with headquarters in Salt Lake most of the time since 1924, arrived Sunday by plane from San Francisco for a few hours stay before proceeding to New York, where he will take a steamer for Liverpool. For several weeks, Mr. Allen has been experimenting, testing and installing radiophones for the Boeing Air Transport. He was taken from his regular pilot schedule some months ago and stationed in Seattle as test pilot for the Boeing airplane factory.

While testing the radiophone, Mr. Allen practically perfected the instruments and system which will be adopted on all Boeing mail and passenger planes, a system which permits ground-to-pilot and pilot-to-ground conversation as well as a separate system for ground-to-pilot only communication. The adoption of the radiophone is a new step toward overcoming the handicap of adverse weather conditions for flying, and Mr. Allen announces the system really meets one of the main requirements, continuous and adequate weather information while ships are in the air. Mr. Allen goes to Europe on a three months mission for the Boeing airmail lines to make an intensive study of air transport in several countries. After a short stay in England, he will proceed to Germany and spend several weeks at various German airports. His next move will be to Russia. His study will end with short stays in Italy and France. — Salt Lake City newspaper

[envelope return address:
American Express Co
New York City N.Y. until May 16
After May 16th   c/o American Express Co. Berlin, Germany]

May 6, 1929

Dear Mother:

My plan is about to go through for the trip to Europe as a wind-up for my education. I am now on my way east where I have to make some connections before sailing. I am hoping nothing will interfere with my plan of getting to DeLand.

Will you please sign and have a Notary Public seal the enclosed document which is necessary for me to get a passport? Please keep it there in readiness for me or for sending to Washington if I write you in a day or two.   Hoping to see you soon,

With love   E.

Dr. Thomas Dyer Allen
1417 Peoples Gas Building   Chicago

May 19, 1929

Dear Aunt Margaret:[35]

Both sister and I have received letters from Mother and, knowing her, we both believe she has written you recently, regarding her dissatisfaction, an urge to be in a different place. We are at our wits end.

It was occasioned by a telegram and letter sister sent when there came a possibility of a job for her in Nashville for the summer. Now that this seems to have been given to someone else, sister is apparently free to do whatever she wants. She feels that this summer she should devote her time entirely to Mother. She can be in Florida in 3 weeks - and expects to be there.

Edmund has just spent a few hours with her - I think this is their first visit since the rather fateful parting in our house some six years ago. He was on his way to Europe. Maybe that had something to do with her restlessness.

I am enclosing some of the letters we have recently received. I had a phone conversation a few hours ago, with sister, and since then Floss, Aunt Pink, Mother See and I have discussed the matter pro and con. We thought of phoning you, but felt that would make you think there was an emergency about the matter. Maybe there is; Floss said "She is apt to move any moment and write about it after". She will not make any big movement alone, but she may go to Daytona Beach, or to some sanitarium.

Well, now, the upshot of all our discussion seems to be something to this effect; that there may be a possibility of her going into the community house in Hamilton. Uncle D.R. and you know the details about the house, the necessity of the applicant, etc. We do not want you to have any fruitless burden in the matter, but a few suggestions especially in regard to the community house would be welcome. I think Mother does not know of money's expectations, and I hesitate to

---

[35] Aunt Margaret is Abby's sister

tell her. Yet her knowledge of this would in a measure help to explain matters.

I am writing Mother to hold the fort while we see what can be done.

We are all well here. Aunt Pink has had a little disturbance which we are investigating but think is not serious.

Florence has taken 10 or 15 steps alone several times today and Bob is talking more and more freely. So we are getting a grown-up family. Thomas had an afternoon with Howard See last Saturday. He came to town Saturday noon. That night Mother See threw a big party at the Palmer House -9 of us at dinner; Mr &Mrs J. Len Young and son of Honolulu have been in town for several days. Mrs Y. you know is Mother See's sister "Val". After supper we all went up to a parlor on the 12th floor and played bridge for a couple of hours. Howard had to leave for Atlanta and Gordon for Albany that night. Uncle Len and Aunt Val are leaving for Cincinnati tomorrow; then they will go on to Washington, N.Y.C. and elsewhere. Mother See has had a big week and is as happy as can be. One night all three sisters were alone together in her own apartment.

Let me know how you are and please give us some advice.

Love from all of us

        Thomas

Hamburg-Amerika Line Stationary
(No date - probably July 1929)
Dear Mother:

I've been having a most instructive and enlightening time here meeting people and seeing the air traffic and management of the European Air Lines. I find I have many friends remaining from my former visit to Europe in 1922 and flying seems ever to produce a comradeship which supersedes nationalistic boundaries and feelings. I am writing this letter on the train enroute back to Germany from a trip I made into Switzerland. Next week I go to Fulda to the old glider camp and from there probably to Paris unless I have a cable to the contrary from America.

I found a letter which I had sent from Alameda but lost. I will enclose it in this. I hope you have gone to the beach or to the sanatorium by this time if it is more comfortable and enjoyable there for you.   Your son   E.

[Saxony is a small German state bordering the Czech Republic's western end. Dresden and Leipzig are its largest cities.]

August 16, 1929
Dear Mother:

Life has been so strenuous during the past weeks that I have hardly had time to digest all that I have taken in. so now I have arranged a rest in the Saxony mountains where I will do some arranging of materials and incidentally rest up a bit.

Your last letter surprised me very much. Life weaves such a strange pattern that most of us who have not found a motif and a means of putting the motif into the pattern are bewildered at its seeming meaninglessness.

I hope Daytona Beach remains for you pleasant during these months. I've been thinking considerably about your statement regarding your condition and outlook and the position taken by medical science concerning it. I still feel that the paralysis is not of the origin of most pareses in which there is a progressive degeneration of the central nervous system due to bacterial or other growths. It may be due to a clot or tiny tumor which will yield to certain types of treatment usually not recommended by the standard authorities. I have recently consulted a physician in Germany about my nose and its functional difficulties, and I asked him then about your paralysis. We talked a long time about it and he asked many questions, some of which I was unable to answer - But the point was that he feels that modern medicine is in a cul-de-sac with reference to certain disorders which he has treated successfully. He is working with reference to the causes of diseases outside of, or beyond, bacterial infection. That is, he says the bacteria do not happen upon the organism and then work their will upon it, but that we all have practically all bacilli in our systems and that only when some cause within us, such as poisoning due to lack of proper elimination, perhaps of a tumor, or blood clot, or of the deposit of waste matter in the tissues, lowers the resistance of the organism and produces local fever, can these bacteria gain predominance over the white blood corpuscles. He treats some diseases quite regardless of the usually considered systems, seeking only to remove the poisoning, or, really, to get elimination through the proper channels again when these have ceased to function. He has cured leprosy by this method alone, absolutely without any medicine or attention to the outward symptoms. He has cured syphilis and paralysis also. One nobleman who was cured of syphilis by him makes an open statement of his case out of deference for the value of the method of treatment.

    The first part of the cure consists in getting the individual to take an active, enthusiastic aggressive part in his treatment. First he sits in a shallow tub of lukewarm water in a warm room with his legs out and the water only covering the hips. He then with a small coarse wash cloth or towel washes from the naval down and on the sides and back from the waist downward and on the legs toward the anus as far as possible. This is continued for half an hour four times a day, or less if the patient is too weak. The continued bathing and stroking starts powerful lymph activity removing deposits of all kinds all over the body and getting them eliminated through the channels where they ought to be: - the intestinal tract and the bladder. A case of gonorrhea [sic] was permanently cured in a week and the leprosy with this and two *other types of treatment disappeared completely in two months. The government took this doctor's cure to their leper colony and used it successfully there. Later the bath is made colder and the patient is kept warm by making the room warm.*

He has some unscientific ideas which are out of date, but that does not influence the fact that here he has hit upon a health giving treatment for troubles like yours. Most doctors have unscientific ideas in some realm of their thought processes, either religious or social or ethical, or very often physiological. If they heal that is the main thing. If you want to get well I think you will have to desert orthodox medicine to do it. They give up too easily.

I still have faith that your paralysis will yield completely to the proper treatment and the wasted limbs will build out again -
          Affectionately
          Eddie.

[Please allow me to share a personal experience with German medicine. When the US Army moved us there in 1964, we hired a German unwed mother, Inge, to help me with housework and care of our four young children including 3-month-old Laura. Inge's baby was not even as old as Laura. Within a couple of weeks, Inge got very ill with a urinary tract infection. At her request, I took her to her German doctor, who prescribed four different teas, which I was to give her on four different schedules, two four times a day, two twice a day. I was making tea all day long and taking care of both babies and Inge, and she wasn't getting better. My husband, then an Army M.D., took charge and gave her an American Sulfa drug (Gantrisin, I think) which had her essentially well in 24 hours. At that time, so surely in 1929, German medicine still had a lot of medieval methods – hot-spring and mud baths, teas, steam rooms – for which they were well-respected in Europe and around the world before modern antibiotics were developed. The fame of the city of Baden-Baden, meaning baths (squared), went back to Roman times, and people came there from all of Europe.   -- Margaret Broussard]

[Poland to Germany]

September 26, 1929

Dear Mother:

Again I must write on the train since that is the only hour I have for such luxuries as writing letters. I hardly expected such a strenuous or continuous assignment as this. What an interesting thing it would be if I were to be given this as a permanent position. The company seems to think I would be useful over in Europe as a permanent fixture. I think I should not be very happy in the Slavic portions but I am quite at home in Germany and France. Last night I saw a brilliant Slavic Review in the Exposition theatre in Posnan. The new Polish nation created by Woodrow Wilson,[36] is expressing itself in sundry ways, but chief among them seems to be a cheap copy of Paris manners and dress and ideas, and a terrific military and aeronautical program for the suppression of every minority and the cultivation of hatred toward their two neighbors, Russia on the East and Germany on the west. The young men (Poles) with whom I talked speak of "the war" meaning the next new war which they expect to fight against?

This is a field rich for the study of Political Economy and Sociology.

The train rocks so that I can hardly write. These trains are not like ours in America and I shall be glad to be back if I can return, as I hope, in November.

I was sorry to hear of Margaret's illness. I thought she was always so strong physically and that especially after the hardening effect of the Minnesota winters she enjoyed perfect health. The summer in Daytona Beach should have left her bubbling over with energy and enthusiasm for the new year to begin, and instead she was all used up and had to go to a sanatorium. Something sounds wrong there. Can't you do something about it. I have a feeling it oughtn't to be that way! If I went to Daytona Beach for a summer I would expect to leave sunburned and having an appetite [sic] for my new work.

---

[36] Woodrow Wilson's "Fourteen Points" address to Congress included Point 13 which spoke of the need to establish an "independent Polish state". Wilson was awarded the Nobel Peace Prize and in 1922 Poland conferred upon him its highest state distinction, the Order of the White Eagle.

What has happened to land values as a result of the Florida panic? I suppose one can buy a beautiful frontage on the ocean at Daytona for half the usual price - with perhaps a house on it at less than cost. Did Palm Beach also suffer from the losses? I remember our drive toward the inlet along that beautiful coast beach. What an ideal place for a country house, especially if one had electricity and water supply. I suppose these do not extend far beyond the limits of the city. I do not remember how the isolated houses farther out obtained their water and electricity. Palm Beach must be delightful in winter. In summer it is too hot, I should think. Somewhat like San Diego California.
    Affectionately   Eddie

Independent Flight Test and Consulting (1930-1939)

Following the European trip sponsored by Boeing, Allen continued his work in support of Boeing. The company became interested in building a metal covered low wing aircraft designed specifically for mail transportation but with some limited passenger capability. The "Monomail" was the company's solution. Allen flew the aircraft on its first flight in May of 1930. His test results caused concern with the propeller selection. They could adjust the blades on the ground to get more "bite" of the air by using a flat pitch to get enough power at low speeds on takeoff, but then that pitch was not good for cruise. The clean airplane needed a prop that could be adjusted in-flight. Allen spent the summer of 1930 making tests, including use of smaller props turning at faster speeds. A new transport evolved called Model 247 which used the new controlled pitch propeller and which became Boeing's first successful all-metal transport.

Following the work on the Monomail, Allen found himself unemployed.* "Why" is not explained, but Philip Johnson steered him to Northrop in Burbank, California, and a new direction in his career. Eddie's letters to his mother had expressed optimism that he would be hired for a permanent position (with Boeing?). That did not happen. Instead he went to southern California and flew tests of Jack Northrop's Alpha, Beta and Gamma aircraft. Then in the spring of 1932 he was contacted by Douglas regarding the DC-1, an aircraft for which he would do extensive testing. ETA's work with W. Bailey Oswald at Douglas produced technical flight procedures that would revolutionize the approach to commercial aviation. These resulted in changes to cruise operation that significantly increased cruise speed and range. These procedures were applied to the DC-2 and DC-3 — the latter became the standard for commercial aviation. These innovations were documented in thirteen articles in Aviation Magazine starting in February of 1934.

*[As a retired Boeing engineer, I am well aware of the " technical arrogance" to which we occasionally would fall victim. I can visualize Eddie from "outside" the company being perceived as a challenge to someone's technical "turf" which could force management into a decision they might not have wanted to make. I have been told ETA experienced it when he joined the company in 1939. Not everyone agreed with the combination of flight test and aerodynamics into the same organization. Also, the fact that he did not report to the chief engineer was a radical organizational change. But by then Eddie Allen was far too dominant a figure and management bypassed the issue by having him report directly to the company president. Note that after ETA's death in 1943, the aerodynamics and flight test disciplines returned to their original organizations. However, the flight test and evaluation procedures established by Eddie Allen were still in active use in 1962.] —— TKO

In the next nine years, Eddie worked for Northrop, Douglas, Consolidated, Chance-Vought, North American, Eastern Air Lines, Panagra Airlines, Sikorsky, Curtiss-Wright, Lockheed, and Boeing.

The list of "first flight" aircraft flown by Edmund T. Allen is extensive. The "first flight" designation certainly reflects the customer's (and sometimes insurance company's) confidence in the test pilot. However, often the following developmental testing is even more demanding.

Allen's "first flight" aircraft include:
- Boeing Model 83 (June 1928)
- Northrop Alpha (March 1930)
- Boeing Monomail (May 1930)
- Northrop Beta (March 1931)
- Northrop Gamma (May 1934)
- Stearman 80 (April 1933)
- Douglas Dolphin amphibian R2D (February 1934)
- Lockheed Electra (February 1934)
- Douglas XO2D-1 (March 1934)
- Douglas DC-2 (May 7, 1934)
- Vought XSBU-1 (June 1934)
- North American GA-15 (O-47A) (May 1935)
- Sikorsky S-43 (June 1935)
- Vought SB2U Vindicator (January 1936)
- Spartan Executive (March 1937)
- Sikorsky XPBS-1 (August 1937)
- Boeing XB-15 (October 1937)
- Boeing 314 Clipper (June 1938)
- Boeing 307 Stratoliner (December 1938)
- Boeing Stearman X-100 (January 1939)
- North American BC-1 (NA-16) (March 1939)
- Curtiss-Wright CW-20 (C-46) (March 1940)
- Boeing XPBB Sea Ranger (June 1942)
- Boeing B-17B (June 1939)   Boeing B-17C (July 1940)
- Boeing B-17E (September 1941)   Boeing B-17F (May 1942)
- Boeing XB-29 Superfortress (September 1942)
- Lockheed L-49 Constellation (January 1943)

Eddie Allen also provided development flight test on many aircraft in addition to the "first flight" aircraft listed above. These aircraft include:
- Boeing Model 95 mail plane
- Boeing Model 80 mail / passenger tri-motor
- Boeing Model 203 trainer
- Fokker FX (F10) February 1933

Northrop Gamma 2C May 1933
Douglas XFD Navy fighter May 1933
Northrop 2C Delta June 1933
Douglas DC-1 July 1933
Sikorsky S-38 December 1934
Vought XSB2U-1 December 1934
Vought XSBU-1 May 1935
Curtiss Design 75 (P-36 Hawk) June 1935
Sikorsky S42B October 1935
Vought XSB3U-1 April 1936
Vought V141 April 1936
Northrop 2A May 1936
Northrop 5A September 1936
Northrop 5B September 1936
Vought V143 September 1936
Douglas DC-3 January to November 1937
Boeing YB-17 May 1937
Douglas XTBD Devastator Torpedo Bomber June 1937
Boeing B-15 September 1937
Consolidated PB2Y Coronado January 1938
Lockheed Model 14 Super Electra May 1938
Boeing Clipper June - December 1938
Boeing School Trainer January 1939
Boeing Stratoliner 307 January 1939
Douglas A-20 Havoc (DB-7A) March 1941
Douglas A-20 Havoc (DB-7B) July 1941

While the B-17, B-29, C-46 and C-47 (military version of DC-3) were produced in large numbers, it is also of note that more than 7400 Douglas A-20 (DB-7B) light bombers were also built and more than 3400 were provided to Russia through lend-lease.

Eddie Allen and the Boeing Monomail 1930

Douglas Dolphin

Vought SB2U   Vindicator

Spartan Executive

Sikorsky XPBS-1

Boeing B-15

North American BC-1 (T-6 Prototype)

Curtiss Wright CW-20 (C-46)

Boeing XPBB Sea Ranger

Hotel Jackson   Medford, Oregon stationary
Dec 19, 1929
Dear Mother:

    I have not yet reached Alameda on my way from Seattle. This country is quite mountainous and very wet. It seems as if it rains all the time. Every day when we get up it is raining and all night it is foggy or misty. Last week 36 inches of rain fell near here. The valleys are all covered with water and the roads washed out. This seems to be a good thing for the northwest because they were having lots of fires here and the cities did not have an adequate water supply. Tacoma, Washington was in great need of electricity because the water level was so low in the hydro-electric plant supplying the city.

  [This is obviously not the whole letter, but it is all that was preserved by his family. There is no closing or signature.] – MRB

[Seattle or near there]　　　　　　　　　　　　March 25, 1930

Dear Mother:

What a long slow jolting trip it was up north. 28 hours! And raining all the way.

Wouldn't you like it better to stay another week or so in DeLand? Especially if it were a little more comfortable? Perhaps I could write to Mrs. Smith asking if she could do something to make you extra-comfortable. Would you like to have me do this? I liked Mrs. Smith so very much. She is a dear. If she could take care of you always instead of the selfish nurses you have had it would be nicer.

I liked also the chiropractor. He has some funny ideas, but then we all have some funny ideas. I don't mind his foolish ones nearly as much as other people's foolish ones.

It was cold when I got back to Washington. I needed my overcoat all day today.

If you would rather stay a week or two longer in DeLand and then go direct to Hamilton with Mrs. Smith, taking a compartment on the train, I think it might be more pleasant for you. What do you think?

　　　　　Edmund.

I just received your letter suggesting that I fit in to the plan of your going to Atlanta. I did not realize that you wished this so much and I am sorry I did not try to fit in better in DeLand. I really had to get back here Monday, however, and if you prefer these weeks in DeLand, it is better this way.

If you would like to try some of the baths I spoke of in my letter from Germany, you might ask Mrs. Smith to help. All it needs is a large basin in which one can sit. It must be in a warm room with no drafts.

Maybe the chiropractor has some hydrotherapy to suggest. The type they often recommend would be too big a shock, I should think.

Several of your letters were forwarded from Alameda and arrived today.

If you have any needs, let me know. And tell Mrs. Smith how much I enjoyed meeting her and my stay there.

　　　　　E.

Please send me my picture out of the frame on your dresser. I'll return it in a short time.   E.

[Approximately 3/30/30]
Dear Mother:

Ever since I've been in Seattle I have been in a mad rush. Not a minute even in the evenings to myself. I've been invited out to dinner with one after another of my old friends and in the daytime it seems we never get a moment free enough to write letters. There are conferences and people to meet and problems to discuss and a great deal of just plain talk - all of which seems to be necessary to kid along one's companions.

It has been very hot here - an unusual state of the weather for Seattle which is usually so cool. It is foggy almost every morning but it clears off about 11 or 12 o'clock and the sun breaks through with its heat. The wind does not bring coolness with it, however. But then the next day the fog will be damp and chill and one will not be able to get warm all day. One day last week we insisted on having the steam heat turned on in the office all day long.

I suppose you are quite interested in the coming election. All my friends are. But it seems like toys which children ever find now. Remember the story of the lady who went back after many years to the toy shop in which as a child she had spent so many delightful hours. She said in surprise, "These are all the same toys. Have you no new ones?" The toymaker said, "No, the toys are always the same. It's the children who are new." This is a very significant saying for the meaning of life, I think.   Yours affectionately,   E.

College Club Seattle, Washington   (stationery)
Dear Mother:

How the days have flown since I have been here. I have been testing some new ships and have had some exciting experiences. It has been, "up early" and a lot of social engagements in the evening with often a lot of liquor which is not always polite to refuse. Yesterday I came back early and slept 15 1/2 hours straight through. Tomorrow I am going to San Francisco and then on to Los Angeles.

Seattle is a very interesting city with lakes and sounds and bays and canals until one sees masts of ships or funnels of steamers whichever way one looks. It is very hilly in between the waterways and beautiful concrete roads..........(missing page).......

Margaret, I presume, is with you by this time and is enjoying the sea breezes.

I would like to be with you in Florida and I hope this can be before long. I do not know where my summer will yet find me. It seems now that I may be sent back to 'Frisco then to New York, then back to Omaha, then to New York again, then to Wichita, Kansas, Salt Lake City, and back east.

My year in school has certainly helped me a great deal. I have gotten in touch with engineering subjects and the engineering point of view and I've enlarged my interests and horizon. Incidentally I have met some wonderful people and (not incidentally) some "friends".

I have been assured that I can get a degree next year and perhaps in half a year.

        Affectionately, E.

July 9, 1930
Dear Margaret,

I am sitting with Mother in Aunt Margaret's cool home, reading and talking. Mother seems better than before when I was here, and she is more interested in reading and talking. She does not walk much but she is alert and more at peace. Aunt Margaret is very wonderful in her kindness and solicitude and care of Mother and DR is his jolly old self again.

I am enclosing a check for one hundred fifty dollars in accordance with your suggestion of buying the bonds. I am facing the possibility of being out of employment for two months this summer but will be able to continue, I expect, as I said in my last letter.

I leave here for Daytona tomorrow but hope to be back occasionally during the summer.

    Your brother E.T.A.

February 3, 1931

Dear Aunt Margaret,

    I have not yet reached Burbank, California and do not know if I shall for another two weeks or more. I may yet have to go to Washington but I do not know if I will get through Hamilton this time.

    There are some letters in Mother's trunk or among her things which I do not want to come into the possession of my brother. They are from my ex-wife Allene, when she was out of her head, and they are very very nasty. I wanted to find them when I was in Hamilton but I did not like to ask at that time to go through Mother's things. They are part of the cause of all the strained relations in our family when Tom, Margaret, and Mother joined in with Allene against me. I am very anxious to get these letters and destroy them before anyone could lay hands on them who might use them again unpleasantly as they were used once before. They are in long envelopes, a whole series of them, either in Allene's hand writing or typewritten on her little Corona.

    Can you send them to me? You may already know where they are or that they are destroyed. I don't want anyone to see them again and especially not Tom or Margaret because of their historical connection with the matter. I am afraid that if they are not removed from Mother's things before Tom returns, he may find and read them.

    I am sorry to write to you like this because it is not right to trouble you about my troubles, past though they now are. And I would ask that you please destroy this letter. I want to get the whole business cleaned up so that no more harm can be done than has been done. I think all my letters to Mother should be destroyed if Mother has not already done so. A letter from Mr. Fittor tells me that mother's will is to be probated shortly.[37] I have written him that I know nothing of the affairs of the Florida property. Margaret knows all about that.

    I expect to be at RR#7 Baraboo, Wisconsin for a week or perhaps two, then better address - Northrop Aircraft Corp. Burbank.

    Love to you and Uncle DR.   Edmund

    P.S. Haven't got to LaCrosse yet because I have no car but I may make it yet in a bus.

---

[37] Dr. Abigail Irene (Dyer) Allen died on January 11, 1931, in Hamilton, Ohio

Most personal letters were no longer preserved after 1931 when Eddie's mother passed away. Some letters were preserved from the 1936 to 1942 time period and these have been included. I have included several stories/letters that reflect Eddie's personality. They are stories written by associates of Eddie or by Eddie himself with respect to his trip to Germany in 1935.

Northrop Alpha with Eddie Allen Pilot

NORTHROP
AIRCRAFT CORPORATION
UNITED AIRPORT
BURBANK, CALIFORNIA

January 21st, 1931

Mr. E. T. Allen
"Peekaboo", Wisconsin

Dear Eddie:

I find myself again owing you several letters and at the same time wondering if you ever received my letter of December 30th sent to you c/o the United office at New York at the time you were dashing madly back and forth between New York and Washington in our behalf. I am enclosing a copy of this letter for your information in case you never received it.

I had the Glendale Sanatorium send full data on their facilities to Dr. Klemperer, as I believe this institution is by far the best and perhaps the only one in this locality suitable for his needs. It is well equipped, has an excellent staff of physicians and, I think, would be satisfactory for his requirements as far as cost is concerned. There are a few other small institutions hereabouts, but the Glendale Sanatorium is the only one with any really worth while reputation as far as I know. They forwarded information by airmail, and, also, telegraphed Dr. Klemperer telling him that this had been done.

Our seaplane tests were practically complete a week ago with the exception of a final okay on longitudinal stability with three passengers all in the rear seat. In this condition, the ship lost its excellent qualities of longitudinal stability, and would climb or dive indefinitely without coming out of either attitude. The removal of one passenger from the rear to the front seat cured the trouble, but the Department demanded that it must be satisfactory with all the weight in the rear so we are having to try to correct the trouble.

I am putting on some new stabilizer tips which slightly increase the area and span of the stabilizer in the hope that the difficulty can be cured without any serious structural changes.

Water taxiing is now satisfactory except in a severe down wind condition and is a great deal better according to Department Inspectors than many seaplanes that have passed previous tests. Take-off time with full load is not as good as we might hope -- being about forty-five seconds at the present time, but I believe this might be helped a bit by some other propeller. Our present blades flutter badly during take-off, and as this loss in efficiency occurs at the time we need the thrust most, I believe it can be helped considerably. The above mentioned time is with full load, dead calm and unruffled water conditions, and is reduced to about thirty seconds with a slight chop and a breeze of five or six miles an hour.

Mr. E.T. Allen   -2-   1/31/71

I have checked carefully into the expected performance of the "Alpha" with the new gear and believe that motor for motor, we will have a faster airplane than any Lockheed model except the retractable geared Sirius. The difference even here will be slight. If the information you gave us is true regarding the Top Speed of the metal fuselage Sirius, we may be very nearly as fast with the pants gear, as this ship will be with complete retraction. I do not believe that two or three miles an hour top speed will ever be of much importance or much of an argument to over come. It seems as though the retractable gear would only be justified in a very long range or extremely high speed airplane. In comparing the high and low wing jobs, there is certainly a definite disadvantage from the efficiency standpoint in the low wing arrangement, but by utilizing the new gear which cannot be applied to a high wing airplane, we more than make up the difference.

The cabin heating and ventilation has worked out fine at last. We get an even and satisfactory application of plenty of hot air around the floor, and there is a comfortable breeze around the top of the cabin with the present system.

The stalling phenomenon is still with us like "the skeleton in the closet", but as it has not been nearly as apparent with the production of ships as it was in #2, we are not worrying an awful lot about it.

We are planning to continue our Cal-Tech wind tunnel tests for lift and drag as soon as their installation of balances is complete which will probably be within the next few weeks, and we may then be able to learn something interesting on filleting of the fuselage and wing so as possibly to eliminate a certain amount of this turbulence.

I hope you are having a nice vacation at "Peekaboo", and that the book is coming along fine.

Please give my best regards to Dr. Rahn.

                         Sincerely,

                         Jack

JKN KK

The preceding letter[38] was written to ETA by Jack Northrop and addresses two issues. The first is a response to a request from ETA about medical facilities sought by Dr. Wolfgang Klemperer. Klemperer probably met with ETA at the glider meet in Germany in 1922 where ETA recognized Klemperer as the "world's record holder". He worked for Hugo Junkers at his glider manufacturing plant in Aachen and participated as a pilot at the Wasser Kuppe. Dr. Klemperer emigrated to the U.S. in 1924 and became associated with the Goodyear-Zeppelin Corporation. Klemperer then became involved with the U.S. Army's Stratospheric Research Project and eventually joined the Douglas Aircraft Company in the development of pressurized cabin technology for civil aircraft. I suspect this request for information was related to the facilities for medical research at high altitudes and not any personal health issue. Klemperer remained with Douglas until his death in 1965.

The second subject brings ETA up to date on Northrop's development activity on a seaplane and the Alpha low wing monoplane. Note the issue of acceptable stability revolves around the location of a single passenger in the seaplane. This plus a 45 second take-off time "is not as good as we might hope". This might certainly be considered an understatement as this is definitely marginal stability and performance and would be avoided from the perspective of a passenger.

For the Alpha, ETA had flown it in March of 1930 and had obviously identified issues to be resolved. Northrop calls the "stalling phenomenon" the "skeleton in the closet". **And Northrop will in the future build the B2 stealth bomber - my how technology will improve.**

---

[38] Jack Northrop misidentified Baraboo, Wisconsin as "Peekabo", Wisconsin

Sikorsky S-40 Reminiscences with Eddie Allen
as told by William H. Cook (about 1940)[39]

Crossed Controls: Eddie was a free-lance test pilot who made many first flights on various make airplanes. He had established a reputation with the insurance companies such that they would give a lower rate for the first flight if Eddie was at the controls, or was the copilot in the right hand seat to assist the company test pilot. This may have been the case on the Sikorsky S-40, which first flew in 1931. It was a large four engined flying boat. The hull was located well below a large upper wing. There were a lot of struts connecting the upper wing with the passenger hull and the outboard stabilizing floats.

In the left hand seat was Mike Gluhareff, who came over from Russia with Sikorsky in 1919. While Mike and Eddie were floating out on the water after the launching of the S40 at the Sikorsky facility at Bridgeport, Conn., and just before take-off, Eddie made a preflight check of the aileron operation. From his side window he could look out and up and see the lower side of the aileron. After moving the control wheel from the left and right stops, he told Mike in his usual calm manner, "The ailerons are reversed". Mike looked out on his side and repeated the check. He replied, in his rough Russian accent, and in no uncertain terms, "the aileron control is all right!" Eddie repeated his check, and reaffirmed his previous statement. Mike did the same. After a couple of rounds of this, they looked at each other and observed that their control wheels rotated in opposite directions."

---

[39] [William H. Cook was Chief Aerodynamicist for Boeing with many years of experience with the Boeing Wind Tunnel]

Lockheed Constellation C-69

Lockheed Constellation First Flight as told to William H. Cook

"….When Eddie visited Lockheed to fly the Connie, on first trying the pilot's controls, he found that (he) could not move them. He asked where the control surface lock handle was located. He was told that there was no lock. The problem was friction. The cable pulleys had bronze bearings rather [than] ball bearings. The cable runs had many bends around small pulleys. Eddie got out of the cockpit and told Lockheed to fix it, and left for Seattle. How an experienced company could make such an error is hard to explain. It took a lot of time for Lockheed to straighten out the cable runs. When Eddie returned for the first flight, the control forces were better, but still too high. The Lockheed airplanes did not have leading edge aerodynamic balancing, as was quite common in the rest of the industry. Lockheed had to go to hydraulic boost controls. Thus Lockheed
became the pioneer in hydraulic servo controls. Eddie probably made a third visit to check the hydraulic controls out. "

Northrop Beta

"Once to Every Pilot" by Captain Frank Hawks

Frank Hawks was known as the "Speed Flying King" and entered contests in the 1920's and 1930's to beat the time of transit between different cities/ countries. He generally flew a Northrop Gamma under sponsorship of the Texaco Corporation.

"We are sitting around the table, Jack Northrop, Ken Jay, Eddie Allen, and myself. Yes, swapping yarns. None boasting or particularly trying to out do the other. Just gabbing, about flying and close calls.

Jack pipes up: "Eddie, how about the time you had to land out of a barrel roll when you were testing our Beta?"

Now Eddie is a reticent type of fellow, and not inclined to talk very much about his experiences unless it is to discuss some technical point concerning an airplane in which he is interested. I doubt if there is any finer test pilot in all the world than Eddie Allen. He is a real technician, a most excellent pilot, and he possesses a cool analytical mind that always overcomes every dangerous spot he has been in.

"Come on, Eddie, loosen up," I add.

"Well slowly and deliberately Eddie begins, "You know that little Beta was a nice little airplane. I was making tests on it for the Department of Commerce. We had gone through everything but a series of tailspin tests, which we were on when that incident which you mention Jack, happened."

We all push our chairs closer and lean forward. We had all seen the thing happen months before, but Eddie had never givens his personal story of the inside of some of his troubles battling with that ship in the skies. It looks as though he is going to talk now, and we are all attention."   Eddie Allen speaking:

"The Department of Commerce inspector wanted to cut the travel of the ailerons during the tests, so he had the mechanics install stoppers on both the ailerons of the Beta to prevent them from going beyond a certain set degree of travel.

When the ship was ready, I took her up. Not knowing just how these restricting lugs were going to act, I thought it best to take the ship up good and high, so I climbed up to 10,000 feet, and kicked it into a spin. Everything seemed to be all right. First I spun to the right and then to the left. After a series of these spins and the recording of the data I wanted to report, I still wasn't exactly satisfied with the results, so I regained the lost altitude, resolving to try some more experiments. I wanted to be a little more rough now with the control to see what would happen.

Reaching 10,000 feet again, I threw the little ship violently into a spin again, and in doing so I accidentally forced the stick so hard to the right that I passed by the aileron stop lugs, and there they were jammed tight! With all the strength I could muster I tried to get the ailerons back but now, since they had gone by the stoppers, they wouldn't come back to normal. It was like pushing a door shut with the catch locking it. The ailerons were locked and there was nothing I could do to remedy the situation.

I didn't want to jump because I wanted to save the ship. There was no doubt that I was in a terrible predicament, but I wanted to experiment. I still had plenty of altitude which insured safety for the moment.

Pushing open the throttle, I found that by giving full opposite rudder, all the opposite aileron I could squeeze, and by pulling up the nose, I could fly straight for a brief moment. Then the violent tugging of the locked ailerons would be more than I could muster strength to hold, and the ship would go into a barrel roll.

Well, I would then let the ship roll around the skies until my arms were rested, and then I would try to see how long I could fly straight. You see, I began to consider the possibility of being able to come down near the ground and perhaps hold the ship level long enough to be able to land safely.

I needed some more altitude, so I proceeded to climb in a series of barrel rolls. Around and around I rolled, and up, up, up I climbed until I had reached my original altitude again. Several times more I tried to hold the ship on a straight level course. There was no doubt that the duration of time my strength would hold out against these locked gyrations was very limited, and at such an altitude I couldn't judge whether I could attempt a landing or not.   I didn't dare come down near the ground to be making any such tests. One false move near the ground and it would be all over for me.

I spied some good looking cumulus clouds while rolling around up there. I picked out the one that most resembled a landing field, and resolved to practice some landings from these rolls on this nice flat cloud.

It may sound funny, all this about barrel rolling, to you fellows, but it was serious to me! I was nearly exhausted from the frequent endeavors to hold the ship level and straight.

Several times I thought I would have to jump and let her go, but then, as I said before, it was such a nice little ship and I wanted to save her if it was at all possible.

I maneuvered into position to land on the cloud by rolling over to it. I came in to land, going "round and round". The first time I cut my motor too soon and found that it had been aiding me a great deal to hold the ship level. I undershot the cloud terribly on the first attempt and fell into a spin before I even got to the edge of it.

Back up I went to try another one. The next time I did a little better by using my motor. Yes, I could land on the cloud out of a barrel roll, but it was obvious to me that the landing was being made over a hundred miles an hour, and I was now worried. I was afraid that I could never come down to the ground and make such a fast landing under those circumstances. I would have to practice more on the cloud and get my landing speed down to a lower figure. I made about eight or ten more of these attempts, and finally felt that I was ready to make the try for earth.

Even though ready and anxious to get out of this mess, I couldn't come down yet, because I had too much gasoline in the tanks. I wanted to land with the very minimum of fuel, so that if I crashed there would be less danger of explosion and fire. So I had to barrel roll around the heavens for another hour. Boys, I was certainly getting tired of these gyrations, but I just let her go, sat there and took it, relaxing as much as I could in order to have all the strength possible when I made the real landing.

As I came down toward the ground, rolling and rolling, I wondered if it were worthwhile or not. I was scared, and I sort of wanted to bail out. Then I got to thinking that after all, I was testing the job and it was up to me to do my best to get it back to earth safely. I did feel there was better than a fifty-fifty chance of doing it all right, but there could be no practicing on this real landing. It had to be made on the first trial, for better or for worse! I maneuvered just as I had done so many times for the cloud, and came rolling in. At a hundred feet, the big test had to come, and I straightened her out, holding that stick over to the left with both hands and my knees to keep level until the Beta set down.

It seemed that I would never get near enough to that ground and my strength was about all gone. I would have to let loose in a minute. Then I cut the motor and gave one final tug to hold her a wee bit longer. Fellows, I was exhausted. As I was beginning to let go and as she began her roll, she set down. naturally the ship hit first on one wheel and then the other, oscillating back and forth, and in addition bouncing up and down. For a minute I thought I would go over on my back. I cut the switch and braced myself. But as you all know, she didn't. With the exception of the landing being vey fast and bouncy and taking up all of the field, everything was all right"

"We are all silent. Maybe Eddie will talk more; but we are disappointed . He crawls back into his shell, and while we are discussing other matters, you can see in those bright eyes of his that perhaps he is sorry he has given out so much. I cannot say that this was his greatest thrill or worst experience, but I am inclined to believe it must have made a formidable impression upon him, or he would never have ventured to narrate so much.

Lockheed Electra

"Kelly   - More Than My Share of it All "
(Clarence L. "Kelly Johnson   1910-1990)

Kelly Johnson became Lockheed's foremost aeronautical engineer and the instigator of the design team that would become the "Skunk Works", the premier design group of high performance aircraft in the U.S. Kelly was the 24 year old flight engineer on this test flight of the Lockheed Electra in 1934.

"When the time came for the Electra's first flight, Gross hired Edmund T. Allen, probably the best and most experienced test pilot of commercial aircraft at that time. The Lockheed pilots had no twin-engine experience. Allen alone flew the plane for the first time on February 23, 1934.".......

"In Eddie Allen, I had an excellent teacher. After the first flight , I flew with him as flight test engineer through the entire initial flight test regime - dive tests, stalls, spins, everything. It was an excellent indoctrination to the art, skill, science, adventure - all that goes into flight testing. He taught me what it was all about, what was important, what to record. And he was unflappable."

"On one occasion we had to boost the Electra to its design dive speed, about 320 miles per hour, to prove that the airplane was free of flutter and control problems. We had the airplane loaded with lead bars to simulate the full gross weight. We took off from the old runway behind the factory in Burbank and climbed to about 12,000 feet. Then Eddie pointed the airplane down in a steep, screaming power-on dive. At 6000 feet when I was expecting he would start to pull out, there was a horrendous bang, and everything was flying around the cockpit. I looked over at Eddie to see what he would do, if we were going to try to jump or not. He was holding the stick with one hand, pulling back on it to bring us out of the dive, and with the other brushing insulating material from his face.

"Got something in my eye" he said matter-of-factly. The windshield on the pilot's side of the cockpit had blown in, pulling some insulation with it and covering Allen's face. Obviously, we redesigned the windshield."

As a result of his experience with Allen, Kelly Johnson wrote: "I have a philosophy that those who design aircraft also should fly them - to keep a proper perspective. The engineer knows where the quarter-inch bolts may be marginal, what the flaps are likely to do or not do. I've shared the concerns of the pilot. I figured I needed to have hell scared out of me once a year in order to keep a proper balance and viewpoint on designing new aircraft."

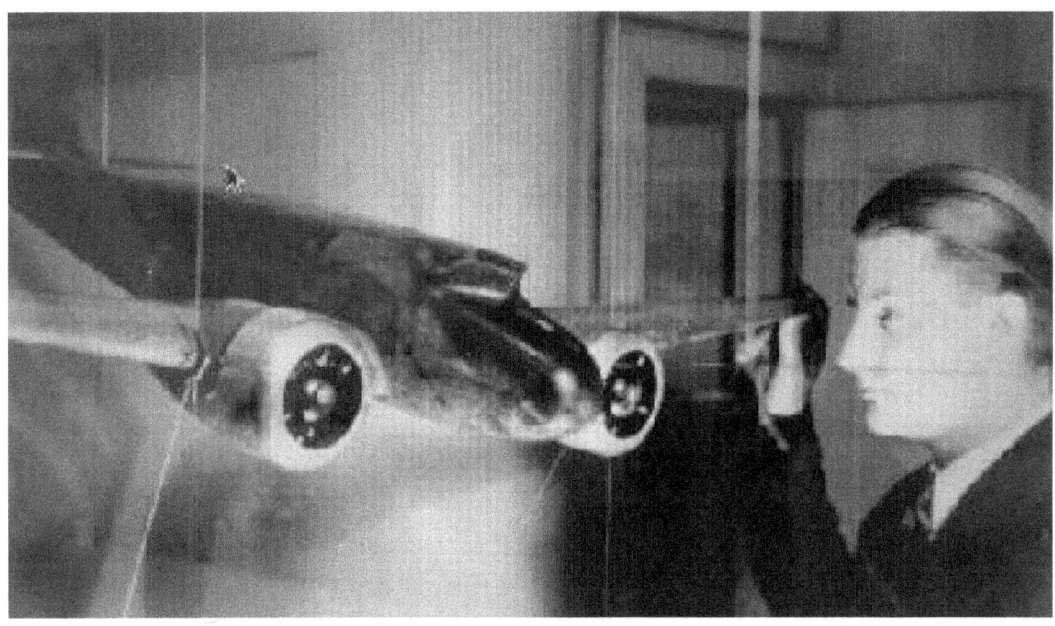

Kelly Johnson and the Lockheed Electra

Douglas Aircraft DC-1, DC-2, and DC-3

In 1931 Dr. Robert Millikan of the California Institute of Technology contacted Donald W. Douglas about providing a weekly course on the practicalities of modern aircraft design. Donald Douglas was an experienced designer himself, educated at MIT. The mutual benefit of this activity was apparent and thus Douglas's chief engineer, Arthur Raymond, would travel over to Cal Tech each week to instruct graduate students. In his lectures he proposed a problem in practical flying that addressed the many flight variables that the pilot must consider. Mr. Raymond was convinced the typical aviator was not knowledgeable of

the optimization of the performance qualities of their aircraft in relation to engine operation, flight profiles, cruise altitude, achievable range, fuel usage, etc. He believed some systematic set of charts could be developed which would enable more efficient flight procedures.

It happened that a young Cal Tech grad student by name of W. Bailey Oswald had just completed 4 years of graduate study and held a fellowship in wind tunnel research. Young Dr. Oswald took the challenge and in a year and a half he developed a system of analytical performance charts, NACA Report 408.

During this same time, Northrop Aircraft Corp., then a subsidiary of Douglas Aircraft, began wind tunnel tests at the Cal Tech tunnel. "Northrop had a pilot by the name of Edmund T. Allen, small, wiry, and modest to the point of shyness, but master nevertheless of breath-taking aeronautical yarns" (Fortune Magazine May 1935). He had discovered during his experience flying the air mail and following test flights, that the optimum aircraft cruise parameters did not follow currently recommended flight procedures. Instead of backing off on throttles as altitude increased, flight speed and fuel efficiency increased by moving the throttles ahead and by increasing the rpm with increasing altitude.

During the Northrop tests, Oswald and Allen discovered each other and became brothers in solving the challenge.

In early 1932, Donald W. Douglas called Eddie Allen into his office in Santa Monica to discuss the design competition for a new passenger airplane for TWA. Eddie had indirectly done flight testing for Douglas through its subsidiary Northrop. Allen had just written an article "Truth and the Test Pilot" published in Aviation magazine in February 1932. Douglas knew well the technical and financial risks associated with a new aircraft and he was anxious to avoid the extensive redesign activity that was nearly always necessary following the first flight tests. The work of Allen and Oswald had been introduced to Douglas by Arthur Raymond. Donald Douglas then hired Oswald as a staff aerodynamics engineer. By the time the DC-1 was ready to fly, Allen and Oswald had sold Douglas on the idea of extended flight tests to develop the performance charts to be provided with each airplane. Douglas had only one DC-1 and the fact that he would turn this airplane over to these two enthusiasts for so much time is really amazing. After flying more than 116 flights and burning more than 9000 gallons in testing the DC-1 in every performance condition, these two flight engineers presented Douglas and the aviation industry findings that totally revolutionized the way commercial aircraft should be operated. For the DC-1 it resulted in an increase in cruise speed of 15 to 20 miles per hour at a much reduced rate of fuel consumption. These charts became the foundation operating procedures for the DC-2 and the DC-3 and in fact became the template for operating procedures for all commercial propeller-driven aircraft.

Clark B. Millikan of Cal Tech (son of Robert Millikan) remarks in his textbook Aerodynamics of the Airplane (John Wiley & Sons, 1941).pp. 131-134. "In 1934 E.T. Allen and W.B Oswald presented a series of fundamental papers on the subject of controlled operation of airplanes which effectively revolutionized certain aspects of airplane operations."

As a direct result of the efforts of ETA and W. Bailey Oswald, the DC-1 would break five recognized world performance records in 1935 and two additional new records. The DC-1 team was recognized with the Collier Trophy presented to Douglas by President Roosevelt on July 1, 1936, the third anniversary of the first flight of the DC-1.

Wheels Up Landing with the DC-1 !!

Even the best can have one of "those days". Allen's flight log dutifully record this test as "Stability - landed gear up" on July 20, 1933. As described in the Journal of American Aviation Historical Society, Summer, 1983:

"Test pilot Eddie Allen (later with Boeing), with Tomlinson as copilot, flight test engineer Frank R. Collbohm and "Doc" Oswald on board, flew over to Mines Field (now LAX) in Inglewood to shoot some landing tests. Oswald would assist with calibrations and handle the gear pump. The main cabin door was removed so that Collbohm lying prone on the floor with head hanging out of the doorway could observe the gear during three-point landings as to which touched first, the main wheels or the tailwheel. Well, the tailwheel hit first and Collbohm got the surprise of his life when pieces of turf began peppering his face as the props began chewing up the sod and the ship caromed down the runway wheels up!

During previous flights, Collbohm had operated the gear hand pump without having to be told each time by the pilots. In this instance, Oswald, while at the pump handle, had not received instructions to pump the gear down. A case of misunderstood signals. And definitely a system redesign was in order for the DC-2's.

The wings were demounted and the airplane trucked back to the assembly hanger where minimal repairs were made, two position props were replaced with the fixed pitch (FP) and tail modifications made. By mid-August the airplane was flying again."

Douglas DC-2

Officially the first flight of the DC-2 was on May 11, 1934, flown by Douglas test pilot Carl Cover. However, according to ETA's log book, Eddie flew the DC-2 on May 7 for two short flights totaling 1:15 hours/minutes checking stability and control. Since ETA had completed a DC-1 flight test program involving 123 hours of testing and had also recently conducted the first flight of the Lockheed Electra in February 1934, it is likely Donald

Douglas asked ETA for an initial check flight prior to the "official" first flight by Carl Cover on May 11, 1934.

On August 29, 1934, ETA initiated flights in support of Pan American-Grace Airways (known more often as Panagra Air Lines)[40]. After the final aircraft checkout and test of several DC-2s consigned to the airline, Allen left the Douglas factory on a delivery flight to South America along with a Panagra crew on-board. Allen then spent three months training Panagra crews on flight operations, nighttime flying and radio procedures, apparently over all the airline routes. He did not return to the U.S. until late December, 1934.

On December 10, 1934 , ETA flew a Panagra DC-2 over the Andes setting a new world's altitude record of 29,800 feet for standard commercial passenger planes. Onboard was Professor Sergea Korff of the California Institute of Technology who conducted "cosmic ray studies" in support of Professor Robert A Millikan's theories on the subject. The flight took 2 hours and 5 minutes according to the Associated Press.

In February to May 1935, ETA conducted similar consulting/training operations with Eastern Airlines for their DC-2 fleet.

In November and December 1937, ETA would nearly repeat the training operations with the DC-3 in South America.

By 1939, the DC-2 and its follow-on the DC-3 would provide 90 per cent of the world's passenger traffic and eventually the DC-3 and its military variant the C-47 would be produced in greater numbers than any passenger aircraft in history.

Eddie Allen   Age 38

---

[40] Pan American and Grace Shipping joined in a joint venture to form Panagra Air Lines in 1929.

Invitation to German Air Ministry (1935)

Eddie was invited by the new German Air Ministry (Deutsche Versuchsanstalt fur Luftfahrt - DVL) in October 1935 to present a paper on transport cruise control. At this time there was very little visibility of the German aviation industry. Germany had repudiated the Treaty of Versailles and both England and France had looked the other way. There was a recognition by some that the terms of the treaty had been unreasonably severe. In fact, England had broken the treaty by negotiating a separate treaty with Germany allowing the German Navy to grow to a level of 35% of the British Navy. German aircraft manufacturers had moved operations to Sweden and Switzerland to work around the treaty constraints. When Hitler repudiated the treaty the firms quickly returned to Germany. German commercial aircraft were flying in South America, Switzerland and Italy. There was no outrage expressed over the German actions. Still the details of the German economic revitalization were not known. Incidents and rumors of anti-semitism, commonly discussed in the western press were denied by the German administration.

Junkers Ju-52

Thus, when E.T. Allen was invited to speak at the new ministry facilities, it was a big deal. He was offered tours of German aviation research and manufacturing facilities and an invitation to fly the Junkers Ju-52 on his trip from Staaken field (outside Berlin) to the Heinkel factory at Warnemunde. The suggestion that Eddie "might like to fly a really good airplane" turned out to be true. While the Ju-52 lacked the comfort and accommodations of a DC-2, Lockheed Electra, or a Boeing 247, it was "superior in stability and maneuverability" according to ETA.

Eddie had previously flown the Ju-52 at Luft Hansa's (the German commercial airline) invitation and took it through many flight test maneuvers, so he knew it was technically an excellent airplane. ETA flew the Ju-52 testing flying characteristics and the operation of the radio landing beam on his return to Berlin.

When ETA was offered a tour of a aircraft facility, he chose the Heinkel factory for its reputation for innovation and technology ("and fast airplanes") and it was the most "secretive" of the German facilities. His host included Dr. Heinkel himself along with key

German aviation engineers. Also in attendance was Lockheed's Chief Engineer, Hall Hibbard who was also presenting a paper.

The invitation was certainly a result of relationships first initiated when he participated in glider events in Germany in 1922 as a member of the MIT glider team. As Eddie noted "all the members of the brilliant 1922 class of Hanover in aeronautical engineering are now famous in German aviation." His trips in 1924 and 1929 and his reputation throughout the industry as a technically oriented test pilot did not go unrecognized. Also the articles published in 1934-35 in Aviation Magazine on Cruise Control were widely recognized.

Heinkel He-70

ETA had the opportunity to closely inspect the He-70 and noted the flush riveting techniques for drag reduction and the elliptical wing planform. This aircraft was first flown in 1932 and was identified as a fast passenger and mail carrier. It was the forerunner of the He-111 bomber used extensively in WW II.

Eddie reconnected with Dr. Georg Hans Madelung, the designer of the "Hanover Vampyr" glider, one of the competitors from the 1922 glider competitions. The Vampyr glider has been claimed to be the model for all modern sailplanes. Its use of the high aspect ratio wing (approximately 10) to achieve superior glide ratios may have been the genesis of the high aspect ratio wings of ETA's personal light aircraft designs as well as ultimately recognized in the design of the B-29 (aspect ratio 11.5). Madelung received his doctorate with the class of 1922 at Hanover University. Eddie visited Madelung's home and commented "Palace in moonlight on hill. Prosperity in evidence". Madelung was one of Germany's leading aeronautical engineers and became the Director of the German Aviation Research Institute in 1937. He had previously worked in the U.S. in 1921 to 1924 and again after the war in 1946-1954.

It is interesting that Eddie initiated the trip to Stuttgart and the phone call to Madelung. The visit with Madelung did not appear to be orchestrated by the DVL. Eddie and Madelung likely had interactions in previous years because of the common interest in gliders and participation at the 1922 Wasser-Kuppe event. Allen also likely visited Madelung in 1929 when he visited the "old glider camp at Fulda".

Hans Martin Antz likely met Allen at Cal Tech in 1934-1935 when he earned a Masters Degree in 1935 before returning to Germany. Hans Martin Antz co-authored a Journal of Aeronautical Sciences paper in September of 1935 at Cal Tech as his masters thesis. Antz became a key engineer in the jet engine airframe design team working with Dr. Alexander Lippisch and Dr. Willi Messerschmitt on the Me 163 Comet and Me 262. It may be that Madelung was with Eddie and Hans Antz the final evening before Eddie left Germany. Both Madelung and Antz were fluent in English and were from the technical world, as contrasted with Nazi politics. Thus, ETA's final evening before his return home may have been a pleasant dinner with technical compatriots.

Georg Hans Madelung and the Hanover Vampyr Glider 1921

Three of the individuals meeting with Allen during his trip had colorful infamous careers later with the Luftwaffe during WW II. Erhard Milch, Ernst Udet, and Wilhelm Wimmer all became high level Luftwaffe generals.

Erhard Milch had been the first managing director of the Deutsch Luft Hansa airline in 1926 following duties in WW I. Milch was never a pilot but he did command a fighter wing during WW I. Milch was active in civilian aviation prior to Hitler's rise to power. He was a favorite of Hermann Goering when the Luftwaffe was being formed in 1933, but the Gestapo refused his admittance to Goering's Luftwaffe because Milch's father was reported to be Jewish. Milch had to have his mother sign an affidavit stating that he and his six siblings had actually been fathered by her deceased non-Jewish uncle. Only then was he allowed to join the Luftwaffe. Even so Goering made the statement that "only he would decide who was Jewish". Working for Goering, Milch was responsible for all aircraft production. He used his position to settle personal issues with key aviation figures such as Messerschmitt and Junkers creating animosities which would hinder his career in the Luftwaffe. He failed to match the aircraft production rates of the Allies and his ineffective management was blamed for the decline of the Luftwaffe later in WW II. Milch tried to convince Hitler to remove Goering as leader of the Luftwaffe in 1944 after the failed invasion of Russia. Hitler refused, supporting Goering who then had Milch demoted to a position working for Albert Speer. Speer successfully implemented reforms that increased the German aircraft production rates but by then it was too late. Following Hitler's suicide, Milch was captured by the British. He was tried as a war criminal at Nuremberg and because of his use of slave labor was convicted of war crimes and crimes against humanity. He was sentenced to life imprisonment. His sentence was commuted to 15 years imprisonment in 1951, but he was released in June 1954. He lived out the remainder of his life in Dusseldorf, where he died in 1972.

Ernst Udet flew in WW I as a member of the Manfred von Richthofen's "Flying Circus" and with Richthofen's death, Udet became Germany's leading surviving WW I "ace" with 62 kills. He flew "barnstormer and stunt" shows during the 1920's. Udet was recruited for the Luftwaffe by Hermann Goering after Hitler came to power in 1933. Following Luftwaffe defeats over Britain and Russia, his mentor Hermann Goering tried to deflect Hitler's ire by blaming Udet for the failures. He would be judged responsible for Hitler's failed air offenses. Udet would commit suicide in 1941, distraught over a girlfriend, alcoholism, problems with the Nazi Party, and his rejection by Goering.

Dr. Wilhelm Wimmer was vice president of the DVL at the time of ETA's visit in 1935. He would rise to Luftwaffe commanding general positions in the forthcoming war responsible for eastern Prussia, then Brussels and northern France. He was captured and released in 1947. He lived his remaining years in Garmisch Partenkirchen and died at the age of 83 in 1973.

Noticeably absent was Wolfgang von Gronau who witnessed ETA's glider crash in 1922 and became famous for his "round-the-world' seaplane flights of 1932. Wolfgang von Gronau seemed to avoid association with the Nazi organization and uniform, yet he was an internationally recognized figure in aviation. Prior to initiation of WW II, he was assigned as air attache to the German Embassy in Japan and lived as a diplomat in Japan until the end of the war. He returned to Bavaria where he lived until his death in 1977.

In 1935 the Nazi Reich was nearing its peak and the negative features of the Reich were kept hidden or rationalized away. Germany was viewed by many as an economic miracle and certainly could be viewed as such compared to the depression plagued economies of the U.S. and other western democracies in the mid 1930's. The Berlin Olympic Games of 1936 were meant to showcase the German accomplishments.

Eddie wrote a summary of his visit with the German aviation industry which was published in the Aviation Magazine in three monthly articles December 1935 thru February 1936. The magazine eventually became the Aviation Week and Space Technology magazine which is published today.

In his summary Eddie noted: "Design and production of aircraft in Germany are practically on a wartime basis. Methods differ radically from those with which we are familiar." — — —ETA

"One of the most interesting features of the flight test branch of the DVL is the school for test pilots where an attempt is being made to bridge the gap between theory and practice —— between the highly skilled test pilot with his invaluable background of instinctive judgement, and the engineer who is thoroughly conversant with theoretical terminology and the manipulation of the differential equations of flight. Here "diploma engineers" and doctors of engineering are inducted into the higher mysteries of the falling leaf, fishtail landings, and the inverted spin." ———ETA

"If a foreigner is offered an opportunity to enter the secret domain of a present-day German aircraft factory and has to name his choice in a hurry he is apt to be torn between Junkers and Heinkel. Heinkel attracts him because of its high speed airplanes and seaplanes (or because it is well known that the Heinkel factory gates are the most difficult for a foreigner to "crash"; Junkers because it is a very great, conservative aircraft factory where we will see 15,000 employees at work, but few radical designs.
My choice was Heinkel". ———ETA

——"Germany is laboring to mend her fences. While much of the world is occupied in other quarters, she is snatching the opportunity to rebuild her "defenses". Ask any German whether the intention of Hitler's government is to force an aggressive war and the answer is invariably "no". But he realizes at the same time that Germany's own security against attack lies in the potential striking power of her own bombers." ———ETA

Allen also wrote a summary on his return from Germany which was not published in Aviation Magazine but was included in his notes which were stored in our family's attic for years. ___TKO

"GENERAL CONDITIONS. Economic recovery has reached very considerable levels as far as the working man is concerned. No idleness, no depression is visible, even in small towns and farms. Wages are very low but people are not discontented. They are emotionally keyed up to the new freedom. The freedom from Versailles is the keynote of the working man as well as the trader and professional man.

Everyone is imbued with the propagandized ideal of duty to one's people rather than "rights". No one has any "rights". The young do not want them; they want only an ideal to which they can give themselves. This is evident in the military chain step of the mechanics who sang (very good harmony by the way), lustily as they marched to their machines. It is not done unwillingly, but enthusiastically. Hitler is to them their savior, a man of keen social insight, a great thinker, a leader out of their misery. Not a trace of the American Press picture of Hitler is discoverable, even under the surface. He is to them dignified, frank, idealistic, a religious man of God. They see him with eyes of devotion. For the thinking, analyzing German, Hitler's principal achievement has been unifying the stormy factions in Germany. He has done one thing never accomplished before, made Bavarians forget everything except "we are Germans".

There are no more Prussians, only Germans; no Saxons, no sectionalists. One people, who must realize their unity and forget all else, can work as a unit, sacrifice as a unit, and _feel_ as a unit. It used to be a saying of the East Prussian Junkers: "German? What is German? We are Prussians." Now this saying would be impossible. The unifying of the sectionalists has led to a defining of the term "German" and has led toward the anti-semitism which puts Jews outside German feeling. Jews are International and therefore "anti-German" for them. Jews are responsible, so they say, for the revolution of 1918, for the economic ruin of 1922-1923 in which Jews cleaned up by buying foreign currency. This was, in the eyes of present-day Germans, "Treason". Now all Jews are regarded as potential traitors.

They cannot think in terms of German racial and judicial instincts. They appeal to Jews <u>outside of Germany</u> and therefore they are anti-German. When these analytical people were asked if this justified pogroms, they said there are no pogroms and have been none except in the minds of the foreign press. I saw Jews taking part in festivals and frequently on streets unmolested.

The second thing Germans praise Hitler for is unifying the fighting religions. Germans had their first loyalty to Protestantism of various shades, or to Catholicism, and a secondary loyalty to government. Now they realized that country comes first and that anyone who tries to come between them and country (or "The German People") is traitorous. It is an intense nationalism.
The desire for freedom is intense and the only bar to freedom seems to them external pressure. The Versailles Treaty, the war guilt bogey." – – –ETA

"The new Air Ministry building in the center of Berlin is nearly ready for occupancy. It comprises an entire city block, a very long one, one quarter mile long. It houses all the central offices of the air arm, land, sea and civil aircraft activities. There is a movement for

unification going on everywhere. They want to unite all similar agencies in order to avoid duplication. The Union for Aircraft

Progress and Research is such a unification. It would be as if our NACA and Wright Field and Anacostia and the Bureau of Air Commerce all got together and eliminated all duplicating departments and duplicating work. All aeronautical universities are also under the control of this group." ----ETA

The Reich Air Ministry received comparatively little damage during the war even though it was the largest building in Europe at the time of its construction in 1935. The maximum damage was sustained in February 1944 when more than 1000 B-17 Flying Fortresses attacked Berlin and the building received 8 direct hits. Goering reportedly had one million bottles of France's best champagne stored in the cellars after the German's captured Paris. Receptions at the Air Ministry were "something like a champagne bath" according to the United Press. Today the building houses the German Finance Ministry — TKO

"ESTIMATES OF AIRCRAFT PRODUCTION: There are approximately fifteen aircraft companies in Germany. Heinkel as a typical example has 600 engineers. They are producing six types on experimental order of 12 each. Then the designs are given to build wings or fuselages or landing gear, etc. requiring that the plant demonstrate its ability to build the entire airplane at a moments notice."—ETA

German Wind Tunnel Technology: ETA was given tours of a variety of wind tunnels and engine test technology. He toured facilities in Berlin and Stuttgart.

"The university aerodynamic laboratories at Gottingen, Hanover, Aachen, Stuttgart, and Berlin are being used effectively in the new program. Envisaging the signifiant value to student designers of contact with practical problems during their university work, the VLF steering committee[1] has designated each university laboratory as a testing station for some particular set of problems. Gottingen with its magnificent wind tunnels and aerodynamic facilities, was chosen to test Junkers designs. Some twenty-odd Junkers project engineers are there all the time to oversee the tests. Probably Dr. Prandtl looks sadly on the interruption of his pure research, although he may find some compensation in having been made Committee Chairman on

---

[41] VLF   Vereinigung fur Luftfahrtforschung - a super committee in control of air development coordinates all related activities toward a common goal.

aerodynamics for the VLF. Aachen University aerodynamics laboratory has been turned over to Heinkel tests. Wind tunnel models are arriving there from Rostock, I am told, at the rate of one a day. The big DVL tunnels at Berlin-Adlershof are rarely used for design-testing and model developments as this is regarded as applied rather than basic research. They are assigned the more fundamental problems...... The automatic balances are ingeniously mounted on a central spindle directly over the throat. They indicate forces and moments in coefficient form, and, at the pressing of a button, they automatically record and plot their value on a chart fed through the recorder. When it is remembered how many months are often required to do a test and work up the results and computations by old-fashioned methods, the great value of this type of recording in speeding up all testing is apparent."

Even Churchill was undecided about Hitler at this time as he wrote: — "We cannot tell whether Hitler will be the man who will once again let loose upon the world another war in which civilization will irretrievably succumb, or whether he will go down in history as the man who restored honour and peace of mind to the great Germanic nation and brought them back serene, helpful and strong, to the European family circle. It is on this mystery of the future that history will pronounce Hitler either a monster or a hero."— —Churchill - November 1935 article in Strand Magazine.

Eddie Allen was impressed by the enthusiasm of the German people and their commitment to what appeared to be a dynamic nation on the rise. He contrasted their unified enthusiasm with the seemingly "constant bickering" and conflicting, divisive nature of the American political system and its difficulty in uniting on common goals. Also, the friends he had made in the German aviation community and the shared "brotherhood" of the German engineers and aviators made it difficult for him to see the negative potential of Hitler and National Socialism. He missed the fact that the "bickering" and divisive nature of the American system is the strength which ensures that a system such as National Socialism would never survive.

The Berlin Olympics are yet to be held in the summer of 1936. The United States along with the world's nations will participate. Unfortunately, Eddie Allen's observations in October 1935 of the wartime design and production activity in the aviation sector and the nationalistic and anti-semitic attitudes of the general population, will foretell Churchill's <u>worse alternative.</u>

Testing / Demonstration in the South American Market 1936

Sikorsky S-43

In early 1936, Eddie Allen spent most of his time on the east coast of the U.S. testing aircraft for Chance Vought, Sikorsky, and North American. His flight logs show that Eddie tested the Vought XSB2U-1 for most of May 1936 at the Navy's ordnance test range in Dahlgren, Virginia. Following this he connected with a Sikorsky S-43 amphibian at the factory in Bridgeport, Connecticut, and flew down to Miami. He then hop-scotched his way down to Lima, Peru.

A letter to Richard Aldrich in June of 1936, briefly summarizes his plans and offers financial support for "Dick's" plan to continue his education at Ann Arbor. Also ETA is in close contact with Carl Rahn at that time.

For the months of June and July, he demonstrated the Sikorsky S-43 throughout South America, including the cities of Cristobal, Medellin, Cali, Quito, Guayaquil, Lima, Arequipa, Arica, and Santiago - all on the western side of the Andes. Some DC-2 flights were done for the Panagra Airlines and also included several Ford tri-motor
flights. Then, in August, he flew over the Andes in a Panagra DC-2 from Santiago to Buenos Aires.

The next five months were spent sales-demonstrating the Vought V-143 fighter and the Northrop 5B attack plane for the Argentine Air Force. No sales were scored for either. His flight log shows he did not leave Argentina until 4 December 1936, and that was aboard a Panagra DC-2. He flew Buenos Aires to Santiago via Cordoba, returning to the U.S. via Panama, Guatemala, and Mexico to Los Angeles where he arrived on December 21, 1936.

Vought V141/14

Northrop 5B

320 E 42nd St   NY City
June - - - 1936

Dear Dick[42]

What have you decided? Are you going to Ann Arbor? Will you need any more money to start out? Or have you decided to stay in Tucson for the summer with the studio work?

Again the Boeing job has been postponed so that now I am going to be able I think to complete my Vought jobs before going west. Probably middle July will see me in Seattle - up to mid August. Then until mid-Sept I probably stay west possibly Encinitas, then Peru for a month and then back to NY. Carl and I were glad to hear of your very pleasant collaborative work with your friend there - The list of projects you have completed sounds impressive.

Eddie

---

[42] Richard Aldrich was a lifelong friend ETA met at the University of Illinois. ETA financially supported Aldrich's education at the University of Arizona and at the University of Michigan. In his will ETA left Aldrich his home in Encinitas, California.

Eddie Allen and Vought Aircraft

Eddie had a unique flight test relationship with Vought Aircraft. While Allen had some recognized design consultant input to the Northrop aircraft and the Douglas DC-1 and DC-2, Vought was the first manufacturer to agree to his request that the test pilot should be involved from the start of the design activity. The following is from Tom Collison's draft biography. Tom was a close friend with Eddie from 1925 until his death.

"When, in 1936, Eugene E. Wilson, then senior vice president of United Aircraft Corporation (now vice-chairman) called Eddie to discuss the SB2U-1, a low-wing monoplane dive-bomber, which Vought planned to build, he realized Vought was moving into an extremely difficult design field. Wilson was anticipating, with no great pleasure, the flight test problem; Eddie was the one man Wilson believed could meet Vought's requirements.

"Eddie said he was entirely willing to undertake the job," Wilson reported later, "which involved terminal velocity dives and unexplored speed ranges with high acceleration pull-outs under heavy bomb loads, but stated he would do so provided he could go on with the design from its inception. This was a new proposal, since the average test pilot came to your factory when the airplane was ready, and after a brief familiarization, went out to see whether it would fall apart or not. The soundness of Allen's proposal appealed to me tremendously, particularly since I had great confidence in Eddie Allen's engineering capacity. The combination of an extraordinary skilled pilot and a highly trained engineer was unusual, but beyond this, Eddie had the ability to cooperate in an engineering organization to a rare degree".

"Eddie was in the design from the beginning, contributing a great deal to the design itself and, therefore, was entitled to a great deal of the credit for the outstanding success of the ship. In the course of the comprehensive flight testing, he contributed still further giving the airplane unusually good flying qualities. One of the problems which went along with the trend in design was the characteristics of the airplane at the stall. We at Vought, as others, ran into this in the SB2U-1. Eddie in his consulting capacity, helped us to solve the problem quickly and so successfully that the NACO gave the airplane a clean bill of health. Service experience with the type later confirmed the NACO result."

"The effectiveness of a test pilot is largely measured by his experience with many types of airplanes, since many of the factors involved in flying qualities are relative. Few of them have as yet been reduced to a quantitative basis. In some measure, Eddie's value lay in his wide experiences."

"Now the relationship of a consulting test pilot to his numerous clients is a delicate one. Eddie Allen's natural honesty and integrity were such that compelling manufacturers had complete confidence in him. As an example, he demonstrated to the Air Corps, both the

Curtiss and Vought Pursuit types. Eddie justifies Vought's confidence that he would safeguard our secrets and our faults. He never told us anything about a competitor's airplane which we would not have been glad to have had him tell a competitor about ours. Eddie was in a unique position....he contributed outstandingly to the success of Vought and Sikorsky airplanes, not only from point of view of safe, economical and effective test flying, but also from viewpoint of fundamental design."

          Eugene E. Wilson Vice Chairman United Aircraft Corporation

In the Spring of 1938, Eddie summed up the significance of his aspirations by saying his objective had been not merely to test airplanes after they had been completed, and find out then what was wrong with them, but rather to discover the defects in flying qualities during initial design stages before the design had reached a point where changes were extremely costly. Emphasizing that his work had contributed to the improvement of stalling characteristics, the correction of bad spinning qualities and the elimination of flutter as these elements were met with and conquered in the air, Eddie said that much effort received neither public recognition nor recognition within the aviation industry, partly because much of the work was confidential Air Corps and Navy business, and partly because the mistakes of his clients had to be kept secret. He was compelled, he said, to comb his articles to prevent illustrations from being recognized. His clients would not tell what he had done for them even after many years. Each design group in the industry wanted to forget its mistakes, especially if those mistakes very nearly cost the life of the test pilot. Very few designers, he said, were proof against this near-Freudian complex, and, as a result, the consulting engineer who discovered the trouble and remedied it, was either forgotten or he was regarded as a particularly trouble-some person, not as the savior of the airplane."

Edmund T. Allen and the Boeing Company

B-17E Flying Fortress

In 1937 and 1938, Eddie Allen continued work for Boeing as a consultant, flight testing the XB-15, the Boeing 314 Clipper, the Boeing Stearman X-100, and the Boeing 307 Stratoliner. He completed the first flight of the Stratoliner on December 31,1938. In the following twenty-one days, ETA flew the B 307 in almost 20 hours of flight test, demonstrating the basic performance, stability, and control of the aircraft. These tests included 3-engine climb, 2 engine ceiling, stalls, dives to 305 mph indicated as well as other stability and control and landing extremes. The final flight on January 20 was a checkout flight for Boeing test pilot Julius Barr.

On March 18, 1939, the Stratoliner crashed during flight demonstrations with key airline personnel, killing all aboard. Victims included KLM and TWA personnel, Boeing's chief test pilot Julius Barr, chief engineer Earl Ferguson, chief aerodynamicist Ralph Cram, and critical flight test personnel; a total of 10 people. The crash of the Boeing 307 had been preceded by the crash of the B-17 prototype in October of 1935, also with key people onboard.

Boeing management had had enough. Both crashes had wiped out critical technical personnel and both crashes had been due to a failure to maintain technical flight disciplines during the demonstration flights. The B-17 prototype crashed because the elevator locks had not been released prior to takeoff. The Boeing 307 apparently crashed because the pilot had been allowed to perform a flight maneuver well beyond the design capability of the aircraft in which both engines on one side were simultaneously shutdown while the opposite side engines were at max performance. The rudder moved hard over to compensate the resulting yaw. The aircraft rudder became locked, and the resulting catastrophic maneuvers tore the aircraft apart.

Eddie Allen was viewed as the man who could restore the technical discipline that was lacking. Claire L. Egvedt, now Boeing president, brought Eddie Allen on as the Director of Flight Test and Aerodynamics in April of 1939. He was given nearly autonomous power of operation and a report chain directly to top management. His team was responsible for the flight test development of all the B-17 modifications to improve the aircraft system effectiveness and survivability. This included the B-17B, C, E, F and G. Aircraft weight was increased by more than 20,000 pounds with additional armament, armor plate and engine horsepower which increased altitude by 10,000 feet. The Boeing Clipper and Stratoliner deliveries continued with reduced priority. The major development activity was the B-29.

Made it Home!!

The loss rate was extremely high. A B-17 bomber crew had only a 40% chance of surviving the required 30 missions. However, many lives were saved because of the survivability features designed into the B-17. The aircraft above (top) returned to its base without the loss or injury to any of its crew. A German aircraft rammed the B-17 completely severing the port side horizontal stabilizer and nearly cutting the aircraft in half. Some B-17s returned riddled with flack and pilots incapacitated (bottom photo). The controls were so forgiving that even the non-pilot trained crew were able to return the aircraft and land successfully.

Howard Hughes with Eddie Allen

Hughes procured his personal Boeing 307 Stratoliner in anticipation of setting a new record for an "around-the-world" flight. Germany's invasion of Poland in September 1939 forced Hughes to cancel his attempt.

Boeing 307 Stratoliner

FUEL DUMPING TEST FLIGHT ON THE BOEING 307B STRATOLINER

Boeing was required to demonstrate the operation of the fuel dumping system to the Civil Aeronautics Authority. One object of the test was to see if the fuel missed the horizontal tail. A large pipe came out of the wing lower surface to expel the fuel so it should be low enough to miss the tail. Instead of fuel, colored water was used for this test. The horizontal tail was painted with white-wash to detect any impingement of the colored water.

Eddie was in the left hand pilot's seat, and Al Reed was in the co-pilot's seat. The CAA observer on the cockpit jump seat was George Haldeman. George was an old time pilot who was well known in aviation......When the 307 was ready to dump fuel, Al lowered the fuel dump pipe. George interrupted and said they were going up wind, and "he wanted to dump downwind" This revolted Al Reed's technical senses. He said emphatically "I won't dump downwind", an equally silly statement. Al imagined himself as a superior scientific test pilot, with all his Cal Tech education. Eddie hushed Al up, and told Al to turn downwind. Eddie then turned to George and said, "George, you can dump now".
 [as told by William H. Cook chief aerodynamicist about 1940]

BOEING 314 CLIPPER

Correcting the Boeing 314 Directional Stability

William H. Cook   June 1, 1992

"I had a conversation with Eddie Allen about his famous first flight of the 314. This was shortly before he was killed in the XB-29 crash. These details have never before been written down, as far as I know.

Eddie was alerted to a probable problem on the 314 by George Schairer, who at that time was working for Consolidated in San Diego. At that time Eddie was a free lance test pilot. Many airplane manufacturers were using his services because the insurance companies would give reduced rates on the experimental first flights if he were at the controls.

Eddie was in San Diego to fly the Consolidated four engined flying boat, the PB2Y-2. This was the second model. It had a twin tail, as was later used on the B-24. The earlier XPB2Y-1 had a single tail, like on the preceding PBY, which was the basis of Consolidated's experience. On the first flight of the XPB2Y-1 the Consolidated pilot, H.A. Sutton, was badly scared due to the lack of directional stability. This experience with directional instability on large flat sided flying boat hulls alerted Schairer to a possible problem on the 314.

George Schairer

Schairer left MIT in 1935 with a Master's degree in aero. He first went to work for Bendix in the middle west. A year of so later he went to work for Consolidated. Before talking to Eddie about the 314 he had accumulated quite a lot of experience in aerodynamics, principally in testing in the Cal Tech 10 foot wind tunnel in Pasadena, and in flight testing Consolidated's flying boats. The Cal Tech tunnel was by far the best in the country for complete airplane model testing. The Douglas DC1, 2, and 3, as well as other designs of the southern California companies, had used this tunnel extensively. There naturally had been a lot of interchange of aerodynamic knowledge between companies. The Cal Tech staff was a catalyst, and was more up to date on advanced airplane design than any other educational institution. The southern California airplane companies had built up substantial production business, and thereby could afford advanced research, much more so than Boeing, which had very little business compared to the southern California companies, and was isolated in Seattle without much money.

Schairer had probably seen a newspaper picture of the 314 with the original small single vertical tail. He told Eddie not to fly it, as it would be directionally unstable. Eddie had been asked by Boeing management to make the first flight on the 314. After talking to Schairer, Eddie went to Seattle and told the Boeing Management that they had to change the tail. At that time Boeing was in very severe financial straights. The founder, W. E. Boeing had bailed out of the company four years before after the so called "air mail scandal". The only business in the Boeing factory was in building 13 B17B's a year. Boeing had lost the bomber production business to Douglas, who had won the competition with the B18 twin engined bomber, that was patterned after the DC2. The Air Force at Wright Field liked the B17 much better than the B18, but was limited by law to buying only 13 service test airplanes of any particular model a year without holding a competition.

When Eddie told the Boeing management that he would not fly it with the small tail, they told him in desperation that they would get somebody else to fly it. I have guessed that the Boeing manager involved was Bob Minshal. He was an old time engineer and was very experienced and capable as a designer, but in later years he did not have the stuff to be a manager. Boeing had only one aerodynamics engineer, Ralph Cram. He was killed a year later in the 307 Stratoliner crash. Boeing did not have a suitable wind tunnel nearby. The University of Washington 8' x 12' Kirsten tunnel first operated in the summer of 1939, about two months after the epic first flight.

When I asked Eddie why he flew it anyway, he said that he thought he could get away with it better than anybody else, as he knew the severity of the problem. Earl Ferguson was in the co-pilots seat (Earl was also killed in the Stratoliner crash). When they took off into the wind in Elliot Bay north, Eddie told me he immediately knew it was a mistake. Probably by this time after un-stick he was approaching Magnolia bluff, and so he could not set back down in Elliot Bay.

As they went north into the wind, Eddie told me that "Earl had to help me on the rudder". I asked "how was this possible?" Eddie answered "it was obvious". I imagine that the nose was diverging from side to side, and that Earl and Eddie had no problem in coordinating on the rudder. The rudder effectiveness was good, but the directional stability was negative. They proceeded north to Everett into the wind. The chop on the salt water ahead into the wind west of Everett must not have been inviting. This was in June 1939, probably in the afternoon when the northerly good weather winds had picked up and caused waves. Landing in salt tidal water would have been a poor location for modification of the tail, but they were probably too busy to bother about this. So they made a 180 deg turn south, probably by a combination of aileron, rudder, and possibly by using unsymmetrical thrust. However the lag in power response might have made this a dangerous last resort. The adverse yaw of the ailerons, with no directional stability, must have been a problem, and a lot of quick thinking based on knowledge and Eddie's very considerable experience, were probably needed to remain in control.

They made a turn up-wind over Lake Washington. As they were approaching Sand Point they cut the power. Eddie told me that immediately he sensed that the 314 was stable with power-off. After landing on Lake Washington the 314 was docked on the west side of the lake just north of Sand Point. There it remained during the modifications and the further testing of the twin tails, and then the final configuration with the fixed center fin.

I saw some wind tunnel data of the 314 that showed very small positive directional stability. However, this was power-off. The power-on slipstream of the inboard engines along side of the hull had sucked the flat sided hull over in yaw, this phenomenon being known as the "Coanda Effect". At that time powered wind tunnel models with running propellers were beginning to be tested in the Cal Tech wind tunnel by Douglas on the DC-4, but Boeing had no such experience.

One interesting side of the fix was the rapidity of the designing process of the twin tails. Ed Wells, who was about number four in the engineering department at that time, and later on the Chief Engineer, and Johnny Sanders, the Project Engineer, designed the twin tails, the attachment to the stabilizer tips, and made the production drawings for the shop, all in a weekend. (Do not expect this ability in the modern generation). Things were still desperate at Boeing, maybe more so. Some months later I was given the assignment of drawing the centerline diagrams for the twin tails, based on Ed's and Johnny's production blue-prints. I asked "what for, since the parts had already been built?" My drawings were for the file."

Thomas Collison added in 1944: "The largest insurance policy ever written on an airplane was placed on the Boeing 314 Clipper during its flight test. Because flight testing of the Clipper, including insurance premiums, cost Boeing $1600. per hour, it was necessary that Eddie achieve his flight test objectives in the minimum number of minutes and yet leave no hidden weaknesses nor dangerous characteristics to the Clipper. He determined, first, the inadequate net fighting moment resulting from the high upsetting moment of the side winds on the water; he discovered the need for an 80 per cent increase in the net righting moment at light load.

He next determined the reversed slope of the yawing moment curve at small yaw angles . He recommended 100 per cent increase in the fin area to the twin rudders. Boeing engineers were obdurate, they added 60 per cent, first, and a new type of yaw stability technique was devised to determine how close the slopes of the yawing and rolling moments were to being satisfactory. ……..

When Eddie predicted the Clipper would possess uncomfortable flying qualities unless a further increase in the slope of the yawing moment was achieved, a central fin was added to the tail. The resulting final stability and controllability proved satisfactory to Pan American Airways pilots, and the control characteristics of the airplane were complemented by CAA inspectors."

Trauma and Celebration - March to July 1939

It has been said that human health and well-being are stressed by major events: death of a family member or close friend, marriage or divorce, change in employment, a move or change in residency, a high stress work environment; all these ETA experienced during this five month interval in 1939.

On March 18, the Boeing Stratoliner crashed killing ten key aviation people. Many of these men were good friends of Eddie who had worked with him on several aircraft programs over the years. Certainly the fact that Eddie had recently completed the initial flight test phase of the Stratoliner and found the airplane to be safe and viable for flight, yet somehow it suffered a cataclysmic failure. Eddie would have had to attend multiple funeral ceremonies in the month of March.

Secondly, as a result of this failure and probably the earlier YB-17 failure, Eddie was now asked in April to step into a radically different role with the Boeing Company, as the leader of a new Flight Test organization which would be unique to Boeing. While the role would allow Eddie to integrate the flight test disciplines into the early phase of aircraft development, the changes he initiated were not universally supported at first, as one might expect in any large technical bureaucracy. He reported directly to top management rather than to the chief engineer, an individual and a technical relationship that was critical in the design process. Eddie had to find a way to work successfully with an engineering design team that might already be considered the best in the aviation world.

It was also true that Eddie had never been in a true management position. He had always been an individual contributor. He may have had natural talents necessary to work with technical personnel, but he certainly had not risen through the technical engineering ranks of a large aviation company nor had he received the management training as had his management peers.

Thirdly, in the personal background to Eddie's professional life, Eddie Allen and Florence Lee Brydon had decided that their marriage would be consummated in the summer of 1939; July 1 would be the date of their marriage.

Finally, It would seem natural that Eddie's best friend since his college years at the University of Illinois, Dr. Carl Leo Rahn, would attend the marriage; perhaps as "best man". Carl did come to Seattle in May, too early for the marriage, and perhaps decided to "see Alaska" and then return in time for the wedding. This is supposition as <u>Carl was not at the wedding.</u> The wedding took place at the Brydon home on Bitter Lake in North Seattle with a reception held at the Arctic Club in Seattle. Eddie and Florence flew to San Francisco that evening for their honeymoon. They spent a few days visiting friends in the Berkley area. The newly married Allens returned to Seattle on the same flight with Howard Hughes who was coming to Seattle to meet with Eddie and Charles Lindberg and other Army and Boeing officials. Allen was scheduled to start flight tests the next day on the modified Stratoliner purchased by Hughes. The aircraft had been extensively modified with internal fuel tanks for his "round the world" venture.

While the timing is not known, Eddie received notice that his friend Carl had drowned at a beach near Ketchikan several weeks earlier and the authorities had not known who to contact. The following article from the Fairbanks Daily News - Miner dated 13 July 1939 summarized the tragedy.

"Strange Case Ended   The case of Dr. Carl Rahn was ended as far as Ketchikan was concerned, when his body was shipped to Seattle. W.C. Stump, his attorney, received a telegram from Edmund Allen, Boeing executive, stating he had communicated with Dr. Rahn's brother, believed to be in California, and instructing Stump to send the body to Seattle. The 57-year old self-styled retired psychiatrist drowned at the Tongass Park beach Memorial Day. Over a week passed without hearing from a friend or relative of Dr. Rahn regarding disposition of the body."

The culmination of these events would be overwhelming for most people. Somehow Eddie and Florence Lee managed to successfully survive and recover for only two months later in September of 1939, while Hitler was busy invading Poland and initiating WWII, my in-laws were busy conceiving my future wife, Turney Allen. — TKO

Wedding Party July 1, 1939

Eddie was married to my mother, Florence Brydon, on July 1, 1939. They met sometime in 1937 while Eddie was photographing my mother's ballet class at the Cornish College of the Arts in Seattle. While Eddie was obviously an accomplished technical engineer, his interest in the liberal arts is expressed throughout his letters with his family. Eddie was an enthusiastic follower of the opera, dance, poetry, was a voracious reader and even became interested in sailing and photography once he moved to Seattle. He wrote a thesis while attending the University of Utah on the psychology of flight and its effect on stress, reflexes and decision-making. Eddie's interest in the arts brought them together as my mother also loved poetry, played the harp and was an accomplished ballet dancer. Her best friend Dorothy Fisher (second from the left), taught ballet at the Cornish College of the Arts and became the Director of the Dorothy Fisher Dance Group which would perform and teach ballet for many years in the Seattle area. —-Turney Oswald

Eddie Allen's final letters are written prior to Pearl Harbor. The letter written in September of 1941 gives an overview of the problems Boeing faced with building a large scale production activity for the B-29.

Even though he is now a full time Boeing employee, Boeing agreed to the request for ETA to flight test the Lockheed Constellation C-69, the Curtiss-Wright CW-20 (C-46), and the Douglas A-20 Havoc. Eddie's desire to ultimately return to his role as an independent test pilot is evident in his letters although it should be noted this particular letter was written before Pearl Harbor which certainly recommitted everyone to the war effort. The B-29 program became the highest priority.

518 N 130th St
Seattle, Wash
April 14, 1940

Dear Margaret:

Just returned this morning from St. Louis and found your letter with the news of Frank's continued difficulty with the bladder ulcers and of his splendid improvement under Tom's treatment. Curiously last night in Chicago, where I changed planes from St. Louis, I phoned Evanston and Tom told me about Frank's arrival in Chicago and his improvement. He said Frank had gone to Marshall for the weekend and was about ready to return to Florida. I am very glad this business is coming out so well. It must have been very difficult for you. Please don't worry at all about the loan. If you are in any need now please let me know and I'll send you some more to tide you over this period until Frank gets going on the teaching job.

Florence is feeling very well and looking forward happily to June 1st or June 15th (the doctor can't make up his mind which)

We have moved out into the country to an old house that used to be a farmhouse. There are about two acres, mostly in flowers, vegetables and "pasture" for two cows and three goats. Florence lives outdoors wandering around among the annuals and flowers. I don't get any time, darn it, to do any of this except on Sunday. Several times I've had to go to St. Louis to carry on a test there going on almost simultaneously with the test here. It means rushing from one to the other as soon as a lull comes.

Spring is so long in coming. Guess I'm in need of a vacation. How is Florida these days?

Sincerely
Edmund

Jan 25, 1941
Dear Margaret:

I was sent to Kansas City two weeks ago to investigate an accident and now I am returning to Seattle. I have a hunch you are in need of money and so I enclose $100. This period when you need to be tided over should cause you no concern in asking for help. I hope Frank is better. I hope even more that he is not worrying about money matters for this keeps one from getting well.

My job in Seattle is very unsatisfactory from many points of view. I have been on the verge of leaving to go east for many months. The advantages and disadvantages seem to cancel at present. Florence and her mother who has become greatly attached to the baby, would not like to leave Seattle. It would mean selling the family place and uprooting everything. We have not decided definitely as yet. I want to return to consulting work which pays better and is more appreciated. When a company gets you on the payroll they forget to be courteous. They forget to be glad you are there.

I wish we could get to visit with you and your family. But right now we could hardly get farther apart within the continental limits of the United States. Florence is much interested in the children and she asks me about you and Frank. If I do leave Boeing I may take a few months off and make a circuit of the industry. It would be nice if we could get to Florida. But I'm afraid this is an idle dream.

Best wishes to all your family.

Sincerely, E.T.A.

518 N. 130th St. Seattle
Sept 1, 1941

Dear Frank and Margaret:

I was glad to get a letter from Citra to hear how things are going for you. It was good news about the job and the improved bladder. I hope improvement continues. There is always a lot of satisfaction comes with the first step of this kind.

This work in Seattle is now getting into its stride. The Boeing Company was one of the last of the aviation companies to get large war orders, owing largely to the lateness of the realization that our need was for heavy - not light - bombers. A year ago there was little realization of this kind. The general public, and some publicity-seeking members of Congress, seem to think that one year after an order is given for a thousand bombers it is possible to get into large scale production. To many, an airplane is an airplane, large or small. 500-a-month seems a small figure compared with the 1000-a-day automobile production. The comparison should be to super-dreadnaughts. They are not produced on a moving belt production line.

The airplanes which we will be building five years from now are now on our drafting boards. We are learning how long they should be from tip to tip for optimum performance, how large should be the tail plane, what the wing-loading should be, how to design the propellers which another company must then make for us. We have to develop mathematical processes to compute the ceiling of the airplane before we even know what it will look like. We balance ceiling against takeoff distance and that result against range in order to determine the dimensions of the wings.

Turney is now 14 months old and quite a problem for two inexperienced parents. We are hoping for a brother in about four months more. Florence says she wants six children, all as close to the same age as possible. If quintuplets should arrive she would have her wish.

I have been taking a course in celestial navigation recently. I hope to get a diploma soon, that will enable me to qualify as a trans-

Atlantic pilot. We can now determine latitude and longitude from two star-sights quite accurately.

Ed

Nov 5, 1941

Dear Sister:

I'm glad to send the enclosed check. Please accept it as our Christmas present.

The property situation cannot continue long to remain as static as at present. I expect a land boom to start soon. It will probably be quite sectional. In fact parts of the country may not feel it for a while. Florida is certain to have another 1928 and then - on the rise - is the time to sell.

I'm glad you are with Frank. The news of his improving health is fine. I've had a strep throat following a blood transfusion I gave Florence's mother who became critically ill with erysipelas last month. I guess I lowered my resistance too much and the strep did the rest. I took sulfanilamide and cleared up the throat but my blood count took a nose dive and so I'm on a liver and iron diet.

Too bad about the flood and the beans. It seems to be always the same story - either a freeze or a drouth or a flood or else extremely low prices!

I would like to have a snapshot of you all.

Sincerely   Edmund

The following letter was written to Florence Lee by Richard Aldrich, a close friend of ETA since 1923. Eddie had designed and built a home on the beach at Encinitas in 1931 to use while he was working on the west coast. Aldrich was staying at the home while he was doing some military training obligation. Following this, Richard would attend the University of Michigan at Ann Arbor and complete his doctoral studies.

[Estimate February 1942]   Camp Roberts, California

My dear Florence

    I stood, like you and Eddy, in the Court. I got home at five o'clock on Sunday morning, and had to leave (to get back to camp) Sunday morning at nine. Thirty-two hours traveling, all told, to see something that perhaps in the whole world only the three of us understand. What, then, with long night marches, and in day-time sitting on the mud firing mortars, I'm pretty tired.

    But I wanted to tell you how relieved I was to get a letter from Eddy. Not hearing from any of you made me surmise you as well as he must be ill again. I'm so glad this anguish was just another gratuitous assumption! I miss so the letters I hope for from you. You cannot know how the humblest doings of home is news to even a quasi soldier, since such a one is no less bound. Something La Turney said or did, the goats, the dogs, or the long days at home that must (at times), feel so far from town, now that friends in town, so many of them, are away, would be a lift.

    The house is in good shape, clean and well looked after. I watch the inching progressions of the vast war, hoping its evil ways, on the other side, of course, will bring themselves to a final abyss, and that after this comes to pass we can go about our comparatively exalted ways. For whatever those ways were they were not bound up with killing someone else. That's all I hear here. Will you help me with a box to Bill? Did you send the book off I left on the piano?

Won't you please write? In a few days I will have been here three months, and in all that time no short note has come from you. But you can't know how I miss all of you; you can't know how I long for the time when we can all be home. I saw Ruby, and they were very kind. We had a good visit.   With love to all of you,

                      Dick[43]

---

[43] Interesting letter from Richard Aldrich "I stood, like you and Eddy, in the Court. …. to see something that perhaps in the whole world only the three of us understand." ? Another puzzle to wonder about   - TKO

Final Flight

Following the first test flight of the #1 XB-29 on September 21, 1942, the test program experienced continued engine related problems. The basic aerodynamics, stability and control of the airplane were validated. The engine problems, however, required the #1 airplane to be grounded for seven months. Testing would continue with the #2 XB-29. The following excerpt was written by Bob Robbins from his book "The Global Twentieth". Bob Robbins was Eddie Allen's co-test pilot and good friend.

"On December 30th the #2 XB-29 (AAF41-003) was ready for this initial flight. It too had engines that were cleared for only 35 hours in positions 1, 3, and 4. It was to be a thorough functional check of the airplane and its extensive instrumentation. The weather was marginal. The functional check proceeded normally until the number 4 propeller would not feather and governing was erratic. Eddie elected to discontinue the flight and immediately headed back to Boeing Field at which time he was advised that the weather was deteriorating rapidly. About 6 minutes out, the # 4 engine caught on fire, the propeller oversped to 3,500 rpm, the propeller would not feather and smoke, sparks and flame were coming from the exhausts. Shutting off the fuel and the use of fire extinguishers were ineffective. The fire continued to get worse. About 2 minutes out the fire was burning fiercely in the accessory compartment. Flames were pouring from the nacelle access door and from the intercooler exit area. Heavy smoke and long fingers of flame were trailing off the wing. In the meantime heavy smoke was pouring from the bomb bay into the cabin making it increasingly difficult to see or breathe. Eddie landed downwind, choking, partially blinded, on the 5200 foot long, 200 foot wide runway. The intense fire was put out by fire equipment on the ground. Eddie later received the Air Medal for his skill and bravery during that harrowing 32 minute flight. Ground inspection showed more trouble. A fire had just started in engine #1 and engine #3 was close to failure, too. Those three 35-hour engines each had less than three hours total ground and flight time. Because of engine shortages two of the three engines had to be replaced with engines cannibalized from the XB-29 which was laid up for some modifications. In addition the fire in #4 had been so severe that the #4 nacelle had to be replaced with the #4 nacelle also cannibalized from the #1 XB-29. At least the #2 XB-29 now had 4 so-called "unlimited" engines.

Unfortunately, engine/nacelle fires similar to the #4 fire continued to occasionally haunt production B-29s and caused at least 19 serious B-29 accidents between February 1943 and September 1944. While Boeing and Wright tried hard to find and correct the cause or causes, there was a natural tendency for each to blame the other. It was 15 months before there was positive proof that the R-3350 was susceptible to induction system fires which could very rapidly get out of hand and become uncontrollable magnesium fires which then destroyed evidence of the fire's origin. That proof came on March 24, 1944, when I had an induction system fire on the #4 engine during a routine test flight on the #1 XB-29. I was fortunate enough to get the engine feathered and the fire out before it broke out of the blower section or the intake pipes and became an external fire. The partially burned magnesium impeller and interior of the blower case were irrefutable evidence. In the face of that evidence Wright developed the fuel injection system to eliminate the potential for induction system fires.

It was almost a month before the #2 XB-29 flew on January 29, 1943. In the next three weeks emphasis was on engine, propeller, governing, and airplane performance testing. Catastrophic engine failures eased up but that was about all. During descent for landing on February 2 there was a strong odor of gasoline emanating from the bomb bay into the cabin. A thorough inspection uncovered nothing conclusive. On a flight on February 17th there was a bad fuel leak over the wing from the #4 fuel filler cap. The leaking cap was fixed.

By February 17, 1943, the #2 XB-29 had made eight flights totaling 7:27 hours — an average of only 56 minutes per flight. In the five months since the first XB-29 flight on September 21, there had been only 31 flights totaling 34:27 hours— a long way from what Eddie had estimated in September could be done. And with an overall average flight of only 1:07 hours the amount of meaningful test data was pretty sparse from that meager 34:27. As hard as everyone was working to solve the problems, the answers were coming painfully slowly. As Eddie and his Project Flight Test Manager left the airplane that afternoon and walked across the ramp to the post-flight conference, Eddie expressed to him the grave reservations he had about continuing flight testing until at least the more serious of the XB-29 problems could be fixed. Unfortunately, the fastest, and maybe the only way to fix some of them was to try out the various fixes in flight—the "try, try, try again" approach that had been so successfully used by Eddie and George Schairer over the years. But now Eddie faced a real dilemma. The B-29 was potentially a fine airplane. It was urgently needed in the Pacific. It was committed to production —1600 B-29s were now on order at four separate plants. Flight test was way behind its expected schedule and the data was badly needed to prove the airplane; quickly find and correct the problems; minimize production disruption, develop training and operating procedures and manuals. But it was currently a dangerous airplane. Major improvements were badly needed. Temporary grounding would be the normal, prudent thing to do. But they were not normal times. The sooner the B-29 could be used in combat, the sooner the war would end and the sooner the casualties and carnage would stop. Eddie concluded that he must continue flight testing as rapidly as possible. His entire crew had also to know the risks—to a man they stayed with him.

The primary objectives of the February 18, 1943 flight were to measure climb and level flight performance and get engine cooling data with 4 and 2 engines operating. Maximum altitude would be limited to 25,000 feet because of the excessive trouble that had been encountered with low engine nose oil pressures above that altitude. The effectiveness of the fixes for some of the past problems would also be evaluated. Takeoff would be at the normal design gross weight of 105,000 pounds with full fuel tanks—5410 gallons of gasoline.

Eight minutes after the 12:09 pm takeoff to the south, while climbing through 5000 feet with rated power, a fire was reported in the #1 engine. Mixture and fuel to #1 were cut off, propeller was feathered, cowl flaps were closed, a CO_2 fire extinguisher bottle was discharged and a descent and return to Boeing Field was initiated. Since the fire appeared to have been put out and everything seemed under control, Eddie elected to make a normal landing pattern and land from the north on runway 13 (128 magnetic) to the SSE into the 5 mph win rather than making a downwind landing on the 5,200 foot runway with a heavy airplane. At 12:24 pm the radio operator routinely reported altitude at 1,500 feet at a point 4 miles NE of the field. They were on the downwind leg, headed NNW and starting a left turn onto base leg. No one suspected the drastic change that would take place in the next 2 minutes. At 12:25 they had just completed turning onto base leg, had just crossed the heavily populated west shore of Lake Washington about 5 miles NNE of the field, were at about 1,200 feet altitude and were heading SW approaching the commercial and industrial south side of downtown Seattle. At that point ground witnesses heard an explosion that sounded like a loud backfire and a piece of metal fell from the airplane. About that time the radio operator, who could see into the forward bomb bay and the wing center section front spar, was overheard by the Boeing tower on an open microphone to say "Allen, better get this thing down in a hurry. The wing spar is burning badly". He told Boeing Radio on a different frequency "Have fire equipment ready. Am coming in with a wing on fire". About a mile down the flight path from the explosion, burned parts of a deicer valve, hose clamps, and instrumentation tubing were later found. They had come from an area normally inside the wing leading edge, ahead of the front spar, and just outboard of the #2 nacelle near the #2 fuel tank filler neck which was rubber like the self sealing fuel cell. The airplane now turned south on an oblique final approach in a desperate effort to reach Boeing Field just 4 miles away. Eddie was about 250 feet high and ground witnesses later reported that part of the wing leading edge between #1 and #2 engines was missing. In the next mile the flight engineers data sheet was found and three of the forward compartment crew members left the airplane — too low for their parachutes to open. At 12:26 pm, only 3 miles from Boeing Field, the #2 XB-29 crashed into the Frye Meat Packing Plant killing pilots Eddie Allen, Bob Mansfield, and the other 6 crew members on board. The crash and resulting fire killed an additional 20 people on the ground and destroyed much of the airplane and the plant. There was clear evidence that fire and dense smoke had gone through the bomb bay into the cockpit in the last moments before impact. Burns on the bodies and clothing of the 3 crew members who bailed out just before impact were a part of that evidence. Eddie Allen and his crew died serving their country the best way they knew how. In one minute the fire had gone from undetectable to catastrophic." (Bob Robbins - The Global Twentieth)

The devastating crash on February 18, 1943, took the lives of Eddie Allen and his entire flight crew, all critical personnel in the development of the B-29. In addition, 20 employees of the Frye Packing Plant and one fire fighter lost their lives.
The Army Air Corps took over the management of the entire B-29 program. The program continued and in April of 1943 the first of 14 YB-29 test aircraft, with modified fire suppression systems, came off the lines in Wichita. In September the first production aircraft were delivered. In total 3895 were built.

Dedication of the B-29 "Eddie Allen" at Boeing Wichita

The "Eddie Allen" was dedicated to his sacrifice. The aircraft was paid for by donations from the employees of Boeing Wichita and given to the USAAF as a gift. The symbols show the aircraft has accomplished seven trips over the Himalayan hump to China and six missions over Japan. The "Eddie Allen" eventually flew ten transport missions "over the hump" and 24 combat missions over Japan before severe flack damage kept the aircraft grounded permanently. The forward section was salvaged and used for training purposes.

"Two months before he died he left his testament to aviation. Chosen to deliver the Wright Brothers Lecture before the Institute of the Aeronautical Sciences, Pilot Allen sat himself down to a codification of test-pilot procedure. Object: to standardize testing, make it result in the same sound, understandable conclusions no matter what pilot is at the controls. The result: a test pilot's bible. Said Edward Pearson Warner of CAB, one time professor in M.I.T.'s Department of Aeronautical Engineering: "More than anybody else, Eddie has made it possible for the performance of aircraft to be determined accurately and scientifically." "[44]

"…..Thus Eddie's vast experience was a big help to Boeing. He had a stature in the company that amounted to being a member of the "Supreme Court"; when it came to safety and technical leadership. He was a leader in technical progress. After his death the influence of Flight Test has gone steadily downhill. Fortunately Eddie was able to lay the cornerstone to the future in his backing of the wind tunnel project, and his decision to go to high sub-sonic speed." — William H. Cook Boeing Chief Aerodynamicist

Eddie Allen left his wife Florence Lee and two children: a daughter Florence Turney Allen (age 2 years, 8 months) and Edmund Carl Allen (age 1 year 2 months)

---

[44] Army & Navy: Test Pilot No. 1 Time Magazine Monday March 01, 1943

Eddie at the helm of the sailboat Tsola

Florence Lee with Eddie and Turney

Boeing Wind Tunnel Blows Strong for Nearly Seventy Years[45]

"On Dec. 17, 1947, the 44th anniversary of the Wright brothers' historic first flight of a powered aircraft at Kitty Hawk, N.C., another flight took place at Seattle's Boeing Field that ranks as one of the most important in aviation. Boeing's B-47 Stratojet bomber flew for the first time that day -- and changed the shape of jet aircraft.

The B-47 was America's, and arguably the world's, first large swept-wing jet. Seemingly forgotten in history, the Stratojet's revolutionary design was the first to pair swept wings with jet engines suspended from the wings in podded nacelles. Discovered in the Boeing High Speed Wind Tunnel in 1945, this basic design is still the model for all jets built today by Boeing, Airbus and others.

For Boeing, the journey to become the pioneer of large swept-wing jets began in April 1939 when the company hired famed test pilot Edmund T. "Eddie" Allen to head its new Flight and Research organization. A respected scientist, Allen was accorded the freedom to do whatever was necessary to advance Boeing's flight research efforts -- and that included building a private wind tunnel. At the time aircraft manufacturers did not have their own wind tunnels and the fierce competition to use the few operated by NACA (forerunner of NASA) and a handful of universities resulted in Boeing falling behind the competition. Allen championed the idea of a company-owned wind tunnel, capable of near-transonic (approaching the speed of sound) speeds. The estimated cost of $1 million represented a huge risk for Boeing at the time. But it also was a great opportunity, and in August 1941 Boeing President, Phil Johnson authorized construction of a high-speed wind tunnel capable of speeds of Mach .9 (625 mph, or 1,000 kilometers per hour).

The B-47 was the first full design tested in the new wind tunnel. The swept-wing concept had first come to Boeing in May 1945 by way of a letter sent from Germany by the company's leading aerodynamicist, George Schairer, who was serving on Air Force Gen. "Hap" Arnold's Scientific Advisory Group. That group was tasked with securing German aircraft and rocket research. Boeing engineers subsequently saw dramatic results during wind tunnel tests of Schairer's swept-wing data, but they also discovered that the wings had to remain "clean" to achieve the high-speed benefits. And this presented a problem since the standard design for multi-engine airplanes at the time was to mount the engines on the wings. As he puzzled over the problem during a train ride back from Wright Field, Ohio (today known as Wright-Patterson AFB), Boeing Chief Engineer Ed Wells came up with the idea of engine pods mounted off the wings. The concept was tested in the Boeing wind tunnel by mounting model engine nacelles on the end of a pole (the "broomstick" test) and moving the nacelles around the wing until the optimal position was discovered -- forward and below the wing.

[45] Boeing Technology Website - October 28, 2013

Eddie Allen and Phil Johnson

These discoveries all came together in the Boeing wind tunnel as the optimal design for a subsonic jet -- and resulted in the revolutionary XB-47 that rolled out of Boeing Plant 2 in September 1947 -- only two years after Schairer sent his note from Germany.

    Just as building their own low-speed wind tunnel was critical to the success of the Wright brothers, so too was the wind tunnel key to success for Boeing and the B-47. Improved over the years, the now-transonic wind tunnel has tested some of the best-known airplanes in aviation history and continues its work today with jets such as the 737 MAX. Boeing was fortunate that a leader arrived at the right time to set a course for success by not only pioneering the organization that continues today as Boeing Test & Evaluation but also insisting the company build its own wind tunnel. The Edmund T. Allen Memorial Aeronautical Laboratories are named in his honor." - Mike Lombardi Boeing Corporate Historian October 2013

"His idea has helped shape many innovations - its a widespread belief without Allen's wind tunnel, Boeing would not be the company it is today. ….. The chances are that one of our competitors — one of them would have built America's first jet bomber and probably have gone on to build America's first commercial jet ….and Boeing would have just faded away.
I firmly believe Boeing would not be here today if not for that wind tunnel."

        Mike Lombardi - Boeing Corporate Historian   October 2013

Eddie was introduced to the benefits of wind tunnel testing in 1920 while working for the NACA at Langley Field. He flew the Curtiss JN-4H "Jenny" to correlate the actual flight characteristics of the aircraft with the wind tunnel predicted performance. While working for Northrop and Douglas in the early 1930's, the value of the Cal Tech tunnel was essential and the inability to get access to the tunnel made the early identification and resolution of aerodynamic issues impossible. Then, when he visited the German wind tunnel facilities in 1935, ETA was even more convinced of the value of the wind tunnel as the wind tunnel complexes across Germany were the foundation of their aviation development program.
At home the late identification of stability and control issues with the tail design of the Boeing 314 Clipper and Boeing 307 Statoliner flight problems just confirmed the need.

This background made Eddie's firm commitment to a dedicated high speed wind tunnel for Boeing.

The Most Unforgettable Character I've Met

by Thomas Collison   The Readers Digest   February 1965

It is a rare privilege to know a great man. I knew one. His name was Eddie Allen, and he was a test pilot. His greatness lay in his vision of aviation's future, and the work he did to make that vision a reality. Now that he is dead, his monument is his contribution to the fact that we can fly safely over continents and oceans; for each time we go into the skies we are cradled in the work of Eddie Allen's hands.

I first met Eddie in the summer of 1925, when I was 18 years old. I had driven my cut-down jalopy from Indianapolis to Cheyenne, Wyo, to visit my brother Hal, who was a pilot in the Mountain Division of the Post Office's airmail service. When I arrived, Hal was off on a flight but had left a note that read, "Go out to the field and meet Eddie Allen. He's a pilot, too, and part of the future , and a kid like you should know him."

At the weed-choked field a man came out of the dispatcher's shack, extended his hand and said, "Hi. You must be Hal's brother. I'm Eddie Allen."

I stared at him, dumbfounded. The pilots of the Mountain Division (the most dangerous leg in the trans-continental route) were the elite of the service and there was a swagger to their walk and their talk and their dress. But this man didn't fit the picture at all. He was short and slight, weighing about 130 pounds   A flattened nose, broken in a plane crash, looked incongruous in a gentle and thoughtful way. He wore ordinary corduroy pants and a heavy sweater. There was no swagger about him - he actually seemed a bit shy.

We walked to my car, and he began asking me questions about it. I was particularly proud of the fuel system I had rigged up. Allen carefully inspected everything, then patted me on the shoulder and said, "Very good. You're not afraid to cut and try. That was the way the Wright brothers invented the airplane; when one thing didn't work they cut a new pattern and tried again. That's the heart of all research, of all knowledge.

He spent an hour talking with me - about fuel systems, engines and aerodynamics. Then it was time for him to take off. As he climbed into the cockpit I saw a small book sticking out of his hip pocket - it was a volume by Nietzsche, the German philosopher. Allen caught my surprised look and said, almost apologetically, "When the air is smooth I wrap my leg around the stick and get a little reading done".

That summer Eddie Allen became my friend. It was my first experience with adult friendship, and I found that with Eddie I could be serious; I could talk about the things important to me without fear of seeming ridiculous. He gave me the courage to be myself, because he set the example.

Search for knowledge. Eddie's mother and father had been medical missionaries to the Indians in the Oklahoma Territory. Theirs was a dedicated and selfless household, but there was always an eager search for knowledge and culture. No matter what the pressures of work and at times of poverty, the evenings were spent reading aloud or playing on the old upright piano the music of Mozart, Schubert and Handel.

While at M.I.T., Eddie designed and built a glider which he took to France and Germany to fly in international competition. In Europe he haunted the great museums and concert halls. Once, speaking about a certain string quartet, he said, "They play really great music without watching public opinion or playing up to public taste."

That was the key to Eddie; in his own life he refused "to watch public opinion and play up to public taste". He was a non-conformist, but not in any aggressive way. He just had to know the why of everything he did.

While the Mountain Division boys flew daringly, their lives sweetened by risk, Eddie set for himself a deliberate program of research to learn what a plane could and could not do, and why. He took time out of flying for an engineering course at M.I.T. so that he might better evaluate his own research. He took a college course in psychology, in order to observe objectively his own role as pilot.

For all his self-imposed hours of research, Eddie never neglected his friends. And if a person was in trouble, Eddie was first to offer help. A school teacher's younger sister developed severe rheumatism, and her only relief game from hot baths full of pine cones. All one summer Eddie used his days off to drive the teacher and the child to the mountains, where he filled the tub with water and pine cones and built a fire under it. While the child bathed, he designed airplane wings in his notebook.

Cut and Try. In 1926 the Post Office Department began phasing out its own mail-flying operation, and awarding airmail route contracts to private operators. This brought a great surge of flight research as companies began to build specific planes for specific jobs. By 1929, Eddie Allen, with his background of flying and engineering, was one of the nation's top consulting test pilots.

Many test pilots of that day were brute-force men; they would sometimes fly a precious experimental plane beyond their power to control it, then bail out, losing hundreds of thousands of dollars and man-hours in an instant. When Eddie tested a plane he always brought it back, and with a detailed analysis. He was there not to murder the plane, but to measure its performance with every instrument he could install.

As his fame and authority grew, he demanded that his work begin with design, long before the first rivet was driven. He insisted on wind-tunnel work to pre-test every aerodynamic feature of each new plane design.

"Cut and try" was his constant battle cry and slowly he convinced the industry. He tested the first helicopters flown in this country, as well as early flying boats and other heavy planes for Boeing, Curtis-Wright and Lockheed. For Douglas he helped design the DC-1, the DC-2 and the DC-3, which greatly advanced commercial air transportation in the 1930's. Eddie's flight-procedure manual for the crews of the DC-3's was the first of its kind in the industry and a model for all other manuals that were to follow.

The shape of his thought and method was so clearly silhouetted in the American skies that when, in 1940, the Institute of the Aeronautical Sciences made its first annual "Chanute Award" for outstanding scientific contribution to aviation, it went to Edmund T. Allen.

After college I went to work in the aviation field as a technical writer, and I saw a good deal of Eddie. I observed that his developing fame and authority made no difference in the man; he continued as unassuming and considerate as he had been that first summer in Cheyenne. And always he pursued his own way, a unique blend of philosopher and engineer.

Designing the Impossible. As World War II approached, there was a crisis in aviation and , fortunately, a man to meet that crisis - Eddie Allen. His whole life suddenly seemed, in retrospect, a preparation for this moment. Aircraft were to play an enormous part in this war ;under conditions never before faced. Potential enemy targets were thousands of miles away, and the Allies needed a bomber that would reach them. The best candidate for the job was a four-engine bomber developed by Boeing in 1934. Eddie was called back to Boeing , given the title of director of aerodynamics and flight research, and told to take the old B-17 and give her 10,000 feet more ceiling and an added 15 tons of armament and accessories.
"Impossible!" was the first cry of Eddie's colleagues.
Eddie didn't agree. He said, "Let's cut and try". With Eddie's help, the B-17 became the fightingest four-engine bomber in military history. Eddie used broader-bladed propellers which took better "bite" in the thin air at 35,000 feet. He increased the payload by changing the center of gravity. He improved the engine power by finding a better way to install turbo-superchargers (high-altitude lungs). Then he said to his staff, "Remember, it is not enough that these planes fight the enemy; they must endure, they must bring their crews home!"
To this end, he and his men labored to make a plane that would fly with many of its controls shot away.
The improved B-17 Flying Fortress, the plane many men said could not be built, *was* built, and it soon darkened the skies of Europe. And it brought many an airman home "on a wing and a prayer".

New Job. One day shortly after Pearl Harbor, I was in Seattle, and Eddie invited me to attend one of his staff meetings.
It turned out to be a very special meeting, for Eddie was to reveal for the first time a new job the company had given him. There were about 30 young men present, all dressed in conservative business suits and looking more like college students than what they were - the greatest pilot and aviation-research staff ever assembled.
Eddie sat on top of a desk in front of them and reviewed the B-17 program. Then he said, quite matter-of-factly, "For the Pacific war, the Air Force is calling for a new bomber, a superfortress. It must have a third more speed, twice the range, and carry double the bomb load".
A stir immediately went through the room, as though he were voicing a goal that was utterly unattainable. And yet I could sense an almost physical bracing by these young men, a mental digging in, not only because they knew that victory in the Pacific might well depend

upon their success, but also because this mild-mannered little man had asked them to do it.

In the hectic months that followed, Eddie worked almost around the clock, turning out a mountain of memos, flight reports and directives about the aerodynamics of the new superfortress - the B-29. Throughout 1942, the plane was put together piece by piece, tested by Eddie and his staff, revised, redesigned, tested again. "Cut and try" became the watchword of the entire Allen staff.

And slowly the plane began to meet the specifications - all except the engines. They just couldn't be cooled, and time after time they caught fire during the test flights.

Eddie could have stayed on the ground and evaluated the reports of his other pilots, but that was not his way. Almost always he was int the pilot's seat when the monster took off for tests. He was determined to make the B-29 the plane he knew it could be. He was equally determined that no soldier or airman would needlessly lose his life when that plane reached the battle front.

"Am Coming In". On February 18, 1943, he scheduled another test for the B-29. The day began with an overcast, but, as the morning wore on, visibility increased to five miles and the cloud ceiling was 10,000 feet. Eddie and his colleagues entered the plane, ran up the engines, roared down the runway and climbed into the low, shifting clouds. It was 11 minutes past noon.

Five minutes later the Seattle control tower heard these words from the plane: "Fire in No. 1 engine. Coming in. Think we have it under control."

Then there was silence for an eternity of five minutes. The plane spoke again: "Twenty-four hundred feet, descending. Request immediate clearance for landing. No. 1 engine fire. Propeller feathered. Order crash equipment to stand by."

The tower replied, "Roger, Cleared to land. Runway 13".

The plane sliced over Seattle's business district at 1200 feet. Smoke poured from engine No. 1, and bits of burning metal began to fall from the plane.

The plane spoke to the tower: "Have fire equipment ready. Am coming in with wing on fire." Flying south for Boeing Field, steadily losing altitude, the plane was burning like a cattail. Pieces of the de-icer system were falling, the landing gear was twisted in heated agony. Gasoline rushed to the leading edge of the wing and exploded there. Flames were flashing into the cockpit. But still the two pilots remained fiercely at their posts, holding the plane in level altitude.

With seconds to live, Eddie Allen and co-pilot Robert Mansfield pulled back on the controls, in an attempt to clear building level. But the left wingtip smashed into a building, spinning the giant flaming plane in a slow and sickening arc. A series of explosions echoed through the distant hills.

I was in the Boeing factory at the time, unaware of the developing tragedy until there came that awful cannonade. It was followed by a moment of stunned silence; then office doors banged open and feet began to run, hundreds of pairs of feet. But not mine. I could not move. I knew with a terrible certainty that I had just heard the death of my friend.

Eddie Allen's work did not stop with his death. He had established criteria for heavy-airplane stability, control and flight procedures, that finally brought the B-29 program to success. And the B-29 brought us victory in the Pacific.

Our debt to Eddie did not end with the war, for every airliner now flying was incubated from those wartime bombers. Whenever I take a flight, jet-propelled and silken-smooth, I seem to see his face before me. He watches as I settle myself in the deeply upholstered seat, as I am swept aloft to be catapulted through the skies at 570 miles an hour. Then he gives me a bemused smile as if to say, "Isn't it surprising what you an accomplish if you're only willing to cut and try?"

Memorable Quotations

"I've decided I might as well have a patron saint. I have a friend whose patron saint is Shelley. Mine shall be Lafayette. I love Lafayette. He represents what I feel we are fighting for. He was the Savior of America. If I can help save France and the same liberty and democracy he loved so, my life shall have been worth while." 1917

"As there was no field big enough to get into with a DH4, I made use of the old axiom "Two fields are bigger than one," and I picked out two fields and used both of them. I didn't know how much damage the barb wire fence in between them would do, but I did not have much time to waste so I took a chance and my gallant DH4 galloped right thru the fence" 1919

"Twenty years ago Father was telling me about the Wright Brothers experiments and now flying is so common that nobody gives a second thought to taking the Transcontinental airplane in New York and getting off in Los Angeles 26 hours later." 1929

"I suppose you are quite interested in the coming election. All my friends are. But it seems like toys which children even find now. Remember the story of the lady who went back after many years to the toy shop in which as a child she had spent so many delightful hours. She said in surprise, "These are all the same toys. Have you no new ones?" The toymaker said, "No, the toys are always the same. It's the children who are new." This is a very significant saying for the meaning of life, I think. "   1930

"I've wanted to stay young in the sense of being able to learn, of taking attitudes characteristic of children toward life, of even being a bit "impractical" because one who is too "practical" in his personal life becomes closed to new possibilities of worth in human living. I've wanted to take piano lessons, even at 33, to study new foreign languages so as to be able to understand and love people quite different from ourselves, to fly, to keep my mind agile enough to consider appreciatively unconventional attitudes. It takes effort to achieve this even at 33 in any considerable measure and I have wanted to put this effort into life." 1929

"Saturday evening we went out to Wellesley to the big Vassar – Wellesley debate. The subject was the restriction of Immigration........ I was much disappointed also in the type of argument. The negative (Vassar) opposed a further restriction on the grounds that (1) not so very many immigrants are coming in now and not many are likely to come (2) our present laws keep out all undesirables – (diseased, criminal and radical) (3) our plans of Americanization are able to assimilate all that get in (4) we need foreigners to do our work for us.

They ought to have lost and I'm glad they did. Not a mention of the great moral issues involved. Where is all the idealism of 1917? – Vanished into thin air." 1921

Honors and Awards

Octave Chanute Flight Award   1939

Wright Brothers Lectureship in Aeronautics -
    Institute of Aeronautical Sciences 1942

Daniel Guggenheim Medal Award 1943

Air Medal - Rarely presented to civilians - Presented
    posthumously at the direction of President
    Truman 1946

The Edmund T. Allen Memorial Research Facility
    1944 — The Boeing High Speed Wind Tunnel

Boeing Pathfinder Award   1983

Eddie Allen 747-400 Flight Simulator 1989

Bibliography

The following is from Richard K. Smith reference materials dated 3 September 1992. "A Preliminary Bibliography" - RKS

"I'm sure that I have most of ETA's bibliography, but there may be as many as a half-dozen items missing, probably published in "odd" or relatively obscure magazines. The titles of a few of his articles are wild, "flashy" and sensational; but here it deserves appreciation that once an author has sold a piece to a publisher he has no control over its title. Editors of popular magazines like "exciting" titles for an article, and when an author doesn't provide it, they will create it; they will even jazz up an author's text."

With F H. Norton. Accelerations in Flight. Report No. 99. *Sixth Annual Report of the NACA* (1920), 481-488.

With F. H. Norton. Control in Circling Flight. Report No. 112. *Seventh Annual Report of the NACA* (1921), 69-90.

Three European Gliding Meets. *Aviation V.21* (Nov. 27, 1922), 712-714

During most of 1924 ETA wrote a regular section in the weekly *Aviation* magazine that was entitled "Light Planes and Gliders"

    (March 31, 1924), 340-341
    (April 7, 1924), 369-370
    (April 24, 1924), 426-427
    (April 28, 1924), 444-445
    (May 5, 1924), 483-484
    (May 12, 1924), 512
    (May 19, 1924), 538-539
    (May 26, 1924), 566
    (June 2, 1924), 593-594)
    (June 9, 1924), None
    (June 16, 1924), 646-674
    (June 23, 1924), 673-674
    (June 30, 1924), 701-702
    (July 28, 1924), 810-811
    (Aug 18, 1924), 810-811
    (Aug 25, 1924), 916
    (Sept 15, 1924), 993-994
    (Sept 22, 1924), 1022
    (Sept 29, 1924), 1059
    (Oct 13, 1924), 1129-1131
    (Oct 27, 1924), 1200-1201
    (Nov 10, 1924), 1252-1283
    (Nov 17, 1924), 1281-1309

(Nov 24, 1924), 1308-1309
(Dec 15, 1924), 1404-1405
(Dec 22, 1924), 1434-1435
(Dec 29, 1924), 1463-1464

Blind Flying. *Slipstream,* V.7 (April 1926), 11-12

Night Flying in Bad Weather, *Aviation,* V.22 (March 7, 1927) 461-463

Flying through Fog and Night, *Aviation,* V.23 (Nov 21, 1927) 1234-1239

The Kick of Blind Flying, In Franklin K Mathews (Editor), *Flying High: A Book of Aviation Stories and Model Airplanes for Boys,* New York: Grosset & Dunlap, 82-85

Safety in Aerial Navigation Through Radio Communication, *ASME Journal* (January 30, 1930) p. 847-848

Truth and the Test Pilot, *Aviation* V.31 (February 1932) 53-55

The Problems of a Modern Test Pilot, *Western Flying,* V.13 (November 1933), 8-9, 22, 29, 31

Breath-Taking Stunts Test New Transport Planes, *Popular Mechanics,* V.124 (March 26-28,1934) 112-113

Then in February 1934 there appeared in *Aviation* magazine (no longer a weekly; now a monthly), the first of a long series of articles (thirteen in sum), by ETA and W. Bailey Oswald relating to "Cruise Control". Their data was drawn from testing experience with the one-and-only Douglas DC-1 airliner. Their product was a classic exposition which established the data base for how the modern commercial airplane should be flown.

Operation at Desired Cruising Conditions, Part 1. *Aviation* V.33 (February 1934), 42-45 Part 2. V. 33 (April 1934) 110-112

Performance Testing and Engine Power, *Aviation* V. 33 (May 1934), 138-141

Power Determination for Cruising Operations, Part 1, *Aviation* V. 33, (July 1934), 215-217, Part 2 V.33 (August 1934) 253-255

Development of Cruising Charts, *Aviation* V. 33 (October 1934) 320-323

Climb and Descent in Cruising Control, *Aviation* V. 33 (November 1934) 361-364

"Trip " Time and Cruising Velocity, *Aviation* V. 33 (December 1934), 394-397

Winds and Control of Scheduled Time *Aviation* V. 34 (January 1935), 15-18

The Future of Cruising Control, *Aviation* V.34 (February 1935) 57-60

Economic Engine Operations for Cruising Reliability, *Aviation* V. 34 (March 1935), 89-92

Power Control and Schedule-Keeping, *Aviation* V. 34, (June 1935) 25-29

This concludes the "Cruise Control" series by Allen & Oswald; eleven titles but thirteen distinct items.

Engine Failure at Takeoff, *Aviation* V.34 (June 1935), 181-183

New Wings for a New Germany, Part 1, *Aviation* V. 34 (December 1935), 11-12 Part 2, V. 35, (January 1936), 14-16. Part 3, V. 35 (February 1936), 23-25

"Here Goes Nothing! —- A Dangerous State of Mind For The Test Pilot or Anyone who Pushes a Stick" *Aviation* V. 36 (December 1937) 24-25, 73

With C. B. Allen, "Tons Aloft; Test Piloting the Transatlantic Clipper", *The Saturday Evening Post,* V. 22 (September 17, 1938), 12-13, 86, 88, 90-91

The Trend of Air Transportation, Institute of Mechanical Engineers, *Journal and Proceedings* V. 142 (December 1939) 127-140

The Testing of Large Aircraft, Society of Automotive Engineers *Transaction* V. 45, (October 1939), 444-448, 456

Flight Testing for Performance and Stability, Sixth Wright Brothers Lecture, the Institute of the Aeronautical Sciences; published in the *Journal of the Aeronautical Sciences,* V. 10 (January 1943), 1-30

Flight Testing is a Sound Business, *Aviation* V. 42 Part 1 (April 1943), 350, 353-355 A posthumous publication.

References

Edmund Turney Allen - Scientific Test Pilot - A Preliminary Chronology by Richard K. Smith   3 Sept 92

Eddie Allen and the Air Mail Service   1925-1927 By Richard K. Smith   28 Nov92

Allen and Dyer Ancestry by Margaret Allen Reynolds 1963

Once to Every Pilot by Captain Frank Hawks , published by Stackpole Sons, New York 1936

Kelly: More Than My Share of it All by Clarence L. Johnson with Maggie Smith 1985

Eddie Allen, Aircraft Designer — Test Pilot by Harry Changnon C 1985 Rev 6-7-86

Reminiscences by William H. Cook   (about 1940 and 1992)

Boeing Historical Archives — Michael Lombardi

Excerpts from "The Global Twentieth" by Bob Robbins

Excerpts from "Boys Life" — March thru June 1931 by Myron Morris

The Most Unforgettable Character I've Met — The Readers Digest February 1965 by Thomas Collison

Test Pilots - by Richard P. Hallion Smithsonian Institution Press 1988

Excerpts from draft biography data by Thomas Collison 1944

Appendix I

Attached are several internal Boeing memoranda. The first two summarize flight test issues experienced by ETA during test flights immediately prior to the fatal February 18, 1943 flight. They are addressed by ETA to Wellwood Beall, the chief engineer on the B-29 program. The final document dated March 10, 1943, documents the findings of the Accident Investigation Committee.

February 3, 1943

To: W. E. Beall

Subject: B-29 Flight Tests

No flights Monday on account of bad weather. One flight Tuesday on #2 for ceiling climb and high altitude performance. At 30,500 feet in climb, one front exhaust manifold on #2 engine failed and started to burn the nose cowling. #2 engine was immediately throttled and then feathered. After landing, it was discovered that quite a little damage had been done by the exhaust flames to dural parts. It is believed that there is something fundamentally wrong about the structural design of the front exhaust. A number of failures had already occurred on this.

Propeller governing, and torquemeter and turbo regulator difficulties were encountered at low temperatures. Steps are being taken to remedy these difficulties.

E. T. Allen

February 10, 1943

To: N. E. [initials]

Subject: XB-29 Flight

An attempt was made yesterday to climb to 35,000 feet in the XB-29. At 26,000 feet, the nose pressure in #3 engine had dropped to 28 pounds and so the climb was discontinued. During descent, an oil leak developed on #3 and this engine was feathered. It was then noticed that #1 had a broken exhaust manifold which might necessitate feathering so #3 was unfeathered for the landing. After landing, it was found that the crack in #1 front exhaust manifold was similar to those four which have already occurred. It was found that the oil leak on #3 was caused by a broken fitting. It was found that a similar broken oil line fitting had also started a serious leak on #4. Both #3 and #4 engines would have failed in flight from this cause had the flight been continued. These things are probably caused by inexperienced help and atrocious working conditions at the hangar.

The oversensitivity of the turbo regulators on this airplane is a very serious handicap to service operation.

The pilot's vision is getting worse on each flight. It is believed that something will have to be done to improve this before the airplanes will be satisfactory in service.

E. T. Allen

March 10, 1943

To: Mr. H. O. West

Subject: Investigation of B-29, Number 1003, Accident, February 18, 1943.

The members of the Committee submit the attached report on their findings and conclusions. In arriving at the conclusions contained in the summary, much data and a great amount of evidence have been investigated, all of which is available in the Engineering Department for further study by interested parties. All remaining parts of the plane and engines are under guard at Warehouse No. 4, undergoing further investigation by Engineering and others.

During the investigation every attempt was made by the Committeemen and their sub-Committees to consider from every angle the cause of the first reported fire. They did not lose sight of the fact that the 3350-13 engines have not been dependable, and close inspection of all remaining engine parts was made to ascertain if an engine failure was responsible for the starting of the fire. Many parts which would have given conclusive evidence were completely destroyed.

During the investigation it was disclosed that many improvements could be made to the plane, (a) to prevent fires, (b) to prevent the spread of fire from one part of the airplane to another, (c) to detect fire more readily, (d) to extinguish fire more positively, (e) to improve means of emergency exit, and (f) to prevent feeding of fuel or oil to any fire which may be started.

Recommendations and further study are under way in the Engineering Department, and the revisions necessary to accomplish such work will be started on Plane No. 1002 as soon as releases are received from Engineering.

Members of the Investigation Committee:

E. C. Wells	G. W. Newton	George Schairer
G. V. Bierst	D. W. Finlay	Chas. Rankin
H. Kent	G. M. Weaver	

SUMMARY REPORT - INVESTIGATING COMMITTEE
ACCIDENT TO XB-29, SERIAL NO. 1003

Based upon evidence examined by the Investigating Committee, it is believed that the following summary outlines the sequence of events during the flight of XB-29 airplane, Serial No. 1003, on February 18, 1943, which flight was terminated when the airplane collided with the Frye Packing plant.

A. Flight Plan and Radio Reports

1. The purpose of the flight, as stated on the flight plan, was as follows:

   "Local Flight - Power Plant Performance and Cooling - Propeller Governing - Airplane Performance - Two-Engine Ceiling."

2. The airplane left the ground at Boeing Field at 12:09 p.m.

3. The airplane reported its position to Boeing Operator at 12:13 p.m. to be over Swan Lake at 2,000 feet altitude, southbound. No other comment in this report.

4. The airplane reported its position to Boeing Operator at 12:16 p.m. to be over Lake Tappé at 5,000 feet altitude, southbound. No other comment in this report.

5. The airplane reported to Boeing Operator at 12:18 p.m. as follows: "Fire in No. 1 engine --- Coming in --- Had fire in engine and used one CO_2 bottle and think we have it under control --- "

6. The airplane reported its position to Boeing Operator at 12:20 p.m. to be south of Kent at 3,500 feet altitude northbound, stating also, "No. 1 feathered."

7. The airplane reported its position to Boeing Operator at 12:21 p.m. to be south of Renton at 2,500 feet altitude northbound. No other comment in this report.

8. The airplane reported its position to Seattle Tower at 12:22 p.m. to be over Renton at 2,400 feet altitude, descending. In addition, this report stated: "Request immediate landing clearance, No. 1 engine on fire, propeller feathered and trouble not serious. Order crash equipment to stand by."

9. The airplane reported its position at 12:24 p.m. to be over the Lake Washington bridge at 1,500 feet altitude. No direction given, and no other comment in this report. However, Boeing Operator states, "The operator did not sound excited until reporting his position over Lake Washington bridge. At that time he first reported an altitude of 2,500 feet and then amended it to 1,500 feet."

At approximately 12:24 p.m., the airplane attempted to communicate with Seattle Tower, but transmission was blocked by other airplanes.

Summary Report
March 10, 1943
Page Two

10. At 12:25 p.m. the airplane reported to Boeing Operator, "Have fire equipment ready as am coming in with a wing on fire." No position reported. Boeing Operator stated, "When radio operator asked for fire equipment, his voice sounded strained and rather hoarse, but perfectly coherent."

Also at approximately 12:25 p.m. the Seattle Tower overheard an inter-phone conversation addressed to Allen by one of the airplane crew members as follows: "Allen, better get this thing down in a hurry, the wing spar is burning badly." There was no further radio contact with either the Boeing Operator or the Seattle Tower.

11. At 12:26 p.m. the lights at the Boeing radio room flickered and they switched to standby power.

At approximately 12:26 p.m. the radio in the Seattle Airways Control Tower went off the air due to power failure.

B. Probable Sequence of Events

1. It is believed that the fire as first reported originated in the nacelle to the rear of the engine and forward of the firewall. This is believed probable since the use of one CO_2 bottle apparently put the fire under control. This is substantiated by the radio reports, Mr. Ainias's log, and by reports from observers of the Interceptor Command, which check with the radio reports as to time and position. From the evidence available it is not possible to determine the exact cause of the first reported fire.

2. It is believed that:

    (a) The nacelle fire spread to the wing leading edge prior to the extinguishing of the nacelle fire and without the immediate knowledge of the crew members; or ---

    (b) The leading edge fire ignited independently of the first reported fire, by contact of gasoline or gasoline fumes and heated nacelle skin in the rear portion of either the inboard or outboard nacelles.

    It is believed that in either case, the leading edge fire began with the ignition of gasoline and/or gasoline fumes. It is probable that the gasoline was supplied to the leading edge by overflow of the fuel tank filler necks (through vent holes) and subsequent leakage around the filler cover plates and their fastenings into the leading edge, and around the junction of the filler neck drain and the lower leading edge surface.

3. The normal gravity flow of any overflow gasoline, and the flow of any ventilating air in the leading edge would both be in an inboard

Summary Report
March 10, 1943
Page Three

direction. It is believed therefore that the leading edge fire progressed through the leading edge to the inboard nacelle (wheel well to rear of firewall) and to the fuselage (bomb bay).

4. Radio reports, testimony of witnesses, and location and condition of parts which fell from the airplane, indicate that the intensity of the fire increased greatly at approximately 12:24 to 12:25, or shortly after the airplane passed over the Lake Washington bridge and turned from a north course to a southwest course, and that the seriousness of the fire was then realized by the crew. It is probable that at this time the fire had traveled to the bomb bay and was readily visible to the crew in the forward compartment. This increase in intensity could have been caused by a failure of the leading edge, a failure of a fuel line or fuel filler neck, ignition of magnesium parts in the leading edge, or failure of oxygen lines in the bomb bay, or a combination of failures due to the heat of the fire. The intensity of the fire could have also been increased by an increased leakage of fuel from the left hand filler necks and drains during the turn from North to Southwest with the left wing down.

5. The hose clamps and the internal parts of a deicer valve, which were found near 17th and Jefferson, indicate that sufficient damage had been done to the leading edge at that time to permit these parts to fall clear of the airplane.

6. With an intense fire burning and with the crew members apparently aware of the seriousness of the situation, three possible alternatives faced the pilot: (1) To attempt to reach the field, risking a further increase in intensity of the fire or risking a structural failure, either of which results would cause loss of control over the city; (2) To maintain altitude and abandon the airplane; (3) To stay with the airplane, attempting to make a crash landing in a lightly inhabited district of the city.

7. It is believed that the third alternative was chosen because the pilot felt that the intensity of the fire would prevent his reaching the field and that required a crash landing in the nearest open area if a landing was to be made before complete loss of control. Although it is possible that power was lost on Number Two engine as well as Number One engine, due to failure of fuel lines or other causes, and such loss of power would require an eventual landing soon after lowering of the landing gear, the rate of descent during the last minute of flight was greater than could be accounted for (at the existing gross weight) by this loss of power or increase in landing gear drag.

8. It is believed that following the increase in intensity of the fire and prior to the time the three crew members left the airplane, the fire found its way to the forward compartment. This could have taken place as a result of burning through of the magnesium tunnel

Summary Report
March 10, 1943
Page Four

wall or as a result of opening the bomb bay access door, or both. It was common practice to open the bomb bay access door to relieve the pressure differential on the nose wheel escape hatch, so that this hatch could be more readily opened.

9. It is believed that the intensity of smoke and heat in the cockpit increased during the last stages of the flight, and that the impairment of vision caused by the smoke may have caused collision with the building rather than continuation of a normal flight path and contact with the ground.

SUMMARY: —

It is believed that the following are the principal factors contributing to the fire:

1. Leakage of fuel and/or oil in the engine nacelle, ignited by contact with exhaust pipe or exhaust shroud, or by operation of electrical equipment, or by failure of the accessory section of the engine or failure of an engine accessory.

2. Leakage of fuel into the leading edge, primarily around the filler neck cover and drain, ignited by contact with heated nacelle skin, or by fire through nacelle firewall openings, or by operation of electrical equipment in or adjacent to leading edge.

3. Spread of fire, following natural course of leading edge ventilation to wheel well and bomb bay, and finally to cockpit.

C. N. Weaver
Assistant to Plant Manager

Glenn V. Dierst
Plant Protection Manager

E. C. Wells
Assistant Chief Engineer

George Schairer
Chief Aerodynamicist

Charles S. Rancin
Chief Inspector

D. W. Newton
Power Plant Unit Chief

INTRA-DEPARTMENTAL ROUTING

Harvey East	D. W. Finlay
Assistant Superintendent	Assistant Project Engineer
3-29	

www.ingramcontent.com/pod-product-compliance
Lightning Source LLC
Chambersburg PA
CBHW062211220526
45471CB00009B/3158